Optimization and Industrial Experimentation

OPTIMIZATION AND INDUSTRIAL EXPERIMENTATION

William E. Biles

The Pennsylvania State University
University Park, Pennsylvania

James J. Swain

Purdue University
West Lafayette, Indiana

A Wiley-Interscience Publication

JOHN WILEY AND SONS
New York · Chichester · Brisbane · Toronto

Library of Congress Cataloging in Publication Data

Biles, William E
Optimization and industrial experimentation.

"A Wiley-Interscience publication."
Includes bibliographies and index.
1. System analysis. 2. Experimental design.
3. Response surfaces (Statistics) 4. Mathematical optimization. I. Swain, James J., joint author.
II. Title.

T57.6.B53 003 79-9516
ISBN 0-471-04244-7

Printed in the United States of America

10 9 8 7 6 5 4 3 2 1

Preface

This book is intended for use both as a textbook in a senior or graduate level course in experimental optimization and as a desk reference for scientists and engineers involved in experimentation. Its focus, the combination of optimization with statistical methodology, is not particularly novel—the field called response surface methodology deals with this combination. This book, however, brings many of the newer developments in numerical optimization to bear on experimentation. The objective is to demonstrate how to optimize systems through experimentation rather than mere computation.

The necessary background for this book is mathematical maturity in calculus and matrix algebra. The required know-how in statistics can be obtained by a careful reading of Chapter 2, which is, in fact, geared to the novice in statistics. Others can use Chapter 2 as a means of becoming familiar with the notation used in the book and reviewing the important concepts of regression and analysis of variance.

The principal theoretical thrust of the book is embodied in Chapters 3 through 5. Although Chapter 3 gives extensive developments in experimental design and response surface methodology, the focus is on the means by which an experimenter would employ experimentation and statistical methodology to develop mathematical models of a process under study. Chapter 4 examines the most important concepts in optimization, stressing search methods for single-variable and multiple-variable optimization, as well as classical and numerical optimization techniques that are directly applicable in response surface methodology. Chapter 5 gives an overview of the combined approach to optimization via experimentation, outlines several general procedures for optimizing systems experimentally, and illustrates and compares some of the proposed procedures.

Chapters 6 and 7 examine a combined experimental optimization approach to various real-world problems. Chapter 6 shows applications in physical systems, including manufacturing machining operations and chemical processes. Chapter 7 demonstrates the concept of "simulation optimization," that is, optimization in simulated systems. It also introduces simulation methodology.

The references at the end of each chapter not only give credit to works from which specific material is taken, but guide the reader to the most important up-to-date developments in the pertinent theory. Thus this book should also serve as a primary research reference in the area of optimization via experimentation.

It gives us great pleasure to acknowledge the many contributions to this effort. For their excellent typing of manuscript materials we thank Mrs. Shirley Wills and Miss Teresa Rankin of the Department of Aerospace and Mechanical Engineering of the University of Notre Dame, and Mrs. Martha Porter, Mrs. Cathi Moyer, and Miss Kim Williams of the Department of Industrial and Management Systems Engineering at the Pennsylvania State University. Our sincere appreciation is also expressed to the National Science Foundation, Engineering Division, for sponsorship of the early stages of this work under grant ENG 74-19313, and to the Office of Naval Research, Naval Analysis Program, for funded support of much of the later development under contract N00014-76-C-1021. Mr. Ming-Lung Lee, a graduate student in mechanical engineering at the University of Notre Dame, and Mr. Michael Davisson, a mechanical engineering graduate of the University of Notre Dame, are hereby thanked for much of the computer programming involved in this work. Dr. Kenneth J. Anselmo and Dr. V. Grant Fox of Air Products and Chemicals, Inc. were very helpful in the development of the material on least-squares methods in Chapter 3 and that on matrices in Appendix B. Their assistance is gratefully acknowledged.

WILLIAM E. BILES
JAMES J. SWAIN

University Park, Pennsylvania
West Lafayette, Indiana
October 1979

Contents

1 Introduction **1**

Fundamental Concepts in Experimentation, 1
Fundamental Concepts in Optimization, 5
Classical Optimization, 7
Functions, 7
Function Continuity, 9
Characteristics of Functions, 9
Conditions for an Optimum, 12
Approaches to Optimization, 14
Analytical Methods, 15
Numerical Methods, 15
Mathematical Programming, 17
Linear Functions, 18
Nonlinear Functions, 18
Optimization through Experimentation, 21
Overview of the Book, 24
References, 26

2 Fundamental Statistical Concepts **28**

Basic Probability, 28
Random Experiment, 29
Sample Space, 29
Probability, 31
Events, 32
Operations with Events, 34
Conditional Probability, 35
Random Variables, 37
Probability Distributions, 38
Expectation, 41

Discrete Probability Distributions, 43
Bernoulli Trial, 44
Binomial Distribution, 45
Poisson Distribution, 47
Continuous Probability Distributions, 49
Uniform Distribution, 49
Exponential Distribution, 50
Gamma Distribution, 51
Normal Distribution, 52
Bivariate Normal Distribution, 54
Sampling and Sampling Distributions, 55
Tabular and Graphical Representation of Samples, 58
Sample Statistics, 59
Sampling Distributions, 61
Normal Distribution, 61
Student's *t* Distribution, 63
Chi-Square Distribution, 64
Snedecor's *F* Distribution, 65
Estimation of Population Parameters, 66
Point Estimation, 67
Interval Estimation, 67
Statistical Tests of Hypothesis, 68
Tests on Means, 71
Tests on Variances, 73
Goodness of Fit Tests, 74
Chi-Square Test, 74
Kolmogorov-Smirnov Test, 77
Regression and Correlation, 77
Linear Regression, 78
Hypothesis Tests on the Slope and Intercept, 80
Curvilinear Regression, 83
Multiple Regression, 86
Correlation, 87
Analysis of Variance, 89
One-Way Analysis of Variance, 90
Two-Way Analysis of Variance, 94
Summary, 97
References, 97

3 **Experimental Design Fundamentals** **99**

Regression Analysis, 99
Matrix Representation, 99
Statistical Properties of the Least-Squares Estimates, 102
Bias and Lack-of-Fit, 106
Residual Analysis, 109
Distribution of the Residuals, 110
Outliers, 111
Graphical Procedures, 112
Properties of Residuals, 113
Some Computational Considerations of Least-Squares Methods, 114
Coding, 116
Solution of the Normal Equations, 117
Linear Dependence and Sensitivity, 117
Curve Fitting, 118
Generating the Polynomials, 121
Experimental Designs for Linear and Quadratic Response Surfaces, 121
Coding Conventions, 123
First Order Designs, 124
Factorial Designs, 124
Simplex Designs, 132
Second-Order Response Surface Models, 135
Form of the Model, 136
Central Composite Design, 137
Other Designs, 147
Design Blocking, 152
Design Criteria, 153
Advanced Regression Topics, 155
Linearizing Transformations, 155
Explicitly Constrained Least Squares, 157
Implicitly Constrained Least Squares, 160
Weighted Least Squares, 166
Maximum Likelihood Methods, 168
Nonlinear Regression and Modeling, 171
Screening Experiments, 175
References, 178

4 Fundamentals of Optimization **181**

Introduction, 181
Single-Variable Optimization, 183
Interval Reduction Methods, 184
Fibonacci Search, 186
Multivariable Optimization, 189
Direct Search Techniques, 189
Random Search, 189
Sequential Simplex Search, 192
Complex Search, 195
Pattern Search, 195
Modifications for Constrained Direct Search, 197
Gradient Based Unconstrained Optimization, 199
Ridge Analysis, 203
Gradient Search: Hill Climbing, 205
Newton Algorithm: Second-Order Methods, 206
Conjugate Gradient, Fletcher Reeves, and Partan Algorithms, 208
The Constrained Optimization Problem, 212
Applications of the Lagrangian Analysis, 216
Algorithms for Solving Least-Squares Problems, 221
Constraint Handling, 225
Multiple Objective Optimization, 229
Weighted Optimization, 230
Threshold Methods, 231
Goal Programming, 232
Geoffrion-Dyer Algorithm, 234
Nondominated Vector Optimal Solutions, 235
References, 235

5 Optimization Via Experimentation **238**

Overview of the Experimental Optimization Problem, 238
Response Surface Methods, 241
Preliminary: Analysis and Screening, 242
Steepest Ascent, 243
Line Searches, 244
Linear Techniques and Constraints, 245
Second-Order Designs, 247

Problem Formulations, 247
Constrained Optimization, 248
Multiple-Objective Optimization, 248
Optimization Techniques, 249
Direct Search Methods, 250
First-Order Response Surface Methods, 252
Second-Order Response Surface Methods, 255
Effect of Bias and Design on Optimization Methodology, 256
Steepest Ascent with a Full Factorial Design, 257
Steepest Ascent with a Simplex Design, 262
Three Second-Order Experiments, 264
Application of Constrained Optimization Methodologies, 270
Gradient Projection Method, 271
Multiple Gradient Summation Technique, 274
Constrained Gradient Approach, 276
A Simplified Zoutendijk Approach, 279
A Nonlinear Zoutendijk Approach, 281
Second-Order Approach, 285
Summary, 286
References, 287

6 Optimization and Experimentation with Physical Processes 289

Example 1. Optimizing Cutting Fluid Pressure in Machining, 289
Golden Section Search, 290
Polynomial Regression, 291
Example 2. Optimizing Yield in a Chemical Process, 293
Augmented 2^3 Factorial Design, 294
Central Composite Design, 296
Box's Complex Search, 297
Example 3. Multiple Independent Variables, Multiple Responses with a Machining Parameters Problem, 299
Central Composite Design, 299
Example 4. Multiple Independent Variables, Multiple Responses with Chemical Processes, 303
Central Composite Design, 305
Summary, 311
References, 311

7 Optimization and Computer Simulation Experiments **313**

Basic Concepts, 314
Introduction, 314
Models, Systems, and Simulation, 315
Design of Simulation Models, 316
Simulation Languages, 317
Time Control, 319
Monte Carlo Sampling, 320
Random Number Generation, 322
Random Variate Generation, 325
Statistical Techniques in Computer Simulation, 327
Input Analysis, 327
Estimation, 328
Output Analysis, 329
Design of Computer Simulation Experiments, 330
Variance Reduction Techniques, 331
Optimization of Simulation Parameters, 333
Simulation and Optimization: A Historical Survey, 334
A General Procedure, 336
Simulation Examples, 336

References, 338

Appendix A
Selected Statistical Tables **341**

Appendix B
Review of Matrix Algebra **355**

Index **365**

Optimization and Industrial Experimentation

Chapter 1

Introduction

Experimentation is a vital function in our unending quest for progress. Although the aim of experimentation is often the discovery of new scientific principles or phenomena, establishing cause and effect relationships, or seeking improved methods and processes, a significant part of the experimental activity conducted in science and industry is devoted to finding the *optimum* conditions for carrying on a process. "Optimum" means *best*, and is a stronger word than "improved."

This book focuses on experimentation as a means of optimizing a process or system. Its objective is to unify the concepts of optimization and experimentation, too frequently viewed as disparate endeavors. It presents techniques for optimization in the experimental domain, rather than the computational realm within which it is most often applied. A primary objective of this book is to motivate experimenters to apply the tools of optimization in their experimental efforts and thus enlarge the catalog of applications of optimization in experimentation.

The purpose of this chapter is to introduce the fundamental concepts of optimization and experimentation, and to propose ways in which these elements can be combined to produce workable experimental methodologies. A brief overview of the book is provided in this chapter to show how these concepts fit together.

FUNDAMENTAL CONCEPTS IN EXPERIMENTATION

If we take a simple "black-box" view of the experimental process, as illustrated in Figure 1.1, we see the important elements involved in experimentation. We can regard the "experiment" block as embodying the physical process within which experimentation is performed, the mechanism which is manipulated, the "pot which is stirred." It possesses the *unknown* aspect, the unanswered questions, the unresolved problem. We may understand the basic principles of the process, or we may not; but there is something about it that remains unknown to us. The "input variables" in the black-box model are those quantities that we hypothesize as having a *causal* relationship to the problem, the factors that we can control. We call them the *independent* variables, and in this book they are denoted x_i, $i=1,\ldots,n$. This notation says that there are n distinct controllable factors or variables in the experiment. The "output responses" in the black-box model represent the quantities that we hypothesize as having an *effect* relationship to the problem. They are *dependent* variables, and are denoted y_j, $j=1,\ldots,m$ in this book. We cannot control the y_j quantities directly, but the results of the experimental effort may tell us how to effect such control indirectly by controlling the x_i quantities. The "uncontrollable factors" that affect the process, which we shall denote z_k, $k=1,\ldots,p$, cannot be ignored. Although these factors lie beyond our direct control—we may not even be able to identify them—we must conduct the experimental process in such a way as to avoid any disruptive effects they may exert. Thus we can describe the experimental process as one in which we perform the experiment at controlled values of the independent variables x_i, $i=1,\ldots,n$ and observe the responses y_j, $j=1,\ldots,m$, while attempting to negate the effects of the uncontrollable factors z_k, $k=1,\ldots,p$.

The experimental variables x_i and y_j require further scrutiny to fully understand their role in experimentation. Suppose, for example, that we are engaged in the manufacture of graphite products for nuclear applications, and we desire to experimentally determine ways to produce an improved graphite material. Let us first focus on the independent variables, x_i, $i=1,\ldots,n$. For the sake of simplicity, suppose that we consider three variables as controllable factors in the mixing stage of the manufacturing process, as follows:

(1) x_1, binder level, %

(2) x_2, mixer temperature, °C

(3) x_3, mixing time, min

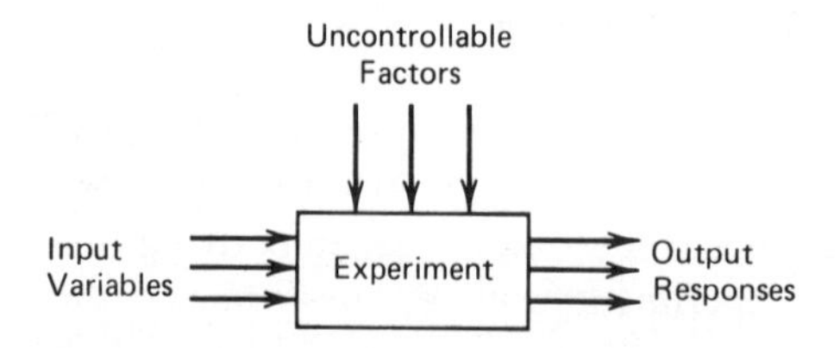

Figure 1.1. A "black-box" view of experimentation.

In addition to these three variables, we desire to evaluate three different types of mixers for their effectiveness in achieving a thoroughly mixed product. The first three variables are clearly *quantitative* in nature; they can be controlled over a continuous range of values. With respect to the type of mixer employed in the experiment, this factor can be viewed as the fourth variable, x_4, which has three distinct, *qualitative* levels. We would expect to have to vary each of the first three factors over each level of the qualitative variable in our experiment.

Suppose that we seek a graphite material that possesses desirable values of such properties as density (kg/m^3), y_1; flexural strength (kilopascals), y_2; and electrical resistivity (ohm-cm), y_3. These properties are the *responses* in the experiment. These are clearly quantitative responses, but many experiments have qualitative responses that determine the preferability of one set of values for the independent variables.

Although we generally consider that we can control the independent variables x_i, $i=1,\ldots,n$ at specified values of levels without error, the responses y_j, $j=1,\ldots,m$ are typically affected by error. This *error* can be viewed as resulting from the combined, random effects of all the uncontrollable quantities z_k, $k=1,\ldots,p$. The manner in which this random error manifests itself is to cause a response y_j to assume a *set* of values y_{jl}, $l=1,\ldots,r$ at a specified set of values of the independent variables x_i, $i=1,\ldots,n$. We call this set of values a *distribution*. Thus statistical methodology is needed to perform an objective analysis of experimental results. This book is based on a statistical approach to experimentation. Chapter 2 reviews the fundamental statistical concepts that are needed for the procedures discussed herein.

Two requirements that derive directly from the influence of random error on the experimental process are *replication* and *randomization*. By replication we mean a repetition of the experiment under precisely the same controllable conditions. In the graphite material experiment, replication would involve repeating a mix at the same binder level (x_1), the same mixer temperature (x_2), the same mixing time (x_3), in the same type of mixer (x_4). If three mixes are prepared under these same conditions, we say that three *replicates* have been obtained. Replication allows the experimenter to estimate experimental error, as we shall discuss in Chapter 2, and to calculate the sample mean $\bar{y}_j$ for each of the responses, or material properties in this case. The sample mean $\bar{y}_j$ is usually a more reliable estimate of the true response, which we shall denote as η_j, $j=1,\ldots,m$, at the particular input conditions x_i, $i=1,\ldots,n$, than the single observation y_j. This is due to the fact that replication tends to average out the effects of the uncontrollable factors z_k, $k=1,\ldots,p$. Randomization, which is the arrangement of the individual runs or trials of the experiment in a random sequence, will also assist in minimizing the effects of any uncontrollable factors that exert a consistent bias. The purpose of randomization is to eliminate any systematic effects or trends due to these extraneous factors, whereas the objective of replication is to enable better estimates of the responses despite the presence of these factors.

If an experiment is to be performed most efficiently, then a scientific approach to planning the experiment must be taken. The *statistical design of experiments*, described in Chapter 3, entails two tasks: (1) the *design* of the experiment, including selecting the values of the independent variables x_i, $i = 1, \ldots, n$, and the randomization of the sequence of experimental trials; and (2) the *statistical analysis* of the data obtained from the experiment. But it is also necessary to approach the experimental process in a systematic manner. The following procedure is recommended:

STEP 1. *Formulate the problem.* Establish the goals and objectives of the investigation. Delimit the study, so that useful results can be obtained with the time and funds available. Identify the key independent variables x_i, $i = 1, \ldots, n$ which are hypothesized to influence the problem, and the principal responses y_j, $j = 1, \ldots, m$ that are to be measured. If possible, postulate a mathematical model relating these variables; that is, hypothesize a particular *form* for the relationship

$$y_j = g_j(x_i, i = 1, \ldots, n) \qquad j = 1, \ldots, m \tag{1.1}$$

The postulated model usually contains one or more unknown quantities.

STEP 2. *Design the experiment.* Select the statistical procedures to be used in analyzing the results of the experiment. Determine the number of experimental trials that will be conducted and establish the set of values x_i, $i = 1, \ldots, n$ for each such trial. Decide what physical procedures will be employed in the experiment and establish a random order in which the trials will be performed.

STEP 3. *Perform the experiment.* Perform the physical process embodied by the experiment in accordance with the design in Step 2. Record all data, including the values of the independent variables x_i, $i = 1, \ldots, n$ and the responses y_j, $j = 1, \ldots, m$ at each experimental trial. Note any unusual occurrences or circumstances that might negate the results obtained and require repeating a given trial of the experiment.

STEP 4. *Tabulate and analyze the data.* Employ established statistical methods to compute appropriate *sample statistics* such a mean and variance. Fit the models postulated in Step 1. Perform any statistical tests of hypotheses that may be necessary to enable conclusions to be drawn from the experiment. Graphical presentations of data are often useful in analyzing the results of an experiment.

STEP 5. *Draw conclusions and offer recommendations.* Once the data have been analyzed, the experimenter may draw conclusions and inferences about the results. A physical interpretation of the statistical results is needed here. Then recommendations are made, either for continued experimentation or for direct implementation of the results.

This five-step procedure forms the cornerstone of a systematic experimental program. We shall discuss techniques by which methods of optimization are merged into the experimental process, requiring slight amendment to this recommended procedure.

FUNDAMENTAL CONCEPTS IN OPTIMIZATION

As stated earlier, the principal objective of this book is to demonstrate how classical optimization theory can be applied to experimentation to derive the most beneficial results from an experimental program. This objective supposes that one of the fundamental reasons for experimentation is to establish the *optimum* conditions for the operation of a system. Optimum conditions are those that produce the best results from a system or activity. "Results" and "conditions" are key words here. The word "results" conveys the idea of a dependent variable or *response*; it suggests an *effect* relationship with the system or activity, and hence corresponds to the response variables $y_j, j=1,\ldots,m$ is our concept of an experimental process. The word "conditions" implies an ability to control selected quantities; it suggests a *causal* relationship to the experimental process, and hence corresponds to the independent variables x_i, $i=1,\ldots,n$.

Optimization is the process of seeking the best solution for a system or activity. If we consider the system responses to be related to the independent variables by a mathematical model of the form (1.1), then optimization entails a manipulation of the x_i's to achieve optimum values of the y_j's. When known mathematical forms exist for (1.1), the well-known procedures of *classical optimization* can be employed to obtain a solution to the problem.

There are several schemes by which optimization techniques can be classified. One means of categorizing methods of optimization is by the number of independent variables. Single-variable methods are used with a function of a single variable $g(x)$, and multivariable methods are used for optimizing a function of several variables $g(x_1,\dots,x_n)$.

A second scheme for classifying methods of optimization depends on whether the method employs derivatives. This is, there are *derivative methods* and *numerical methods*. Derivative methods seek to determine those values of $x_1,\dots,x_n$ for which the first derivative vanishes; that is, for a single variable, the value of x for which

$$y' = \frac{d[\,g(x)\,]}{dx} = g'(x) = 0 \tag{1.2}$$

We illustrate this concept later in this chapter for functions of a single variable. When the derivative either does not exist or is too complicated to manipulate, we resort to numerical methods. Numerical methods typically use algorithms, and they are therefore well suited for implementation of a digital computer.

A third way to classify optimization methods is to distinguish between unconstrained and constrained problems. In constrained problems, the solution must satisfy special conditions. Equation (1.1) gives the unconstrained case; that is, we would seek to optimize a function $g(x_1,\dots,x_n)$. The general form of the constrained problem is as follows:

Optimize
$$y_0 = g_0(x_1,\dots,x_n) \tag{1.3}$$
subject to

$$a_i \leqslant x_i \leqslant b_i \qquad i = 1,\dots,n \tag{1.4}$$

$$y_j = g_j(x_1,\dots,x_n)\left\{\begin{matrix} \leqslant \\ = \\ \geqslant \end{matrix}\right\} d_j \qquad j = 1,\dots,m \tag{1.5}$$

Equation (1.3) is exactly like (1.1); it states that we want to find conditions $x_1^*,\dots,x_n^*$ that optimize a primary criterion to yield y_0^*. However, (1.4) limits the range of values that the ith independent variable can assume; that is, x_i can range between a lower bound a_i and an upper bound b_i. These bounds might be due to limitations in the equipment's capability, to costs in a competitive marketplace, to specifications of the product's performance, or to legal restrictions established by federal, state, and local codes. The expression given in (1.5) states that there are other criteria that must be controlled. These criteria also depend on the independent variables $x_1,\dots,x_n$. Thus changes in the variables $x_1,\dots,x_n$ that are favorable to the primary criterion y_0 might be unfavorable to one or more of the other criteria $y_1,\dots,y_m$. As with (1.4), these restrictions might be due to any number of technical, economic, social, or legal factors.

Implicit in the constrained versus unconstrained classification of optimization procedures is yet another basis for categorization—the number of dependent variables or responses, y_j, $j = 1,\ldots,m$. We can refer to a *single-response* or *multiple-response* problem. (In the optimization sense, these are often called single-objective and multiple-objective problems.)

Classical Optimization

Although the principal orientation of this book is optimization through experimentation, it is essential that we develop the basic principles of optimization theory. These basic principles will not only enhance our understanding of the science of optimization, but they will also afford a body of computational techniques that can be applied in the case where we have formed approximating functions by fitting equations to experimental data and wish to find an optimum for those functions. We shall make use of these basic concepts even in Chapter 2 in developing certain basic statistical concepts.

Functions Let us denote a *function* as $y(X)$, where y represents the dependent variable and X the n-vector of independent variables x_i, $i = 1,\ldots,n$. Figure 1.2 shows a one-dimensional representation $y(x_1)$, and Figure 1.3 illustrates a two-dimensional function $y(x_1,x_2)$. In general, $y(X)$ is said to be a *function* of X, defined in some region with respect to X, if there exists a rule that allows us to determine y for any set X in this region. Such a rule may be represented by an equation, a graph, a table, or an experiment. Figure 1.2, suitably scaled, would allow us to estimate the value of y corresponding to a specific value for x_1. Such estimation would be almost impossible for the function $y(x_1,x_2)$ shown in Figure 1.3, but a

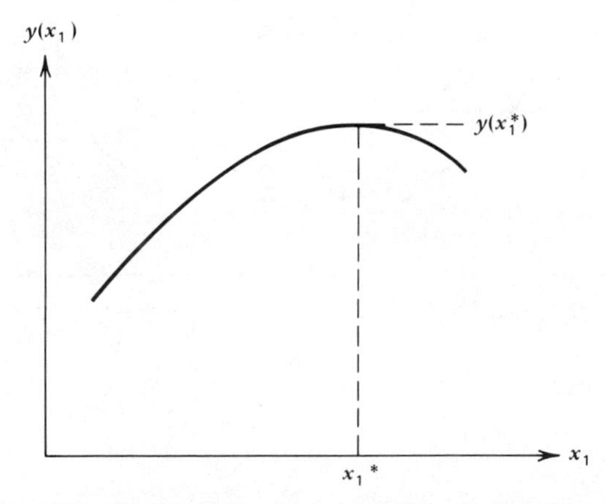

Figure 1.2. One-dimensional representation of a function $y(x_1)$.

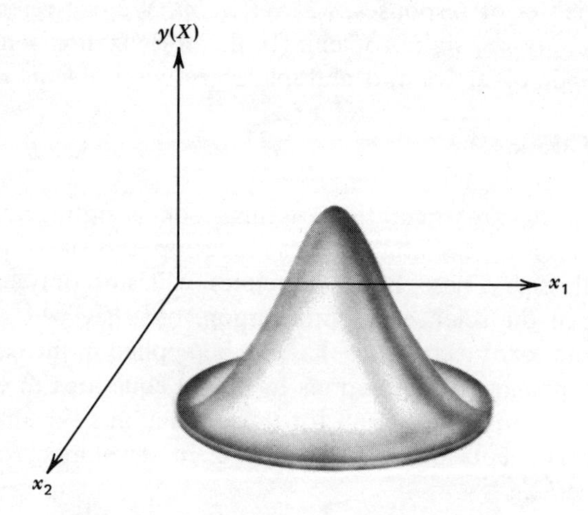

Figure 1.3. A two-dimensional function $y(x_1, x_2)$.

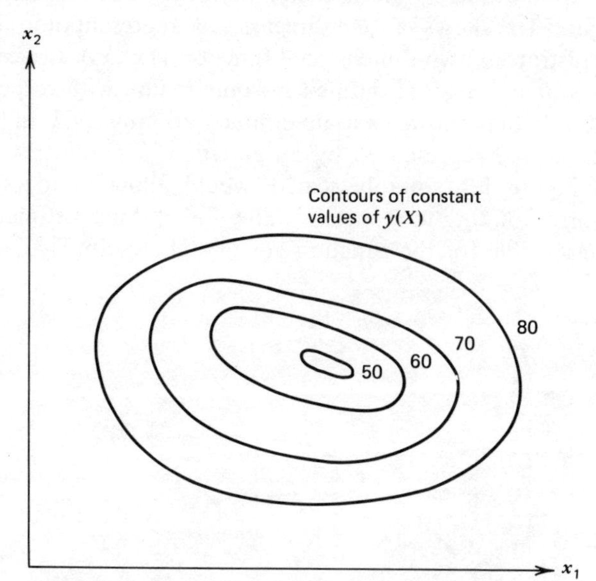

Figure 1.4. Contour representation of a function $y(x_1, x_2)$.

somewhat different graphical representation, such as the two-dimensional contour representation of $y(x_1, x_2)$ shown in Figure 1.4, does enable accurate estimation.

Function Continuity A function $y(X)$ may be termed *continuous* at some point X if

$$y(X) = \lim y(X+\delta) \qquad \delta \to 0 \tag{1.6}$$

for all possible ways that δ can approach zero. A function is said to be *discontinuous* if the relation (1.6) does not hold. Figure 1.5 illustrates a discontinuity for a function $y(x)$. An example of a discontinuous function is the function $y(x)$ in which y represents the pressure drop in a given length of pipe having diameter x. Since pipe is only available in standard diameters, the function $y(x)$ is a discrete-valued function of x, and hence discontinuous.

Characteristics of Functions There are certain characteristics of functions that are useful in optimization. A function $y(x)$ is said to be *monotonically increasing* when, for $x_2 > x_1$, $y(x_2) > y(x_1)$. Here x_1 and x_2 refer to successive values of a single variable x, not to several variables as seen earlier. A function $y(x)$ is called *monotonically decreasing* if, for $x_2 > x_1$, $y(x_2) < y(x_1)$. The function $y(x)$ is *monotonically nondecreasing* and *monotonically nonincreasing* if, for $x_2 > x_1$, $y(x_2) \geqslant y(x_1)$ and $y(x_2) \leqslant y(x_1)$, respectively. These concepts are graphically depicted in Figure 1.6.

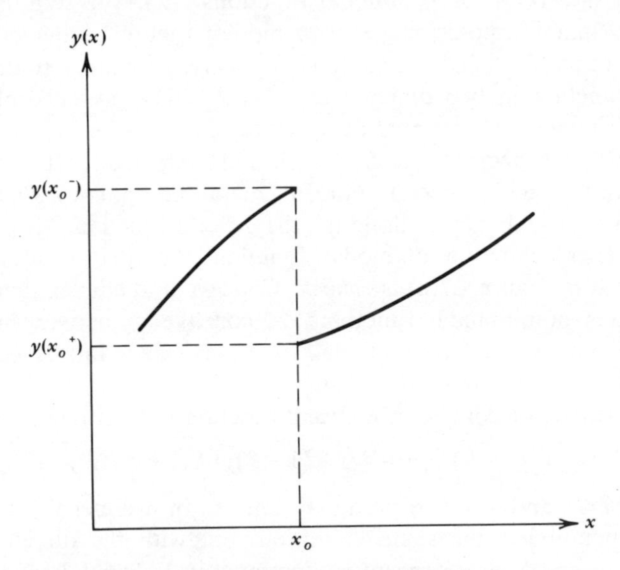

Figure 1.5. A function $y(x)$ with a discontinuity at x_0.

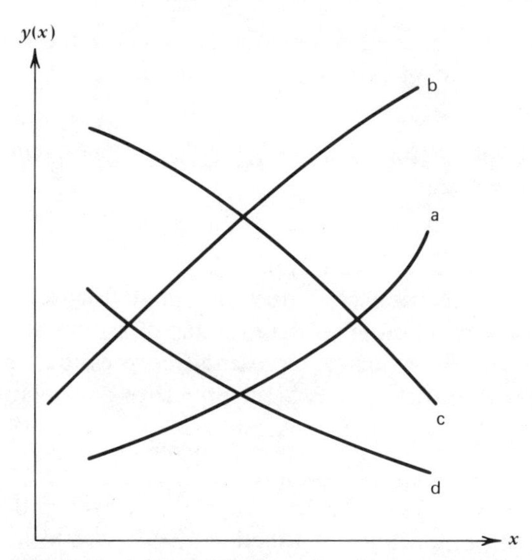

Figure 1.6. Examples of (a) monotonically increasing, (b) monotonically nondecreasing, (c) monotonically decreasing, and (d) monotonically nonincreasing functions $y(x)$.

The *modality* of a function is particularly important in the optimization of a function $y(X)$. The term *unimodal* refers to a function having a *single* mode, which can be either a peak (maximum) or a trough (minimum). Figure 1.7 illustrates four unimodal functions $y(x)$. The term *multimodal* refers to a function possessing *several* modes. Figure 1.9 depicts a multimodal function of a single variable $y(x)$, whereas Figure 1.10 illustrates a bimodal function in two dimensions, $y(x_1, x_2)$. The presence of multiple modes confounds our attempts at optimization, because to determine the best solution it is necessary to find each and every mode. It is difficult to know if every mode has been found. In general, no optimization procedure guarantees that such can be done for the general function $y(X)$.

Figure 1.7(a) shows a unimodal function $y(x)$ that is also *concave*, whereas that in Figure 1.7(b) is *convex*. Concave and convex functions are special cases of unimodal functions. A concave or convex function is necessarily unimodal, but unimodal functions need not be concave or convex—Figures 1.7 (c and d) illustrate unimodal functions that are neither concave nor convex. A concave function $y(X)$ is one in which

$$y[(1-\theta)X_1+\theta X_2] \geqslant (1-\theta)y(X_1)+\theta y(X_2) \tag{1.7}$$

for all $0 \leqslant \theta \leqslant 1$ and any two points X_1 and X_2 in n-dimensional space. A convex function has the same definition, but with the direction of the inequality reversed. This concept is illustrated in Figure 1.8 for a function of a single variable $y(x)$.

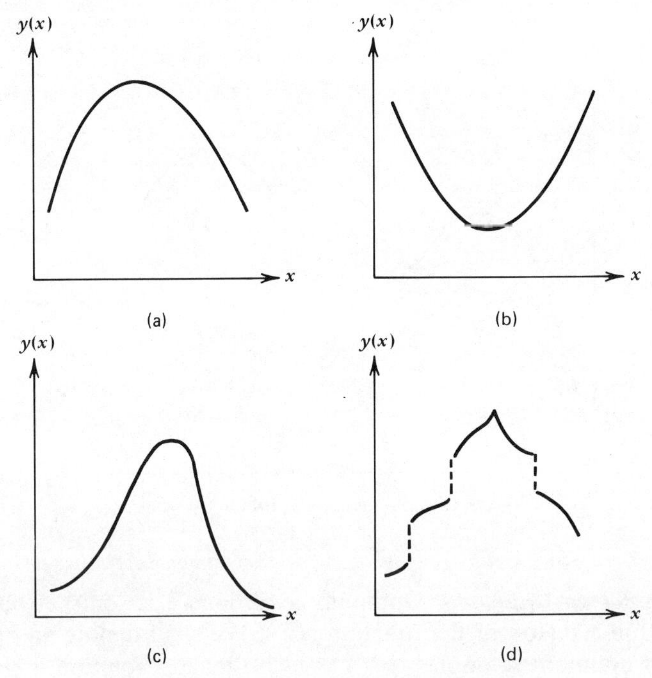

Figure 1.7. Examples of unimodal functions $y(x)$.

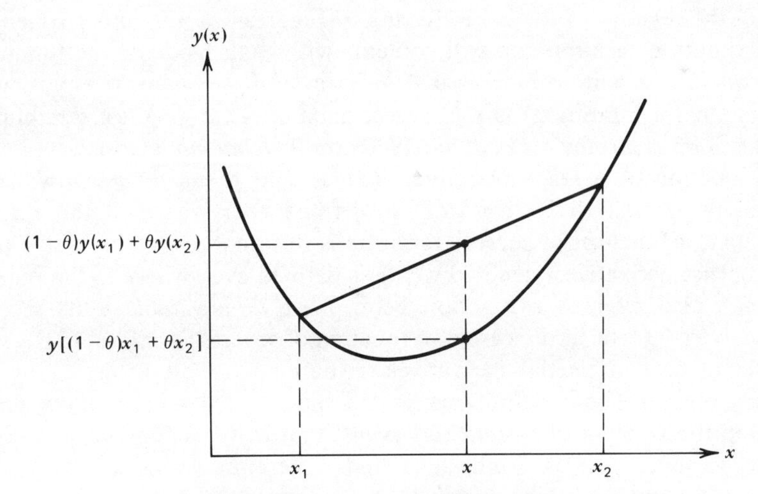

Figure 1.8. A convex function $y(x)$.

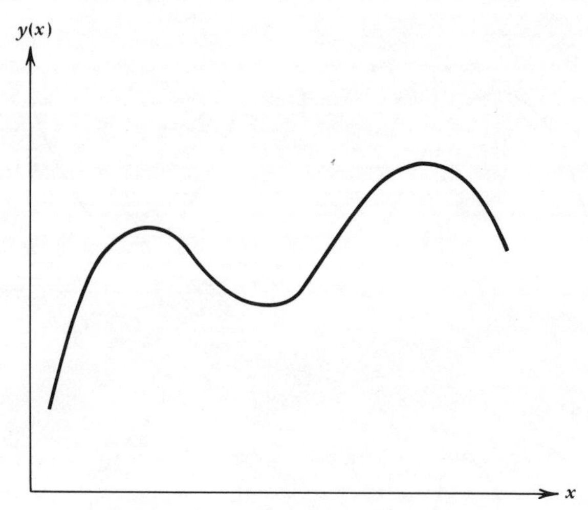

Figure 1.9. A multimodal function $y(x)$.

Conditions for an Optimum Optimum conditions exist only at particular points within a region of the function $y(X)$. We shall denote an optimum as X^*. An optimum solution is $y^*(X^*)$ such that this solution is preferred to any other $y(X)$.

A function $y(X)$ has one or more maximum values and minimum values within the region of interest. Referring to Figures 1.7, 1.9, and 1.10, either a maximum or a minimum will coincide with each mode of the function. We call such a point a *local maximum* or a *local minimum*. If a particular maximum (or minimum) is the largest maximum (or smallest minimum), such a point is termed a *global maximum* (or a *global minimum*).

A function of a single variable, $y(x)$, is said to be *differentiable* in a region if its first derivative $y'(x)$ is defined everywhere in the region. Similarly, a function of several variables is differentiable if its vector of n first partial derivatives, denoted $y'(X)$, is defined everywhere in the region. We can also refer to a function being *twice differentiable* if its second derivative $y''(x)$, or in the case of a function of several variables its Hessian matrix of second partial derivatives, is defined for all values X in the chosen region. These definitions, stated only casually here, allow us to develop the concept of a *stationary point*; that is, for a function $y(x)$ of a single variable that is continuous and differentiable in a region, the stationary point is a point x such that

$$y'(x) = \frac{d[y(x)]}{dx} = 0 \tag{1.8}$$

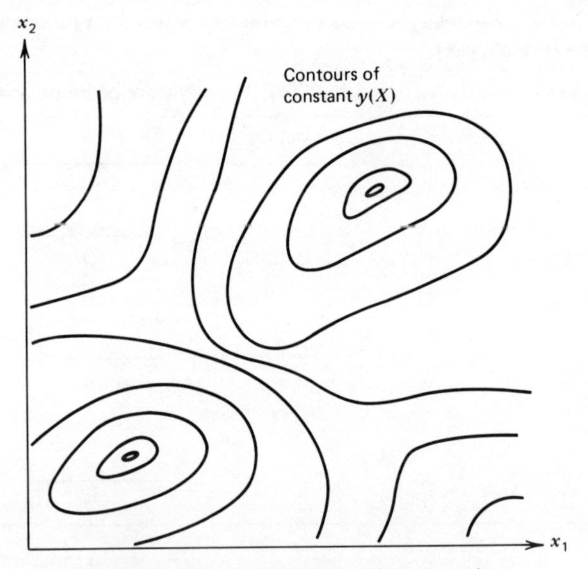

Figure 1.10. A bimodal function $y(x_1, x_2)$.

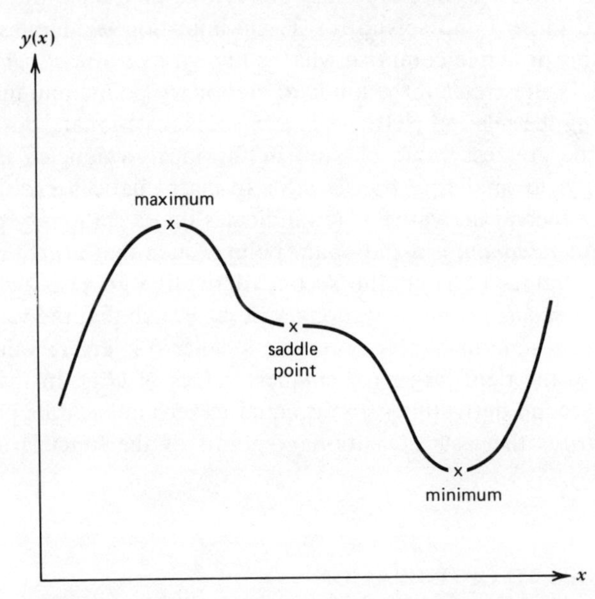

Figure 1.11. Stationary points in a one-dimensional function.

Table 1.1. Summary of tests of stationary points for a function of a single variable $y(x)$

$y'(a)$	$y''(a)$	$y'''(a)$	Nature of Point, $x=a$
0	−	Exists	a 
0	+	Exists	 a
0	0	+	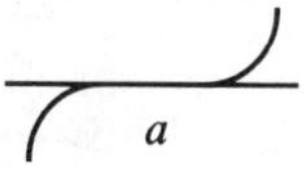 a
0	0	−	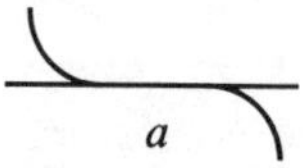 a

For a function of several variables, a corresponding condition must be met; that is, the set of n first partial derivatives of $y(X)$ with respect to X are equated to zero and solved for X. Optimization techniques which are based on this principle comprise what is known as *classical optimization.*

Figure 1.11 illustrates three kinds of stationary points one might obtain by exercising the relation (1.10). A *maximum* is a stationary point x, which produces the greatest value of $y(x)$ in the local region, so that nearby values of x yield smaller responses $y(x)$. In mathematical terms, a *negative* value of the second derivative $y''(x)$ indicates that a stationary point x is a maximum. A *minimum* is a stationary point x such that nearby values of x yield larger values of $y(x)$; the second derivative $y''(x)$ is *positive* for a minimum. A *saddle point* is a stationary point x such that the values of x to the left of the stationary point produce smaller (or larger) values of $y(x)$ and those to the right larger (or smaller) values of $y(x)$. In mathematical terms, the second derivative $y''(x)$ is equal to zero at a saddle point. Table 1.1 summarizes the tests of stationary points of the function of a single variable $y(x)$.

APPROACHES TO OPTIMIZATION

As stated earlier, three schemes by which optimization procedures can be classified are (1) the number of independent variables x_i, $i=1,\ldots,n$, (2) whether *derivative methods* can be employed or *numerical methods* must be applied, and (3) whether the problem is constrained or unconstrained. We said that the third scheme usually implied another, whether a single response or multiple responses are measured. These and other considerations give rise to several different approaches to optimization. Although the complete development of these techniques is described in Chapter 4, some introductory remarks about them will afford a clearer perspective from which to consider how they might be incorporated into experimentation.

Analytical Methods

Analytical methods are those based on the concept of the *derivative* from calculus. We have already seen some of the basic concepts relating to these in our previous discussion. Our approach to the optimization of a function $y(x)$ of a single variable is (1) to equate the derivative $y'(x)$ to zero as in (1.2) and solve for the stationary point x, and (2) to determine the kind of stationary point at x by evaluating the second derivative $y''(x)$. For multidimensional functions $y(X)$, the equivalent optimization procedure involves (1) equating the gradient vector $\nabla y(x)$ to zero and solving for a stationary point, and (2) evaluating the Hessian matrix $H(X)$ to determine the nature of the stationary point X. Of course, the gradient vector $\nabla y(X)$ is the n-vector of first partial derivatives $\partial y/\partial x_i$, $i=1,\ldots,n$, and the Hessian matrix $H(X)$ is the n-square matrix of second partial derivatives $\partial^2 y/\partial x_{ij}^2$, $i=1,\ldots,n$, $j=1,\ldots,n$. These procedures are fully described in Chapter 4.

Often we are confronted with the optimization of a function whose independent variables must satisfy one or more *constraints*. This problem leads into the domain of the Lagrangian function, in which each of the constraining function $g_j(X)$, $j=1,\ldots,m$ is associated with a *Lagrange multiplier* λ_j and single function is formed which must then be optimized relative to both the independent variables x_i, $i=1,\ldots,n$ and the Lagrange multipliers λ_j, $j=1,\ldots,m$. The solution procedures for such problems also entail the use of classical optimization. These procedures are discussed in Chapter 4.

Numerical Methods

A variety of numerical techniques are available for optimizing both unidimensional and multidimensional functions. Certain of these methods borrow from classical theory to the extent that they involve functions that are continuous and differentiable. These methods, known as *gradient techniques*, require the computation of certain partial derivatives. They are numerical in that we typically terminate computation at an "approximately" optimal solution, one at which the optimality criterion is not satisfied exactly. We shall exploit these procedures in two ways: (1) by experimentally estimating the gradient vector $\nabla y(X)$ using so-called *first-order response surface methods*, and (2) by applying the gradient procedures computationally to quadratic models obtained by employing least-squares multiple regression with experimental data from a *second-order response surface design*.

Other methods, called *direct search techniques*, are applied either computationally or experimentally in situations in which few, if any, assumptions are made about the function being estimated. Certain of these techniques can be applied to functions that are not even continuous. One standard assumption that does apply to these techniques is the unimodality of the function, but we will often attempt to apply them to each modal region of the function, if indeed we can identify such regions.

Most of the numerical methods described in this book are *sequential* search procedures. That is, we perform a set of computations or experiments at one or more X values, and determine the next set of X values that will be evaluated. For the most part, the numerical methods applied here involve applying a set of steps called an *algorithm*. We shall refer to an algorithm as a stepwise procedure for optimization. An algorithm is such that it applies to an entire class of problems, rather than to only one specific problem.

Search techniques are classified as either unidimensional or multidimensional, depending on the number of independent variables x_i, $i=1,\ldots,n$. They are basically structured for the optimization of a single response y, but as we shall see in Chapter 5 that they can be easily adapted to handle several responses, y_j, $j=1,\ldots,m$.

Some of the more popular unidimensional search techniques are exhaustive search, dichotomous search, Fibonacci search, and golden section search. Exhaustive search simply evaluates all possible points in an interval $a \leqslant x \leqslant b$; hence, it is not very efficient and is not usually a worthwhile experimental procedure. It is typically applied to situations in which the response function is multimodal. The other unidimensional search techniques are both computationally and experimentally efficient, but require that the response function be unimodal. They share the feature that one or two trials are evaluated and an unfavorable interval (one in which, due to the assumption of unimodality of the function $y(x)$, the optimal solution x^* *cannot* occur) is eliminated. The search continues in this way until the remaining interval is sufficiently small, and the best value of x obtained is accepted as x^*. Beveridge and Schechter [3], Wilde and Beightler [28], and Gottfried and Weisman [14] all give detailed treatments of the unidimensional search procedures. We shall examine several of them in Chapter 4.

The multidimensional search procedures are of two types: (1) *gradient methods*, which are based on the gradient vector $\nabla y(X)$, evaluated at the current search point X^k. This direction is normal to the tangent of the response contour at X^k. Various procedures, including unidimensional searches, are then employed to establish the *step length* along the gradient direction. This two-phase procedure is repeated until a near-stationary point is obtained. Gradient search is commonly called the "method of steepest ascent."

Direct search procedures determine the search direction by evaluating the objective function $y(X)$ at several points rather than calculating derivatives. One of the most popular direct search procedures is *pattern search*, developed by Hooke and Jeeves [18]. Pattern search makes a set of *exploratory moves* to establish a favorable direction, and then accelerates along this direction. Pattern search exploits successful directions with large moves, but shrinks and adjusts when turning maneuvers are needed. Hence, it has good ridge-following properties. It is computationally efficient, but uses too many function evaluations to be a successful experimental procedure. The *sequential simplex* procedure by Spendley et al. [26] is very useful in an experimental environment. It starts by evaluating $n+1$ points located at the vertices of an n simplex, discards the worst point, replaces it with its "mirror image" point, and continues in this fashion until it surrounds an optimum. Sequential simplex search progresses laboriously, however, in the presence of constraints. The *complex search* procedure by Box [8] remedies this shortcoming, however, and provides an experimentally efficient multidimensional search procedure. These and other useful search procedures are described in Beveridge and Schechter [3] and Wilde and Beightler [28]. They are discussed in Chapter 4.

Mathematical Programming

The computational procedures by which we systematically seek optimal solutions to multivariable problems are generally called *mathematical programming*. We shall use this term to describe a variety of optimization problems involving several *independent variables* x_i, $i=1,\ldots,n$ and several dependent variables or *criteria* y_j, $j=1,\ldots,m$. These problems can be classified according to several schemes, including the following:

1. Linear versus nonlinear criterion functions $y_j(X)$.
2. Constrained versus multiple-objective problem formulations.

We must also include the dichotomy between computational and experimental procedures.

Linear Functions Two well-known classes of optimization techniques apply to problems in which all functions $y_j(X)$ are linear functions of the independent (or *decision*) variables x_i, $i=1,\dots,n$:

1. Linear programming.
2. Goal programming.

Linear programming was first developed by Leontief over 50 years ago, but widespread use followed the development in 1947 of the *simplex algorithm* by Dantzig [11]. This algorithm, which has been extensively refined since its initial development, allows for efficient computer solution and postoptimal analysis of a problem having the following basic structure:

$$\text{Maximize (or minimize) } y_1 = \sum_{i=1}^{n} c_i x_i$$

subject to

$$y_j = \sum_{i=1}^{n} a_{ij} x_i \begin{Bmatrix} \leqslant \\ = \\ \geqslant \end{Bmatrix} b_j \qquad j=2,\dots,m \tag{1.9}$$

and the condition that $x_i \geqslant 0$, $i=1,\dots,n$. Each of the m functions is linear.

In the linear programming formulation expressed in (1.9), we isolate a *single, linear objective function* $y_1(X)$, which is to be maximized or minimized, subject to specifications or bounds on the performance of several other linear functions $y_j(X)$, $j=2,\dots,m$. It is sometimes not practical to frame a problem involving a single objective function. In the early 1960's, Charnes and Cooper [10] presented an approach to the solution of linear decision models having more than one objective. Called *goal programming*, this methodology involves multiple-objective formulations wherein preemptive priorities and weightings are associated with the several objectives. Ignizio [19] presents a particularly lucid development of the linear goal programming problem and a computerized computational procedure for its solution.

Nonlinear Functions We have already referred to the Lagrange Multiplier approach to obtaining a solution to a problem involving several functions $y_j(X)$, $j=1,\dots,m$. Several classes of computational procedures exist for approaching *nonlinear programming* problems. One powerful class of techniques rests on solving linearized forms of the original problem. These include the following:

1. Using linear programming to solve linear *approximations* to the original problem.

2. Using extended linear programming techniques for problems in which the only nonlinearities are quadratic (*quadratic programming*).
3. Using *separable programming* to approach problems that can be expressed as the sum of n nonlinear functions, each of a single variable x_i.
4. Using a *cutting plane* procedure in which a series of linear programming problems, each step increasing the number of rows in the simplex matrix, is solved.
5. Using *approximation programming* in which the objective function and all constraints are approximated by their tangent planes at a given starting point X^0, to produce a linear programming problem.

Gottfried and Weisman [14] present a well-organized treatment of these nonlinear programming procedures. Of these linearization methods, the most powerful appears to be that of Griffith and Stewart [15].

A second general class of nonlinear programming techniques are *the methods of feasible directions*. The best known of these include the following:

1. Klingman and Himmelblau's [20] multiple-gradient-summation technique.
2. Zoutendijk's [29] feasible direction methods.
3. Rosen's [24] gradient projection method.

Experience has shown that feasible direction methods are less powerful than linearization techniques, due (1) to their requirement for a feasible starting point X^0 and (2) to their being limited to inequality constraints.

Penalty-function methods form a single, augmented objective function by adding each of the constraints, suitably transformed and multiplied by a penalty factor, to the original objective function and solving a series of unconstrained problems. As the solution nears one or more bounds, the penalty factor becomes small and the value of the transformed constraining function [such as $\ln y_j(X)$] becomes large. Penalty function methods include the following:

1. *Interior-point methods*, such as those by Carroll [9] and Fiacco and McCormick [13], which originate within the feasible region and approach the bounds from inside.
2. *Exterior-point methods* which are initiated outside the feasible region, and approach a solution at one or more bounds.

Penalty-function techniques are perhaps the most powerful of the nonlinear programming techniques.

Geometric programming is a very useful computational technique for solving optimization problems in which the functions $y_j(X)$, $j=1,\ldots,m$ have a very special structure. Consider a function $y(X)$, which is the sum of terms that involve the products of the independent variables x_i, $i=1,\ldots,n$. For example, the surface area of a closed cylindrical tank, expressed as

$$y(X)=\pi r^2+2\pi rh$$

where $x_1=r$ and $x_2=h$, is such a function. Beightler and Phillips [2] present an excellent development of the entire theory of geometric programming, along with numerous applications in engineering design and operations research. Importantly, geometric programming allows the transformation of a nonconvex optimization problem to a convex problem; hence it often affords a "dual" approach to solving a nonconvex problem.

Most of the nonlinear programming techniques mentioned thus far apply to a constrained formulation of the optimization problem. Ignizio [19] describes several computational procedures, including one based on the method of Griffith and Stewart [15], for solving the *nonlinear goal programming* problem in which, not one, but several objectives are considered at once.

There are other formulations to both the linear and nonlinear programming problems that deserve mention here. They are as follows:

1. *Integer programming* in which some or all of the independent variables x_i, $i=1,\ldots,n$ are restricted to the nonnegative integer set. Integer programming applies to the linear, nonlinear, goal and geometric programming problems.
2. *Zero-one programming* in which some or all of the independent variables x_i, $i=1,\ldots,n$ are restricted to zero or one values. A one indicates that the variable is "in solution," whereas a zero signifies that the variable is not in solution.
3. *Dynamic programming*, which is actually a misnomer for what should be termed *recursive optimization*, involves problems that can be considered to be made up of a sequence of stages, with each stage consisting of an input, a decision, a return, and an output. Nemhauser [23] has presented a very readable treatment of dynamic programming techniques.

Chapter 4 presents detailed developments of some of the mathematical programming procedures mentioned here. Others, although not described in this book, afford useful techniques in experimental situations. The next section introduces certain basic concepts relating to the application of selected optimization methods in the experimental process.

OPTIMIZATION THROUGH EXPERIMENTATION

In previous sections we have separately discussed some of the basic concepts associated with experimentation and optimization. However, we have heretofore viewed experimentation as a process by which we perform and statistically analyze physical experiments and optimization chiefly as a mathematical exercise for computing optimal solutions to problems. Although we have often alluded to the experimental evaluation of a function $y(X)$, we have not yet begun to focus on optimization as an intrinsically *experimental* procedure. In this section we begin to form some of the fundamental framework, to be enlarged upon in later chapters, for experimentally determining optimum conditions for a physical process. This is the *raison d'etre* for this book.

Earlier, in Figure 1.1, we considered a "black-box" view of the experimental process. In Figure 1.12 we see a similar view of the interface between optimization and experimentation. The primary function of the *optimization procedure*, having taken the current experimental responses y_j, $j=1,\dots,m$ into account, is to provide the values of the independent or controllable input values x_i, $i=1,\dots,n$ to be introduced into the next set of trials with the *experimental process*. This interactive process is initiated at some starting point X^0, which is often the current operating conditions, and repeated until an optimum solution X^* is obtained. The experimenter is an integral part of this interactive process. The premise of this book is that an optimization focus, as suggested by Figure 1.12, is a worthwhile approach to experimentation.

Before setting about the process of examining how optimization and experimentation can be combined, let us consider how others have viewed this combination. Bartee [1] takes the approach that the methodology for designing experiments can itself serve as an optimization technique, since, by its proper use, cost and effort can be minimized and effective results can be maximized. The principal thrust of Bartee's work is to provide the

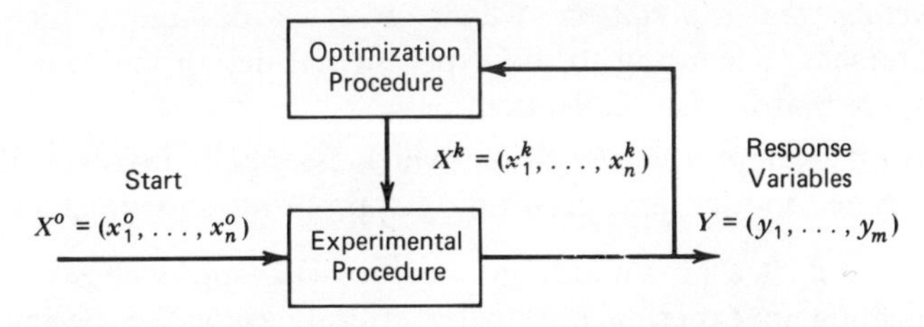

Figure 1.12. Interface between optimization and experimentation.

fundamental concepts that are necessary to accomplish effective planning, design, and analysis of engineering experiments, so that the use of experimental equipment and resources may be optimized. Thus, Bartee is primarily concerned with the optimization of the experimental process. He advocates a highly structured methodology for experimentation following a three-phase approach that involves eight steps, as follows:

Analysis Phase

1. *Formulating the experimental problem*, including establishing the *need* for the investigation and analyzing that need.
2. *Analysis of the experimental mechanism*, consisting of measurement of the physical variables, selecting the dependent or "response" variables, and selecting the independent or control variables.

Synthesis Phase

3. *Designing the structural model*, which involves selecting the number of *factors* (control variables) and the levels at which they can be appropriately measured.
4. *Designing the functional model*, which includes selecting the precise *levels* of the independent variables at which response measurements will be obtained.
5. *Designing the analytical model*, which includes hypothesizing a mathematical model that relates the dependent variable or response to the independent or control variables.
6. *Designing the experimental model*, which determines the number of times the functional model is repeated to provide the desired precision in estimating the response.

Evaluation Phase

7. *Conducting the experiment*, which involves obtaining the physical measurements according to the experimental design laid out in steps 3 through 6, that is, data collection.
8. *Deriving a solution from the model*, which essentially involves fitting any hypothesized models and performing appropriate statistical analyses.

In contrast to Bartee's methodology, Fedorov [12] approaches the problem of optimization in experimentation by offering *optimal* experimental designs. Fedorov's approach is to establish a *criterion function* relating some property of the experimental design to the allocation of its design points across the experimental region and to choose the design points to obtain the optimum value of the criterion function. His presentation includes extensive mathematical developments based on the theorems of mathematical statistics and experimental design. Thus, as compared to Bartee's procedural emphasis, Fedorov offers a theoretical approach to optimally designed experiments.

The presentation in this book differs substantially from either Bartee's work [1] or Fedorov's [12]. This book combines the theory and practice of optimization with that of statistical experimental design to form procedures for optimizing a physical process via experimentation. Thus it concentrates more on achieving an optimal system or process rather than optimizing the experimental program (as in Bartee [1]) or the experimental design (as in Fedorov [12]). Some of the procedures presented in this book are closely akin to those described by G. E. P. Box [5–7] in the context of *response surface methodology* and *evolutionary operation*. Each of these procedures attempts to model the unknown functional relationship $\eta(X)$ between the response y and the n-vector of independent variables, x_i, $i=1,\dots,n$ via designed experiments. Other techniques presented here represent the application of search techniques, including the *Complex* method by M. J. Box [8], the *pattern search* technique by Hooke and Jeeves [18], and the *sequential simplex* procedure by Spendley et al. [26], to the optimization of systems via experimentation. This book also describes how the methods of *mathematical programming* such as linear programming, nonlinear programming, and goal programming, can be employed with response surface models estimated via experimentation to find "optimal" solutions. Of course, the word "optimal" will not apply in the classical sense due to the presence of random error in the experimental milieu.

In this book we shall consider several possible approaches to *formulating* the experimentation/optimization problem, as well as various procedures by which the experimentation/optimization process is conducted. The basic formulations that we will consider are as follows:

1. The optimization of a single response function $y(X)$.
2. A constrained optimization formulation such as represented by relations (1.3) through (1.5).
3. A multiobjective optimization in which we form a prioritized function $W(y_j(X), j=1,\dots,m)$ and seek the best results for that function.

We shall treat three basic optimization approaches that apply to any one of the problem formulations above. These are as follows:

1. *Direct search procedures* in which an experimental trial is conducted at each point X^k in the search, and in which the m-dimensional response vector is measured.

2. *First-order response surface methods* in which $N \geqslant n+1$ experimental trials are arrayed in a local region, the observations $y_{jl}(X), j=1,\ldots,m, l=1,\ldots,N$ are used to fit m linear models from which the *gradient direction* $\nabla y(X)$ is estimated, and a line search is performed along $\nabla y(X)$ to an optimum point. This procedure is repeated until $\nabla y(X^*)$ is equal to zero.
3. *Second-order response surface methods* in which $K \geqslant (n+1)(n+1)/2$ experimental trials are arrayed over an experimental region $a_i \leqslant x_i \leqslant b_i$, $i=1,\ldots,n$ and the observations $y_{jl}(X), j=1,\ldots,m, l=1,\ldots,K$ used to fit m quadratic models. These models are then employed as known mathematical functions $y_j(X)$, $j=1,\ldots,m$ in a standard optimization procedure. Note that here experimentation and optimization are indeed separate, *sequential* processes.

Chapter 5 is devoted to a detailed description of these combined experimentation/optimization procedures.

OVERVIEW OF THE BOOK

This chapter has presented some introductory remarks concerning experimentation and optimization, and briefly proposed how these methodologies can be combined to provide several rather novel and "experimental" approaches to conducting experimentation. The remainder of the book is devoted to quite basic descriptions of the fundamental concepts needed to understand the methodologies proposed here, to systematic development of the methodologies themselves, and to various examples of their implementation in physical and simulation experiments.

This book is predicated on a statistical approach to experimentation, because statistical methodology is needed to perform an objective analysis of data that are affected by random error. Chapter 2 presents the fundamental statistical concepts that are the underpinnings of experimentation. It reviews the basic ideas in probability, including random experiments, sample spaces, probabilities, events, random variables, and probability distributions. It describes several of the most common discrete and continuous probability distributions that arise in the real world, and which form the basis of experimental data. It examines the basic structure of sampling and sampling distributions. It describes the very important principles of hypothesis testing that allow us to draw conclusions about experimental results. It discusses the basic procedures of regression and analysis of variance as a prelude to Chapter 3. Chapter 2 is certainly not an exhaustive treatment of statistics, but it is our intention that all of the basic statistical concepts needed for this book are presented there. The presentation in Chapter 2 is kept very basic and illustrated with numerous examples.

Chapter 3 presents the fundamental principles involved in the statistical design and analysis of experiments. It concentrates on the developments needed for first-order and second-order response surface design and analysis. The 2^n factorial and n-simplex first-order experimental designs are emphasized for estimating linear approximations of the functions $y(X)$. The central composite designs, based on 2^n factorial designs, and certain extensions of the n-simplex designs are described for use in estimating quadratic functions. Attention is also given in Chapter 3 to designs that are used when specific functional forms, such as exponentials, etc., are to be fitted. Thus, rather than being a "stand alone" treatment of experimental design, Chapter 3 concentrates on the designs that are most often used in connection with optimization. The techniques known as *response surface methodology* are discussed in Chapter 3.

Chapter 4 discusses the most important search methods and optimization procedures from the standpoint of their usefulness in the experimental, rather than purely computational, domain. Emphasis is given to the golden section technique for unidimensional search, Box's complex search method for multidimensional search, gradient search, and gradient-based nonlinear programming procedures. Whereas many other optimization techniques would deserve discussion if computational approaches to optimization were the focus of this book, its concentration on experimental technology distills the array of pertinent optimization methods to these few.

Chapter 5 develops procedures by which one would combine optimization and experimentation to seek the optimum conditions in either a physical or a simulation environment. As stated previously, the various ways of formulating an optimization problem are considered, as are the most pertinent and successful optimization procedures. The experimentation/optimization procedures are described in an algorithmic fashion in such a way that they will apply to any one of the formulations that the experimenter might adopt.

Chapter 6 presents several examples illustrating the application of optimization in physical experimentation. For instance, one example dealing with the optimization of the cutting fluid jet pressure in machining illustrates how to optimize a function $y(x)$ of a single variable x using golden section search. This procedure is compared with another method involving polynomial regression with the same example problem. Another example, dealing with maximizing the yield from a chemical process, compares various procedures for optimizing a single response as a function of several variables. Other examples treat problems involving multiple responses. Each of the example problems discussed in Chapter 6 are taken from realistic industrial technologies.

Chapter 7 describes the application of selected optimization methods to experimentation with computer simulation models. Discussion is given to the special characteristics of computer simulation as an experimental tool. Techniques are described by which optimization techniques can be interfaced with computer simulation models to obtain estimated optimal responses. Examples of optimization in simulation are presented. These include the optimization of stock control level, reorder point, and time between reviews in a simulation of a stochastic inventory system, and the optimal selection of fire control procedures in a tank duel.

REFERENCES

1. Bartee, E. M., *Engineering Experimental Design Fundamentals*, Prentice-Hall, Englewood Cliffs, NJ, 1960.
2. Beightler, C. S., and D. T. Phillips, *Applied Geometric Programming*, Wiley, New York, 1976.
3. Beveridge, G. S. G., and R. S. Schechter, *Optimization: Theory and Practice*, McGraw-Hill, New York, 1970.
4. Biles, W. E., "A Response Surface Method for Experimental Optimization of Multi-Response Processes," *Industrial and Engineering Chemistry: Process Design and Development*, vol. 1, no. 2, April 1975, pp. 152–158.
5. Box, G. E. P., "The Exploration and Expolitation of Response Surfaces," *Biometrics*, vol. 10, 1954, pp. 16–61.
6. Box, G. E. P., and K. P. Wilson, "On the Experimental Attainment of Optimum Conditions," *Journal of the Royal Statistical Society*, Series B, vol. 13, 1951, pp. 1–45.
7. Box, G. E. P., and N. R. Draper, *Evolutionary Operation*, Wiley, New York, 1969.
8. Box, M. J., "A New Method of Constrained Optimization and a Comparison with Other Methods," *Computer Journal*, vol. 8, 1965, pp. 42–52.
9. Carroll, G. W., "The Created Response Surface Technique for Optimizing Nonlinear Restrained Systems," *Operations Research*, vol. 9, 1961, pp. 169–184.
10. Charnes, A., and W. W. Cooper, *Management Models and Industrial Applications of Linear Programming*, Vols. I and II, Wiley, New York, 1961.
11. Dantzig, G. B. *Linear Programming and Extensions*, Princeton University Press, Princeton, NJ, 1963.
12. Fedorov, V. ., *Theory of Optimal Experiments*, Academic Press, New York, 1972.
13. Fiacco, A. V., and G. P. McCormick, "Computational Algorithm for Sequential Unconstrained Minimization Technique for Non-Linear Programming," *Management Science*, vol. 10, 1964, pp. 601–617.
14. Gottfried, B. S., and J. Weisman, *Introduction to Optimization Theory*, Prentice-Hall, Englewood Cliffs, NJ, 1973.

15. Griffith, R. E., and R. A. Stewart, "A Nonlinear Programming Technique for the Optimization of Continuous Processing Systems," *Management Science*, vol. 7, 1961, pp. 379–392.
16. Hicks, C. R., *Fundamental Concepts in the Design of Experiments*, Holt, Rinehart, and Winston, New York, 1961.
17. Himmelblau, D. M., "Process Optimization by Search Techniques," *Industrial and Engineering Chemistry: Process Design and Development*, vol. 2, 1963, pp. 296–300.
18. Hooke, R., and T. A. Jeeves, "Direct Search Solution of Numerical and Statistical Problems," *Journal of the Association of Computing Machines*, vol. 8, 1961, pp. 212–229.
19. Ignizio, J. P., *Goal Programming and Extensions*, Heath, Lexington, MA, 1976.
20. Klingman, W. R., and D. M. Himmelblau, "Nonlinear Programming with the Aid of Multiple-Gradient Summation Technique," *Journal of the Association of Computing Machines*, vol. 11, 1964, pp. 400–415.
21. Montgomery, D. C., *Design and Analysis of Experiments*, Wiley, New York, 1976.
22. Myers, R. H., *Response Surface Methodology*, Allyn and Bacon, Boston, MA, 1971.
23. Nemhauser, G. L., *Introduction to Dynamic Programming*, Wiley, New York, 1966.
24. Rosen, J. B., "The Gradient Projection Method for Nonlinear Programming, Part II, Nonlinear Constraints," *Journal of the Society of Industrial and Applied Mathematics*, vol. 9, 1961, pp. 514–532.
25. Shannon, R. E., *Systems Simulation: The Art and Science*, Prentice-Hall, Englewood Cliffs, NJ, 1975.
26. Spendley, W., G. R. Hext, and F. R. Himsworth, "Sequential Application of Simplex Designs in Optimization and Evolutionary Operation," *Technometrics*, vol. 4, 1962, pp. 441–461.
27. Wilde, D. J., *Optimum Seeking Methods*, Prentice-Hall, Englewood Cliffs, NJ, 1964.
28. Wilde, D. J., and C. S. Beightler, *Foundations of Optimization*, Prentice-Hall, Englewood Cliffs, NJ, 1967.
29. Zoutendiijk, G., *Methods of Feasible Directions*, Elsevier Publishing Company, Amsterdam, and Van Nostrand Company, Princeton, NJ, 1960.

Chapter 2

Fundamental Statistical Concepts

The purpose of this chapter is twofold: (1) to present certain fundamental concepts of statistical theory that are important to the understanding of experimental methods; and (2) to describe the statistical procedures and techniques that are essential to a sound, cost-effective approach to designing, conducting, and analyzing experiments.

In this chapter we consider the basic concepts of probability, the science upon which statistics is based. The fundamental discrete and continuous probability distributions are discussed. The theory of sampling and the key sampling distributions are presented to provide the necessary background for such topics as statistical estimation, hypothesis testing, regression and correlation, and analysis of variance.

The presentation in this chapter foresakes mathematical rigor in favor of a procedural approach to the fundamental topics of statistics. Simple examples are given to illustrate each of the statistical procedures outlined in this chapter. Since a single chapter cannot possibly provide an exhaustive treatment of the science of statistics, references are cited for some of the most authoritative texts on the subjects presented here.

BASIC PROBABILITY

One means of classifying experiments is according to a *deterministic* versus *probabilistic* scheme. In deterministic experiments, specifying the values of the independent variables x_i, $i=1,\ldots,n$ yields a precisely predictable outcome y. For example, specifying the current I and resistance R in a simple electrical circuit yields a unique voltage E according to Ohm's Law $E=IR$. Thus Ohm's Law is a deterministic model. Although we readily accept this basic law of physics in theory, we must realize that the experiment is only deterministic if the uncontrollable factors z_k, $k=1,\ldots,p$ in the experimental setting produce a negligible effect. If these factors are not negligible, we could observe that repetitive measurements of E at constant values of I and R produce a *set* of values. Such a collection of repetitive measurements is called *distribution*. A real-world situation is said to be probabilistic when a distribution of values for the dependent variable y occurs with repetitive trials of the experiment at constant values of the independent variables x_i, $i=1,\ldots,n$.

The science that enables us to objectively analyze the results of probabilistic experiments is called *statistics*. Statistics is based on the laws of *probability*, and any basic study of statistics requires that we have a fundamental understanding of probability. This section examines the fundamental concepts of probability.

Random Experiment

Fundamental to the basic concepts of probability is the notion of a *random experiment*. A random experiment is a process that, in a given trial, yields any one of several possible outcomes. Chance completely determines the outcome that actually occurs on any given trial; we cannot predict it with certainty before the trial. Common examples of random experiments include tossing a coin and rolling a die. Examples of random experiments in industrial technology include inspecting a manufactured assembly for defects, observing the status of several channels of communication at a specific instant in time, and carrying on a chemical production process. In each of these instances, we cannot state with certainty what the outcome of a trial will be until we have observed it.

Sample Space

Although we cannot predict with certainty the precise outcome of a single trial of a random experiment, we can list all possible *outcomes*. This set of n outcomes, called the *sample space*, is written as

$$S=\{a_i\}=\{a_1,a_2,\ldots,a_n\} \tag{2.1}$$

where a_i, $i=1,\ldots,n$ represents the ith possible outcome of the random experiment. For the toss of a coin, the outcomes are a head H or a tail T. The sample space for this simple experiment is

$$S=\{H,T\}$$

Rolling a single die produces an outcome in which the die shows 1, 2, 3, 4, 5, or 6 spots on the top face. The sample space for this experiment is

$$S=\{1,2,3,4,5,6\}$$

The inspection of a manufactured assembly might be regarded as having just two possible outcomes, defective D or nondefective N. The sample space for the inspection of a single assembly is

$$S=\{D,N\}$$

For the random experiment involving multiple channels of communication, consider the situation in which just three channels are available. If a channel is not in use it is assigned the value 0, whereas a 1 indicates that the channel is in operation. The sample space for this random experiment is therefore

$$S=\{(0,0,0),(1,0,0),(0,1,0),(0,0,1),(1,1,0),(1,0,1),(0,1,1),(1,1,1)\}$$

This example points out that the outcomes in the sample space show all possible ways that the channels can be either idle or operational. For instance, the outcome $(0,1,0)$ shows that only channel 2 is operational at a given time and that channels 1 and 3 are idle. This is physically different from the outcomes $(1,0,0)$ and $(0,0,1)$. This sample space has two important properties. The outcomes are mutually exclusive, and they are exhaustive. *Mutually exclusive* simply means that the outcomes are distinct; if one occurs, another cannot occur. *Exhaustive* simply means the only possible outcomes of the experiment are those delineated in the sample space S.

Consider the operation of an electrical circuit for a fixed period of time T. If the circuit survives the entire period T, we denote the outcome by a 1; if it fails at some time $t \leqslant T$, we represent the outcome by a 0. This sample space is simply

$$S=\{0,1\}$$

Now let's take a more complicated view. Suppose that the circuit operates for some random time $t \leqslant T$. Then the sample space contains an infinite number of possible values of t. We would represent the sample space for this experiment as

$$S=\{t_i : 0 \leqslant t_i \leqslant T\} \tag{2.2a}$$

As another example, consider the yield (%) from a chemical process. This random experiment has the sample space.

$$S=\{y_i : 0 \leqslant y_i \leqslant Y\} \tag{2.2b}$$

where Y is the maximum theoretical yield for the process. We say that (2.1) is a *discrete* sample space, and that (2.2a) and (2.2b) are *continuous* sample spaces.

Probability

For a given trial of a random experiment, one does not know precisely which outcome in the same space will result. But repeating the experiment many times lets us estimate the frequency of a given outcome. That is, if we repeat the random experiment M times and the outcome a_i results m_i times, the outcome a_i's frequency of occurrence is m_i. If we divide the frequency m_i by the total number of trials M, we get the probability p_i of outcome a_i,

$$p_i = \frac{m_i}{M} \tag{2.3}$$

The sum of the frequencies of the n outcomes must equal the total number of trials of the experiment; that is,

$$m_1 + m_2 + \cdots + m_i + \cdots + m_n = M \tag{2.4}$$

Dividing both sides of (2.4) by the total number of trials M shows that

$$p_1 + p_2 + \cdots + p_i + \cdots + p_n = 1 \tag{2.5}$$

Associated with each outcome in the sample space is a probability, so that we can write the set of probabilities as

$$P = \{p_1, p_2, \ldots, p_n\} \tag{2.6}$$

If the outcome a_i never occurred in M trials, its probability would be

$$p_i = \frac{0}{M} = 0$$

We would say that outcome a_i is *impossible*. Similarly, if the outcome a_i occurred with each and every trial, its probability would be

$$p_i = \frac{M}{M} = 1$$

and outcome a_i would be called *certain*. These observations enable us to state two important laws of probability:

$$0 \leqslant p_i \leqslant 1 \qquad i = 1, \ldots, n \tag{2.7}$$

$$\sum_{i=1}^{n} p_i = 1 \tag{2.8}$$

Let's see what these laws mean. In the toss of a fair coin, for example, the head is just as likely to occur as the tail, so that the probabilities are

$$p(H) = \tfrac{1}{2} \qquad p(T) = \tfrac{1}{2} \qquad P = \{\tfrac{1}{2}, \tfrac{1}{2}\}$$

In the roll of a fair die, each of the six outcomes is equally likely to occur, so that the set of probabilities is

$$P = \left\{ \frac{1}{6}, \frac{1}{6}, \frac{1}{6}, \frac{1}{6}, \frac{1}{6}, \frac{1}{6} \right\}$$

These two examples illustrate the motion of *equally likely outcomes*. This concept is simple: Each of the n outcomes a_i in the sample space S has just as good a chance of occurring as any other outcome. The probability of the outcome a_i is thus simply $1/n$. Not all random experiments produce such convenient results. The inspection of a manufactured part, for instance, will usually have a much lower probability for the D outcome than for the N outcome.

The concept of probability with a continuous sample space is somewhat more complicated, since we can no longer define a finite set of outcomes. Indeed, there is an infinite number of outcomes y_i in the continuum of values, say $Y_{\min} \leqslant y_i \leqslant Y_{\max}$; hence, the value of the denominator M in expression (2.3) is infinity, so that the probability p_i is zero. Instead of the probability of a specific value y_i, we refer to the probability of an outcome of the experiment falling within some range of values $r \leqslant y_i \leqslant s$. A more precise statement of this concept will be presented after the notions of an *event* and a *probability function* have been developed.

Events

Another fundamental concept in probability is that of an *event*. An event is a collection of those outcomes in the sample space that have an attribute or characteristic of interest. To illustrate this, let us consider the problem of rolling a pair of dice. Each die can show 1, 2, 3, 4, 5, or 6 spots on the top face. Hence the pair of dice produces the following outcomes:

$$\begin{aligned} S = \{ &(1,1),(1,2),(1,3),(1,4),(1,5),(1,6),(2,1),(2,2),(2,3),(2,4),\\ &(2,5),(2,6),(3,1),(3,2),(3,3),(3,4),(3,5),(3,6),(4,1),(4,2),\\ &(4,3),(4,4),(4,5),(4,6),(5,1),(5,2),(5,3),(5,4),(5,5),(5,6),\\ &(6,1),(6,2),(6,3),(6,4),(6,5),(6,6)\} \end{aligned}$$

Structuring the sample space in this fashion implies that we can distinguish the outcome (1,3), say, from (3,1). It is simple to make this distinction if we have one red die and one green die, for example. Then these outcomes are (Red 1, Green 3), and (Red 3, Green 1). Thus there are 36 distinct outcomes in this sample space, and each outcome is equally likely to occur. Hence the probability of each outcome is $1/36$.

Suppose we are interested in the event R that exactly one 3 appears. Collecting the outcomes in the sample space that produce exactly one 3, we have the smaller space

$$R=\{(3,1),(3,2),(3,4),(3,5),(3,6),(1,3),(2,3),(4,3),(5,3),(6,3)\}$$

Ten outcomes yield the event R. How do we find the probability of the event R? Just as we aggregate the set of outcomes that define the event R, we must similarly aggregate the probabilities of these outcomes to determine the probability of the event, that is, $P(R)$. Since 10 of the 36 equally likely outcomes yield event R, we quickly establish that

$$\begin{aligned}P(R)&=p(3,1)+p(3,2)+p(3,4)+p(3,5)+p(3,6)+p(1,3)+p(2,3)\\&\quad+p(4,3)+p(5,3)+p(6,3)\\&=\frac{1}{36}+\frac{1}{36}+\frac{1}{36}+\frac{1}{36}+\frac{1}{36}+\frac{1}{36}+\frac{1}{36}+\frac{1}{36}+\frac{1}{36}+\frac{1}{36}\\&=\frac{10}{36}\end{aligned}$$

Thus we sum the probabilities for those outcomes in the sample space S that yield the event R. That is, if r of the n outcomes in the sample space S yield the event R, then the probability of event R is

$$P(R)=\sum_{j=1}^{r} p_j \tag{2.9}$$

where we sum over all outcomes in the event R. This notation suggests that the first r outcomes in the sample S produce the event R, but we can reorder the outcomes in the sample space without changing the nature of the space.

For a continuous sample space, we define an event as the occurrence of an outcome within some interval of interest, $r \leqslant y_i \leqslant s$. For example, we might be interested in the event that an electrical circuit survives at most 200 hours; hence, the event is $0 \leqslant t \leqslant 200$. Or we might be interested in the event that process yield y will exceed 55%, so that the event is $55<y<Y_{\max}$, where $Y_{\max}>55$ is the maximum theoretical yield. To compute the probability of such an event, we must know the *probability function* $f(t)$ or $f(y)$. Then the probabilities of the events are

$$P(0 \leqslant t \leqslant 200)=\int_0^{200} f(t)\,dt \tag{2.10a}$$

and

$$P(y>55)=\int_{55}^{Y_{\max}} f(y)\,dy \tag{2.10b}$$

Operations With Events

Considers two events B and C. Two operations with events are of fundamental importance in the study of probability: union and intersection. The symbol $B+C$ denotes the *union* operation, shown in Figure 2.1(a); that is, at least one of the two events B *or* C occurs. The symbol BC represents the *intersection* operation, depicted in Figure 2.1(b); that is, both events B *and* C occur. Suppose that b of the n outcomes in the sample space S form the event B, c outcomes form C, and d outcomes yield both events B *and* C. Then the probability of the intersection BC is

$$P(BC)=\sum_{k=1}^{d} p_k \tag{2.11}$$

The probability of the union $B+C$ is

$$P(B+C)=\sum_{i=1}^{b} p_i+\sum_{j=1}^{c} p_j-\sum_{k=1}^{d} p_k \tag{2.12}$$

Since d of the b outcomes in event B are also in event C, the probabilities of these d outcomes appear in both the first and second terms in (2.12). Hence we subtract them out in the third term. This leads the additive law of probability; that is,

$$P(B+C)=P(B)+P(C)-P(BC) \tag{2.13}$$

Now suppose that event C cannot occur if event B does, or vice versa. Then we say that events B and C are *mutually exclusive*. That is, the probability $P(BC)$ is zero, and (2.13) reduces to

$$P(B+C)=P(B)+P(C) \tag{2.14}$$

As an example of the additive law of probability, consider a manufactured lot containing 1000 parts. Of this lot, 84 parts have a type B defect, 53 parts have a type C defect, and 13 parts have both types of defects.

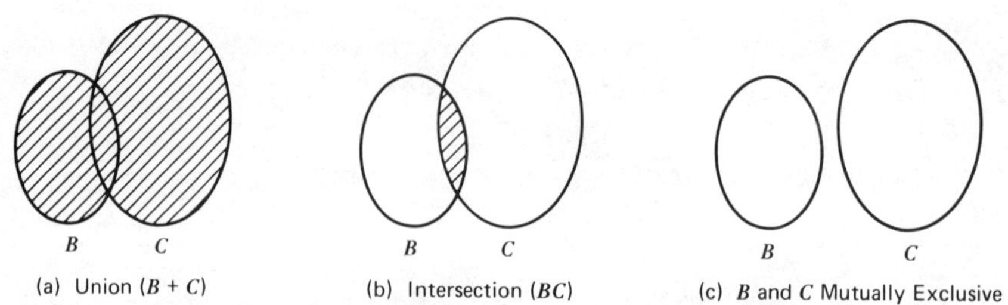

Figure 2.1. Operations with events—a Venn diagram approach.

Therefore the probabilities of these events are

$$P(B)=\frac{84}{1000} \qquad P(C)=\frac{53}{1000} \qquad P(BC)=\frac{13}{1000}$$

The probability of a defect is then the probability of a type B defect or type C defect; that is, the probability of the union of B and C:

$$P(B+C)=\frac{84}{1000}+\frac{53}{1000}-\frac{13}{1000}=\frac{124}{1000}=0.124$$

As an example of mutually exclusive events, consider a machine that automatically fills boxes with 28 g of breakfast cereal. According to manufacturing specifications, the range of acceptable net weights of cereal per box is from 27.0 to 29.0 g. Suppose that the probability of an underweight box (less than 27.0 g) of cereal is 1/100, and that of an overweight box (more than 29.0 g) is 2/100. Representing these as events B and C, respectively, we see that

$$P(B+C)=\frac{1}{100}+\frac{2}{100}=\frac{3}{100}$$

It is clear that events B and C are mutually exclusive, because an underweight box cannot also be an overweight box. Observe that, in this example, $P(B+C)$ is the probability that a given box of cereal does not meet manufacturing specifications.

Frequently we are interested in the composite event BC. If the occurrence of event B in no way affects the occurrence of event C, and vice versa, we say that they are *independent*. For independent events B and C, the probability of the composite event is simply

$$P(BC)=P(B)\cdot P(C) \tag{2.15}$$

We can generalize this relationship to any number of events $B,C,D,E,\ldots$

$$P(BCDE\ldots)=P(B)\cdot P(C)\cdot P(D)\cdot P(E)\cdots \tag{2.16}$$

For example, consider the receipt of a transmitted signal from each of two satellites. The probability of receiving the signal from satellite B is 0.9, and from satellite C it is 0.8. These are independent events, so that the probability of receiving both signals is

$$P(BC)=(0.9)(0.8)=0.72$$

Conditional Probability

One sometimes faces the problem of determining the probability that an event C will occur, given that another event B occurs. This situation implies that the occurrence of event B somehow affects the occurrence of event C; that is, events B and C are dependent,

$$P(BC) = P(B) \cdot P(C|B) \tag{2.17}$$

The symbol $P(C|B)$ is read "the conditional probability of event C, given that event B occurs." By rearranging, we can determine this conditional probability $P(C|B)$ from the equation

$$P(C|B) = \frac{P(BC)}{P(B)} \tag{2.18}$$

for $P(B) \neq 0$.

We can extend the principle of conditional probability given in (2.18) to establish the probability that an event A, which we know has occurred, might have occurred in a particular manner. That is, suppose that a random experiment has n mutually exclusive outcomes $A_1, A_2, \ldots, A_n$, and that there is an event B such that $P(B) \neq 0$. Then the conditional probability of outcome A_i given that event B has occurred is

$$P(A_i|B) = \frac{P(B|A_i) \cdot P(A_i)}{\sum_{j=1}^{n} P(B|A_j) \cdot P(A_j)} \tag{2.19}$$

Equation (2.19), called *Bayes's theorem*, is a very useful tool in experimental analysis. For example, suppose that a part is produced on three machines: A, B, and C. Machine A yields 40% of the output, machine B 35%, and machine C 25%. The percentage of defective output from the three machines is 8%, 5%, and 3%, respectively. We select one part at random from the total output of the three machines and find that it is defective. What is the probability that it was produced on machine A? Let D denote a defective part; let A denote the event that the part came from machine A, B from machine B, and C from machine C. Then the probabilities of interest are

$$P(A) = 0.40 \qquad P(D|A) = 0.08$$

$$P(B) = 0.35 \qquad P(D|B) = 0.05$$

$$P(C) = 0.25 \qquad P(D|C) = 0.03$$

The statement $P(A) = 0.40$ says that the probability that a given part is produced on machine A is 0.4, since machine A produces 40% of all parts. The statement $P(D|A) = 0.08$ says that the conditional probability that a part is defective, given that it was produced on machine A, is 0.08, since 8% of all parts produced on machine A are defective. From (2.19).

$$
\begin{aligned}
P(A|D) &= \frac{P(D|A)\cdot P(A)}{P(D|A)\cdot P(A)+P(D|B)\cdot P(B)+P(D|C)\cdot P(C)} \\
&= \frac{(0.08)(0.40)}{(0.08)(0.40)+(0.05)(0.35)+(0.03)(0.25)} \\
&= \frac{0.0320}{0.0320+0.0175+0.0075} \\
&= \frac{0.0320}{0.0570} \\
&= 0.561
\end{aligned}
$$

This result says that, given a defective part, the probability that it was produced on machine A is 0.561. Similarly, the probabilities that the defective part was produced on machines B and C are found to be 0.307 and 0.132, respectively.

Random Variables

An important concept associated with the notion of the sample space is the *random variable*. It is a real-valued function $g(Y)$ of the outcomes in the sample space S. For the discrete sample space S, the n outcomes produce $k \leqslant n$ possible values of the random variable. Denoting the discrete random variable as Y and its possible values as $y_1, y_2, \ldots, y_k$, we see that one or more outcomes a_i aggregated yield a possible value y_j. Similarly, the probability of value y_j, denoted $f_Y(y_j)$, is the sum of the probabilities for those outcomes a_i that combine to produce the possible value y_j.

For example, consider the system of three channels of communication and a random experiment that consists of checking which channels are in use at randomly selected times. The sample space is

$$S=\{(0,0,0),(1,0,0),(0,1,0),(0,0,1),(1,1,0),(1,0,1),(0,1,1),(1,1,1)\}$$

Suppose the probabilities for this random experiment are

$$P=\{0.25,0.15,0.15,0.15,0.09,0.09,0.09,0.03\}$$

Define the random variable Y as the *number* of channels in use. Table 2.1 shows the possible values of Y, the outcomes in S that produce these values, and the associated probabilities of the values y_j.

Table 2.1. Probabilities of values y_j of random variable Y

Possible value y_j	Outcomes a_i in S that yield value y_j	$f_Y(y_j)=\Sigma p_i$
0	(0,0,0)	0.25
1	(1,0,0),(0,1,0),(0,0,1)	0.45
2	(1,1,0),(1,0,1),(0,1,1)	0.27
3	(1,1,1)	0.03

For the continuous sample space, the definition of a random variable can be considered to be practically the same as that for the outcome; that is, the outcome y_j of a random experiment is typically a value of the random variable of interest, Y. In our two examples, we can refer to the random variables

$$t = \text{time to circuit failure, hours}$$
$$y = \text{process yield, \%}$$

which are in fact the measured outcomes of the experiment. We must remember, however, that the probability of occurrence of specific values of t or y is zero due to (2.10a) and (2.10b).

Probability Distributions

The probabilities $f_Y(y_j)$ for possible values of Y form the *probability distributions* of the discrete random variable Y. We denote this distribution by $f_Y(y)$. Note that these probabilities satisfy the basic laws of probability stated in (2.7) and (2.8); that is

$$0 \leqslant f_Y(y_j) \leqslant 1 \qquad j=1,\ldots,k \tag{2.20}$$

$$\sum_{j=1}^{k} f_Y(y_j) = 1 \tag{2.21}$$

Thus the probability distribution $f_Y(y)$ is often called the *probability function* of the random variable Y. For the discrete random variable, $f_Y(y)$ is usually called the *probability mass function*. Figure 2.2 shows the probability mass function for the example of the communications channels.

In many instances, we are interested in the probability that the random variable Y has a value less than or equal to y. Denoting this probability by $F_Y(y)$, we see that

$$P(Y \leqslant y_j) = F_Y(y_j) = \sum_{l=1}^{j} f_Y(y_l) \tag{2.22}$$

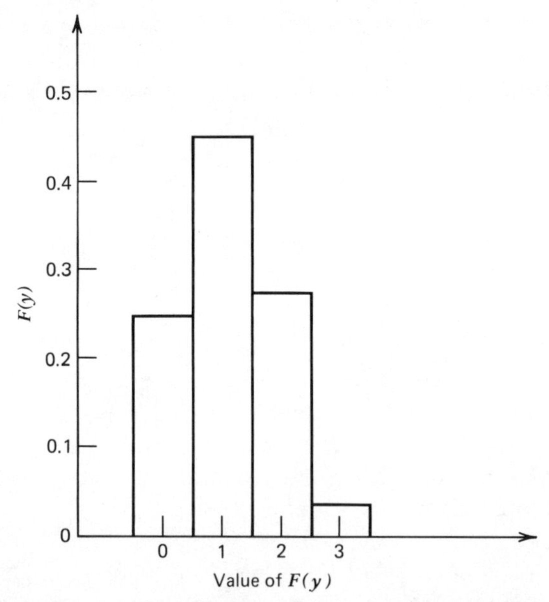

Figure 2.2. Probability mass function of a random variable y.

In our example concerning channels of communication, the probability that two or fewer channels are in use is

$$P(Y \leqslant 2) = F_Y(2) = f_Y(0) + f_Y(1) + f_Y(2)$$
$$= 0.25 + 0.45 + 0.27 = 0.97$$

Often we want to know the probability that the random variable Y has a value greater than y_j; that is, we want to known $P(Y > y_j)$. This is given by

$$P(Y > y_j) = 1 - P(Y \leqslant y_j) = 1 - F_Y(y_j) = 1 - \sum_{l=1}^{j} f_Y(y_l) \qquad (2.23)$$

In our example, the probability that more than one channel is in use is

$$P(Y > 1) = 1 - P(Y \leqslant 1) = 1 - F_Y(1)$$
$$= 1 - [f_Y(0) + f_Y(1)] = 1 - [0.25 + 0.45]$$
$$= 0.30$$

A probability $f_Y(y_j)$ is associated with each possible value y_j of the discrete random variable Y; likewise a cumulative probability $F_Y(y_j)$ associated with each y_j. Thus the cumulative probability function gives the probabilities $F_Y(y_j)$ expressed by (2.22). Table 2.2 shows the cumulative probabilities for the problem about the channels of communication. Figure 2.3

Table 2.2. Cumulative probabilities $F_Y(y_j)$

Possible value y_j	Probability $f_Y(y_k)$	Cumulative probability $F_Y(y_j)$
0	0.25	0.25
1	0.45	0.70
2	0.27	0.97
3	0.03	1.00

shows the cumulative probability function $F_Y(y_j)$ for the random variable Y, the number of communications channels in operation.

For the continuous random variable, we have already seen that the probability of a specific value of a random variable Y is zero. This result can be confirmed by considering the definition of the *cumulative distribution* of the continuous random variable Y; that is,

$$F_Y(s) = P(Y \leqslant s) = \int_{-\infty}^{s} f_Y(y)\,dy \tag{2.24}$$

Here $f_Y(y)$ is the probability function of Y; we call it the *probability density function* in the case of a continuous random variable Y. By definition,

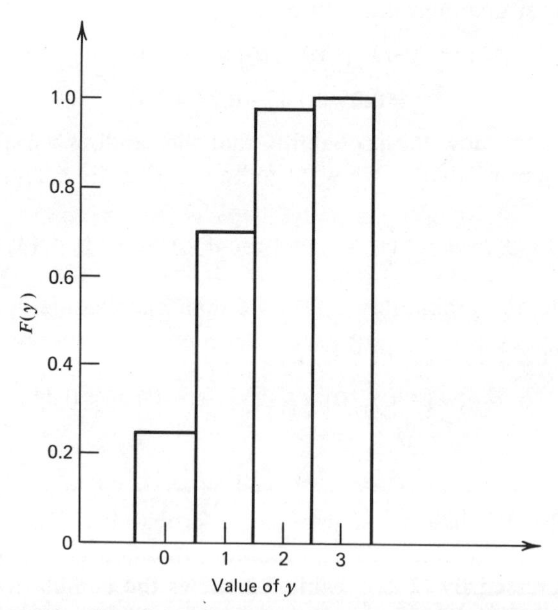

Figure 2.3. Cumulative distribution function of a random variable y.

then,

$$f_Y(y) = \frac{dF_Y(y)}{dy} \tag{2.25}$$

As seen in equations (2.10a) and (2.10b), the probability of an event $(r \leqslant Y \leqslant s)$ for the continuous random variable Y is

$$\begin{aligned} P(r \leqslant Y \leqslant s) &= F_Y(s) - F_Y(r) \\ &= \int_{-\infty}^{s} f_Y(y)\,dy - \int_{-\infty}^{r} f_Y(y)\,dy \\ &= \int_{r}^{s} f_Y(y)\,dy \end{aligned} \tag{2.26}$$

We shall make extensive use of (2.26) in computing probabilities for continuous random variables as described by *continuous probability distributions*. For the most part, the kinds of probability distributions exploited in analyzing the results of experiments are continuous distributions. Since experimentation basically involves *sampling* of values from unknown distributions, the science of sampling and *sampling distributions*, such as the *normal*, *Student's* t, *Chi-Square* and F *distributions* are extremely important in experimental analysis. We shall consider them in detail in a later section.

Expectation

The concept of *expected value* or *expectation* is important in computing certain quantities in a distribution. The expectation of a function $g(Y)$ of the random variable Y is

$$E[g(Y)] = \begin{cases} \sum_{j=1}^{k} g(y_j) \cdot f_Y(y_j) & Y \text{ discrete} \\ \int_{-\infty}^{\infty} g(y) f_Y(y)\,dy & Y \text{ continuous} \end{cases} \tag{2.27}$$

Certain very important results can be obtained by using this concept of expectation. For instance, the average value or *mean* of a random variable Y is simply the expectation of Y, or

$$\mu = E[Y] = \begin{cases} \sum_{j=1}^{k} y_j f_Y(y_j) & Y \text{ discrete} \\ \int_{-\infty}^{\infty} y f_Y(y)\,dy & Y \text{ continuous} \end{cases} \tag{2.28}$$

Another very important characteristic of a distribution of the random variable is its *variance*, found by

$$V(y)=\sigma^2=\begin{cases}\sum_{j=1}^{k}(y_j-\mu)^2 f_Y(y_j) & Y \text{ discrete}\\ \int_{-\infty}^{\infty}(y-\mu)^2 f_Y(y)\,dy & Y \text{ continuous}\end{cases} \tag{2.29}$$

A more convenient form for the variance computation can be obtained by expanding the $(y-\mu)^2$ term to yield

$$\sigma^2=V(y)=E\left[(y-\mu)^2\right]=\begin{cases}\sum_{j=1}^{k}y_j^2 f_Y(y_j)-\mu^2 & Y \text{ discrete}\\ \int_{-\infty}^{\infty}y^2 f_Y(y)\,dy-\mu^2 & Y \text{ continuous}\end{cases} \tag{2.30}$$

The concepts of expectation, mean and variance will find general usefulness throughout this book, and it may be worthwhile to review several elementary results that ensue from them. If Y is a random variable and c a constant, then

1. $E[c]=c$ (2.31)
2. $E[Y]=\mu$ (2.32)
3. $E[cY]=cE[Y]=c\mu$ (2.33)
4. $E[Y_1+Y_2]=E[Y_1]+E[Y_2]=\mu_1+\mu_2$ (2.34)
5. $V[c]=0$ (2.35)
6. $V[Y]=\sigma^2$ (2.36)
7. $V[cY]=c^2V[Y]=c^2\sigma^2$ (2.37)
8. $V[Y_1\pm Y_2]=V[Y_1]+V[Y_2]\pm 2\,\mathrm{Cov}(Y_1,Y_2)$ (2.38)

where

$$\mathrm{Cov}\left[Y_1,Y_2\right]=E\left[(Y_1-\mu_1)(Y_2-\mu_2)\right] \tag{2.39}$$

is called the *covariance* of the random variables Y_1 and Y_2. To illustrate the concept of the mean and variance of a discrete random variable, consider the example of the communications channels for which the probability mass function $f_Y(Y)$ is given in Table 2.1. The mean number of channels in use is

$$\mu = \sum_{j=1}^{4} y_j \cdot f_Y(y_j)$$
$$= (0)(0.25) + (1)(0.45) + (2)(0.27) + (3)(0.03)$$
$$= 1.08 \text{ channels}$$

The variance is

$$\sigma^2 = \sum_{j=1}^{4} y_j^2 \cdot f_Y(y_j) - \mu^2$$
$$= \left[(0)^2(0.25) + (1)^2(0.45) + (2)^2(0.27) + (3)^2(0.03) \right] - (1.08)^2$$
$$= 0.6336$$

To illustrate the concept of the mean and variance of a continuous random variable, consider a random variable Y for which the probability density function is

$$f_Y(y) = \begin{cases} \dfrac{1}{a} & 0 \leqslant y \leqslant a \\ 0 & \text{otherwise} \end{cases}$$

The mean of this random variable is

$$\mu = \int_0^a \frac{y}{a}\, dy = \left[\frac{y^2}{2a} \right]_0^a = \frac{a}{2}$$

The variance of Y is

$$\sigma^2 = \int_0^a \frac{y^2}{a}\, dy - \mu^2 = \left[\frac{y^3}{3a} \right]_0^a - \left(\frac{a}{2} \right)^2 = \frac{a^2}{12}$$

The preceding discussion of probability is by no means exhaustive, but it will facilitate an understanding of discrete and continuous probability distributions in subsequent sections.

DISCRETE PROBABILITY DISTRIBUTIONS

In our previous discussion of discrete probability functions, we defined a random variable Y, enumerated those outcomes in the sample space S that gave a particular value of y_j of Y and summed the probabilities of those outcomes to find the probability that Y has the value y_j. We can often approach this problem more directly by writing a mathematical model that accomplishes what we have heretofore done by enumeration. In the following sections, we shall describe several such models.

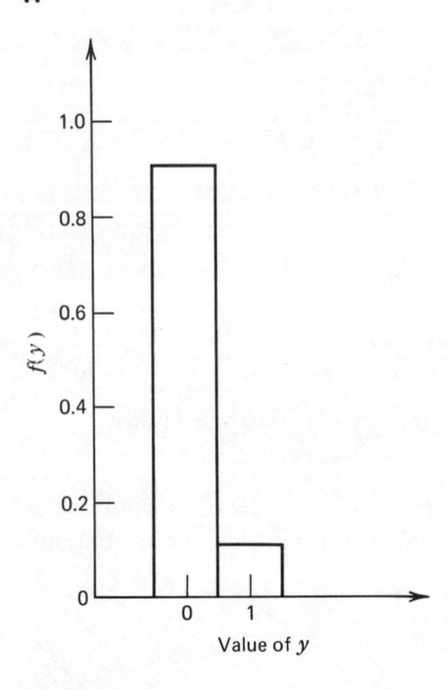

Figure 2.4. A Bernoulli trial for $p=0.1$.

Bernoulli Trial

A Bernoulli trial is a random experiment that has exactly two outcomes. These outcomes are generally termed success and failure; 1 denotes success and 0 denotes failure. Tossing a coin, pulling the starter rope on a lawnmower, and inspecting a manufactured part are familiar examples of a Bernoulli trial. We can represent the probability model for such a random experiment by

$$f_Y(y) = p^y q^{1-y}, \qquad y = 0, 1 \tag{2.40}$$

where p is the probability of success and $q = 1 - p$ is the probability of failure. Thus the random variable y is exactly the same as the outcome y.

For example, suppose that 10% of the parts produced in a manufacturing operation are defective. If discovery of a defective part denotes success, the probability of success is $p = 0.1$. Then the probabilities of the values 0 and 1 are

$$f_Y(0) = (0.1)^0(0.9)^1 = 0.9 \qquad f(1) = (0.1)^1(0.9)^0 = 0.1$$

The set of values $f_Y(y)$ is commonly called the *probability distribution* of Y. Figure 2.4 shows a histogram for this probability distribution.

Binomial Distribution

The binomial random variable Y is the number of successes in n independent Bernoulli trials. By *independent* we mean that the outcome of a given trial of the random experiment in no ways affects the outcome of any other trial. The probability model is represented by

$$f_Y(y)=\binom{n}{y}p^y q^{n-y} \qquad y=0,1,\ldots,n \tag{2.41}$$

The expression $\binom{n}{y}$ denotes the combination of n things taken y at a time. It is given by

$$\binom{n}{y}=\frac{n!}{y!(n-y)!} \tag{2.42}$$

Hence we can rewrite (2.41) as

$$f_Y(y)=\frac{n!}{y!(n-y)!}p^y q^{n-y} \qquad y=0,1,\ldots,n \tag{2.43}$$

which defines the *binomial distribution*. For example, suppose that in the previous example, we inspect exactly four manufactured parts. If N denotes a nondefective part and D denotes the defective part, the sample space is

$$\begin{aligned}S=\{&(N,N,N,N),(D,N,N,N),(N,D,N,N),(N,N,D,N),\\ \times\,&(N,N,N,D),(D,D,N,N),(D,N,D,N),(D,N,N,D),\\ \times\,&(N,D,D,N),(N,D,N,D),(N,N,D,D),(D,D,D,N),\\ \times\,&(D,D,N,D),(D,N,D,D),(N,D,D,D),(D,D,D,D)\}\end{aligned}$$

There are 16 outcomes in this sample space. The number of outcomes in S in which exactly one of the four parts ($n=4$) is defective is

$$\frac{n!}{y!(n-y)!}=\frac{4!}{1!3!}=4$$

Examining the sample space, we see that these four outcomes are (D,N,N,N), (N,D,N,N), (N,N,D,N), (N,N,N,D). Thus the term $[n!/y!(n-y)!]$ in the binomial distribution model (2.43) gives the *number of outcomes* in S that produce the value y. The term $p^y q^{n-y}$ yields the *probability* of each of these $[n!/y!(n-y)!]$ outcomes. Thus, in computing the probability $f(y)$ of the value y, we are essentially summing $[n!/y!(n-y)!]$ separate but equal probabilities. The binomial probability distribution model for our example is thus

$$f_Y(y)=\frac{4!}{y!(4-y)!}(0.1)^y(0.9)^{4-y} \qquad y=0,1,2,3,4$$

Table 2.3. Binomial probability distribution of inspection of manufactured parts, $n=4, p=0.1$

Possible value y_i	Probability $f_Y(y_j)$	Cumulative probability $F_Y(y_i)$
0	0.6561	0.6561
1	0.2916	0.9477
2	0.0486	0.9963
3	0.0036	0.9999
4	0.0001	1.0000

where y is the number of defective parts. Evaluating this model at each of the five possible values of y gives the probabilities and cumulative probabilities in Table 2.3. Figure 2.5 shows the histogram for this binomial process.

The mean of the binomial distribution is

$$\mu = np \tag{2.44}$$

and the variance is

$$\sigma^2 = npq \tag{2.45}$$

For our example, the mean and variance are

$$\mu = (4)(0.1) = 0.4, \qquad \sigma^2 = (4)(0.1)(0.9) = 0.36$$

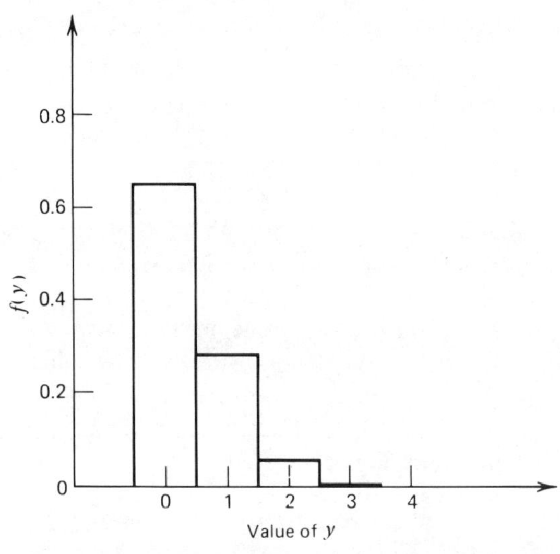

Figure 2.5. A binomial distribution with $n=4$, $p=0.1$.

From the binomial distribution model given by (2.43), as well as from its mean (2.44) and variance (2.45), we can see that this distribution is completely specified if we know the number of independent trials n and the probability of success p in each trial. These quantities are thus known as the *parameters* of the binomial distribution. Each probability distribution can be characterized by one or more parameters.

Poisson Distribution

An important probability distribution in analyzing experiments is the Poisson distribution. The model for this distribution is

$$f_Y(y)=\frac{e^{-\lambda}\lambda^y}{y!} \qquad y=0,1,2,\ldots \tag{2.46}$$

This distribution models random processes, such as the rate at which vehicles arrive at a highway toll booth, the number of defects per manufactured assembly, the number of telephone calls per minute at a telephone exchange, and the number of defects per 1000 feet of electrical cable. That is, we typically use the Poisson probability distribution to describe the *rate* at which an event occurs in a given time, length, area, volume, assembly, and so forth. We use it when the opportunity for occurrence of the event is large, but the probability of its occurrence in a given interval is relatively small. The mean of the Poisson distribution is

$$\mu=\lambda \tag{2.47}$$

and its variance is

$$\sigma^2=\lambda \tag{2.48}$$

As an example of a Poisson process, consider measuring radioactive disintegrations of a radioisotope with a Geiger counter. The Geiger counter measures the number of disintegrations per minute. Suppose the mean rate of disintegration is 2.8 counts per minute. Thus the Poisson model for this process is

$$f_Y(y)=\frac{e^{-2.8}2.8y}{y!} \qquad y=0,1,2,\ldots$$

The mean and variance of this distribution are both 2.8. Table 2.4 gives the probability distribution and the cumulative probabilities. Figure 2.6 shows the histogram for this Poisson process.

Table 2.5 summarizes these discrete probability distributions and their properties.

Table 2.4. Probability distribution of radioisotope disintegration

Possible value y_j	Probability $f(y_j)$	$F(y_j)$
0	0.061	0.061
1	0.170	0.231
2	0.238	0.469
3	0.222	0.691
4	0.156	0.847
5	0.087	0.934
6	0.041	0.975
7	0.016	0.991
8	0.006	0.997
9	0.002	0.999

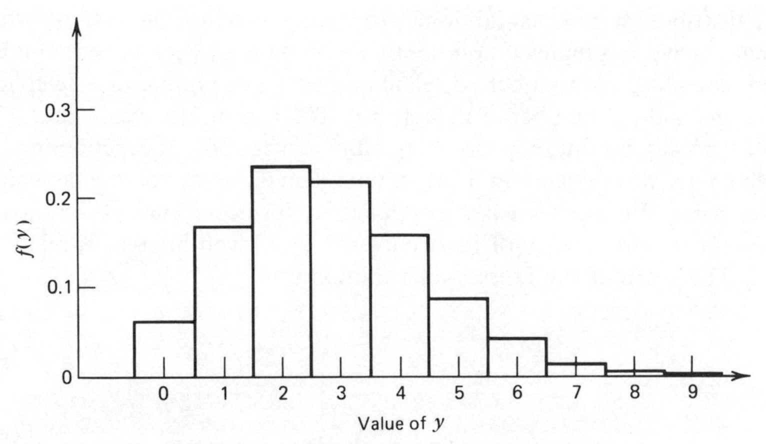

Figure 2.6. A Poisson distribution with $\lambda = 2.8$.

Table 2.5. Some discrete probability distributions and their properties

Distribution	Probability mass function	Parameters and conditions	Mean μ	Variance σ^2
Bernoulli Trial	$f_Y(y) = p^y q^{n-y}$	$y = 0, 1$ $p + q = 1$	p	pq
Binomial	$f_Y(y) = \dfrac{n!}{y!(n-y)!} p^y q^{n-y}$	n and p $y = 0, 1, 2, \ldots, n$ $p + q = 1$	np	npq
Poisson	$f_Y(y) = \dfrac{e^{-\lambda}\lambda^y}{y!}$	$\lambda > 0$ $y = 0, 1, \ldots$	λ	λ

CONTINUOUS PROBABILITY DISTRIBUTIONS

There are several important continuous distributions that arise from physical processes, and hence are of interest to us from an experimental standpoint. These include the uniform, exponential, gamma, and normal and bivariate normal distributions. There are other continuous distributions that are fundamental to the analysis of experimental results, due mainly to the fact that they arise from certain *statistics* involving samples. These include the Chi-Square, Student's t, and Snedecor's F distributions. We shall discuss the principal characteristics of each of these distributions in ensuing sections.

Uniform Distribution

Suppose we have a random variable Y that is equally likely to occur anywhere in the interval $a \leqslant Y \leqslant b$. Then Y is said to have the *uniform distribution* given by

$$f_Y(y) = \begin{cases} \dfrac{1}{b-a} & a \leqslant y \leqslant b \\ 0 & \text{otherwise} \end{cases} \tag{2.49}$$

The parameters of the uniform distribution are a and b. Its mean is

$$\mu = \frac{(a+b)}{2} \tag{2.50}$$

and its variance is

$$\sigma^2 = \frac{(b-a)^2}{12} \tag{2.51}$$

Figure 2.7 illustrates the uniform probability density function.

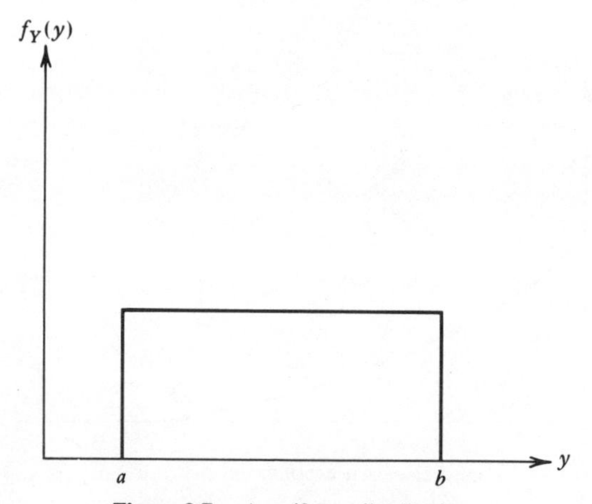

Figure 2.7. A uniform distribution.

Of special interest from the standpoint of experimentation is the uniform distribution in the interval $0 \leqslant Y \leqslant 1$. This distribution defines the set of *random numbers* that are so important in Monte Carlo simulation. This concept is discussed further in Chapter 7, which deals with simulation experimentation.

Exponential Distribution

Earlier we saw how a discrete random variable that derives from the number of occurrences of an event over a specified period of time forms a Poisson distribution with parameter λ. If we focus on a new random variable y which is the time between successive occurrences of this event, then y has the *exponential distribution*, given by

$$f_Y(y) = \begin{cases} \frac{1}{\lambda} e^{-y/\lambda} & \text{for } y \geqslant 0 \\ 0 & y < 0 \end{cases} \tag{2.52}$$

Figure 2.8 shows the probability density function for the exponential distribution.

The cumulative distribution function of an exponentially distributed random variable is given by

$$\begin{aligned} F_y(b) &= \int_{-\infty}^{b} f_Y(y)\,dy \\ &= \int_0^b \frac{1}{\lambda} e^{-y/\lambda}\,dy \\ &= 1 - e^{-y/\lambda} \end{aligned} \tag{2.53}$$

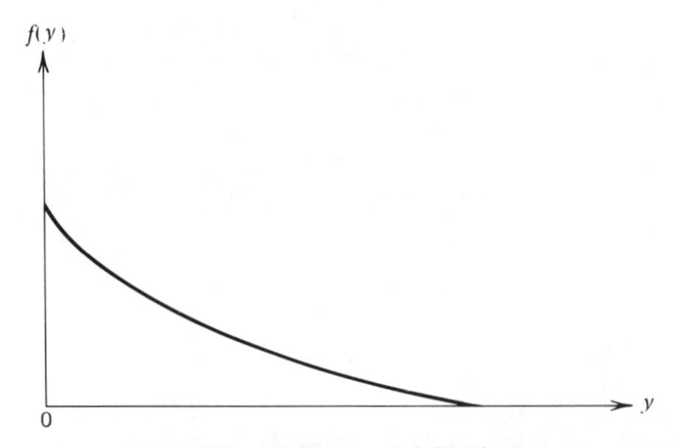

Figure 2.8. An exponential distribution.

The quantity λ is the lone parameter of the exponential distribution, just as it was for the Poisson distribution. The mean of the exponential distribution is

$$\mu = \lambda \tag{2.54}$$

and its variance is

$$\sigma^2 = \lambda^2 \tag{2.55}$$

To illustrate the use of the exponential distribution, consider the example of an electric generator, which has a mean life of 10,000 hours of operation. We might ask: What is the probability that the generator will last at least 4000 hours? This probability is

$$\begin{aligned} P(Y > 4000) &= 1 - P(Y \leqslant 4000) = 1 - F_Y(4000) \\ &= e^{-4000/10{,}000} = e^{-0.4} \\ &= 0.67 \end{aligned}$$

The exponential distribution is very important in the theory of reliability, because it often describes the "life" characteristics of electronic components.

An important aspect of the exponential distribution is its "memoryless" property; that is, the probability that a component will survive Y time units longer is independent of its age. In other words a brand new component is no better than one that has lasted 1000 hours. This implication of an exponential distribution is too frequently overlooked in practice.

Gamma Distribution

A continuous random variable whose probability density function is given by

$$f_Y(y) = \begin{cases} \dfrac{1}{\Gamma(\alpha)\beta^{\alpha}} y^{(\alpha-1)} e^{-y/\beta} & y \geqslant 0 \\ 0 & y < 0 \end{cases} \tag{2.56}$$

is known as a *gamma distribution*. Its importance lies in the fact that it describes the distribution of a random variable Y, which is the sum of α independent and identically distributed exponential random variables with parameter β.

The two parameters of the gamma distribution are α and β, both of which are positive constants. The term $\Gamma(\alpha)$ is called the *gamma function* and is defined as

$$\Gamma(\alpha) = \int_0^{\infty} t^{\alpha-1} e^{-t}\, dt \qquad \alpha > 0 \tag{2.57}$$

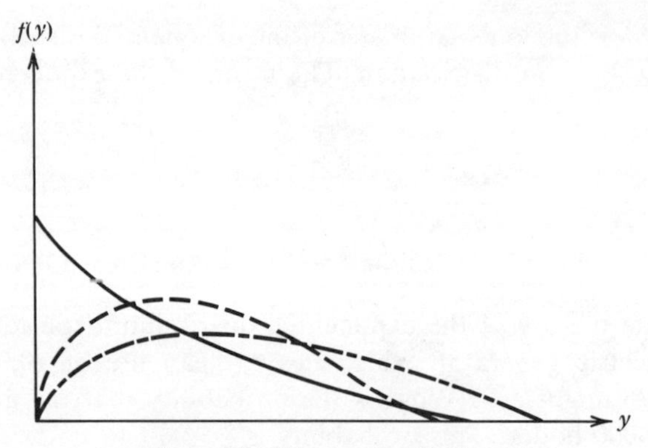

Figure 2.9. Gamma distributions.

If α is an integer, as it is for many useful applications of the gamma distribution, then repeated integration by parts yields

$$\Gamma(\alpha)=(\alpha-1)! \tag{2.58}$$

A gamma distribution with an integer parameter α is called the *Erlang distribution.* A graph of typical gamma distribution functions is illustrated in Figure 2.9.

The mean and variance of the gamma distribution are

$$\mu=\alpha\beta \tag{2.59}$$

and

$$\sigma^2=\alpha\beta^2 \tag{2.60}$$

respectively.

Normal Distribution

Probably the most important continuous probability distribution, from the standpoint of statistical analysis of experimental data, is the *normal distribution*. The probability density function for a normally distributed random variable Y is given by

$$f_Y(y)=\frac{1}{\sqrt{2\pi}\,\sigma}\exp\frac{-(y-\mu)^2}{2\sigma^2} \qquad -\infty<y<\infty \tag{2.61}$$

A typical normal density function is depicted in Figure 2.10. This figure reveals that the normal density function is bell-shaped and symmetric about the mean μ.

The normal distribution has mean μ and variance σ^2. A more convenient form of the normal distribution is obtained by using its parameters μ and σ

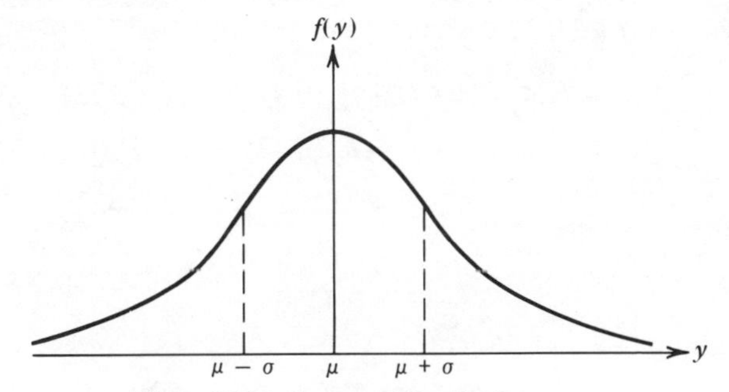

Figure 2.10. A normal distribution.

(σ is the standard deviation, the square root of the variance) to effect the transformation from the random variable Y to a new random variable Z according to relation

$$z=\frac{y-\mu}{\sigma} \tag{2.62}$$

This new distribution has mean $\mu=0$ and variance $\sigma^2=1$. The new probability density function, called *the standard normal distribution*, has the form

$$f_Z(z)=\frac{1}{\sqrt{2\pi}}e^{-z^2/2} \tag{2.63}$$

Its cumulative distribution function is

$$F_Z(b)=\frac{1}{\sqrt{2\pi}}\int_{-\infty}^{b} e^{-z^2/2}\,dz \tag{2.64}$$

The values of the cumulative distribution function for the standard normal distribution are given in Table I in Appendix A. To find any probability $P(a \leqslant z \leqslant b)$, we use the equation

$$P(a \leqslant z \leqslant b)=F_Z(b)-F_Z(a) \tag{2.65}$$

If either a or b is negative, it becomes necessary to use the identity

$$F_Z(-z)=1-F_Z(z) \tag{2.66}$$

To illustrate the use of the table, consider the probability $P(0.91 \leqslant z \leqslant 1.23)$.

$$\begin{aligned} P(0.91 \leqslant z \leqslant 1.23) &= F_Z(1.23)-F_Z(0.91) \\ &= 0.8907-0.8186 \\ &= 0.0721 \end{aligned}$$

The probability $P(-0.72 \leqslant z \leqslant 0.47)$ is

$$\begin{aligned}P(-0.72 \leqslant z \leqslant 0.47) &= F_Z(0.47) - F_Z(-0.72)\\ &= F_Z(0.47) - [1 - F_Z(0.72)]\\ &= 0.6808 - [1 - 0.7642]\\ &= 0.6808 - 0.2358\\ &= 0.4450\end{aligned}$$

The probability $P(z > 1.06)$ is

$$\begin{aligned}P(z > 1.06) &= 1 - P(z \leqslant 1.06)\\ &= 1 - F_Z(1.06)\\ &= 1 - 0.8554\\ &= 0.1446\end{aligned}$$

To use Table I (Appendix A) for calculations involving the normally distributed random variable Y with mean μ and variance σ^2, it is necessary to first transform to the random variable z. For example, if the burning rate of a solid propellant is normally distributed with mean $\mu = 2.3$ in./sec and standard deviation $\sigma = 0.02$ in./sec, what is the probability that it will burn at a rate between 2.28 and 2.34 in./sec? Using the relation

$$P(a \leqslant y \leqslant b) = F_Z\left(\frac{b-\mu}{\sigma}\right) - F_Z\left(\frac{a-\mu}{\sigma}\right) \tag{2.67}$$

yields the result

$$\begin{aligned}P(2.28 \leqslant y \leqslant 2.34) &= F\left(\frac{2.34 - 2.30}{0.02}\right) - F\left(\frac{2.28 - 2.30}{0.02}\right)\\ &= F_Z(2.0) - F_Z(-1.0) = F_Z(2.0) - [1 - F_Z(1.0)]\\ &= 0.9772 - [1 - 0.8413] = 0.9772 - 0.1587\\ &= 0.8185\end{aligned}$$

Thus 81.85% of the missile motors using this solid propellant will experience burning rates between 2.28 and 2.34 in./sec.

Bivariate Normal Distribution

The *bivariate normal distribution* is a two-dimensional normal distribution of the random variables Y_1 and Y_2. It is a generalization of the normal distribution for a single random variable Y. The probability density

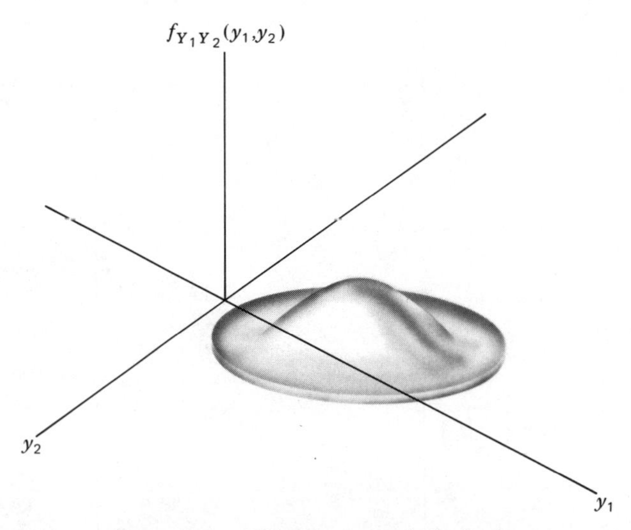

Figure 2.11. A bivariate normal distribution.

function for the bivariate normal distribution is given by

$$f_{Y_1Y_2}(y_1,y_2)=\frac{1}{2\pi\sigma_1\sigma_2\sqrt{1-\rho^2}}\exp-\left\{\frac{1}{2(1-\rho^2)}\left[\left(\frac{y_1-\mu_1}{\sigma_1}\right)^2-\frac{2\rho(y_1-\mu_1)(y_2-\mu_2)}{\sigma_1\sigma_2}+\left(\frac{y_2-\mu_2}{\sigma_2}\right)^2\right]\right\} \tag{2.68}$$

where μ_1, μ_2, σ_1, σ_2 and ρ are constants. The parameter ρ is called the correlation coefficient of Y_1 and Y_2. The covariance of Y_1 and Y_2 is

$$\text{Cov}(Y_1, Y_2)=\rho\sigma_1\sigma_2 \tag{2.69}$$

Figure 2.11 shows the probability density function for the bivariate normal distribution. Table 2.6 summarizes the important characteristics of the continuous distributions treated in this section.

Sampling and Sampling Distributions

When we perform an experiment, we observe and record one or more *measurements* of some physical quantity Y. Because of the influence of numerous uncontrollable factors, the quantity Y is affected by random variation; hence, it is a random variable.

Table 2.6. Some continuous distributions and their properties

Distribution	Density function	Parameters and conditions	Mean μ	Variance σ_y^2
Uniform	$f_Y(y)=\frac{1}{b-a}\quad a \leqslant y \leqslant b$	$-\infty < a < b < \infty$	$\frac{a+b}{2}$	$\frac{(b-a)^2}{12}$
	$=0,\quad$ otherwise			
Exponential	$f_Y(y)=\frac{1}{\lambda}e^{-y/\lambda}\quad y \geqslant 0$	$\lambda > 0$	λ	λ^2
	$=0,\quad y<0$			
Gamma	$f_Y(y)=\frac{1}{\Gamma(\alpha)\beta^{\alpha}}y^{(\alpha-1)}e^{-y/\beta}, y \geqslant 0$	$\alpha > 0$	$\alpha\beta$	$\alpha\beta^2$
	$=0,\quad y<0$	$\beta > 0$		
Normal	$f_Y(y)=\frac{1}{\sigma\sqrt{2\pi}}e^{-(y-\mu)^2/2\sigma^2},$	$-\infty < y < \infty$	μ	σ^2

In most cases, measuring every item in a great number of items is prohibitively expensive and completely impractical. It may even involve destroying the item, as in a strength test. Hence, we usually want to obtain useful information from just a few measurements. The set of items measured is called a *sample*. The totality of items from which the sample is taken is called a *population*. The objective of the experiment is to infer information about the population on the basis of results obtained from the sample.

Most of the methods presented in this book are predicated on the *randomness* of the sampling process employed in experimentation. If a population is finite with N items and a sample of r items is selected, the number of possible samples that could be selected is the combination of N things taken r at a time, or $N!/r!(N-r)!$ A *random sample* is obtained when each of the $N!/r!(N-r)!$ possible samples has an equal probability of being selected. The procedures by which we draw conclusions about a population based on results obtained from a random sample is called *statistical inference*. The following sections describe procedures that are used to extract information from samples. Also discussed are several sampling distributions based on certain important statistics derived from samples.

Table 2.7. Sample of 100 observations of tensile strength (kilonewtons/m²) of concrete cylinders

60	63	56	68	61
56	57	58	60	63
62	59	65	61	61
59	60	60	54	59
61	58	57	58	64
57	61	56	60	63
62	59	64	63	58
59	58	61	57	61
64	62	59	60	61
59	54	58	59	60
58	63	62	58	58
66	59	61	64	62
63	60	64	61	64
57	63	57	66	59
58	61	63	56	60
61	59	60	62	61
60	66	59	61	63
61	61	60	65	62
62	60	60	59	63
59	60	64	60	58

Table 2.8. Frequency histogram of a sample of 100 values of tensile strength (kilonewtons/m²) of concrete cylinders

y	Absolute frequency		Relative frequency	Cumulative frequency	Cumulative relative frequency
54	11	2	0.02	2	0.02
55		0	0.00	2	0.02
56	1111	4	0.04	6	0.06
57	HHT 1	6	0.06	12	0.12
58	HHT HHT 1	11	0.11	23	0.23
59	HHT HHT 1111	14	0.14	37	0.37
60	HHT HHT HHT 1	16	0.16	53	0.53
61	HHT HHT HHT	15	0.15	8	0.68
62	HHT 111	8	0.08	76	0.76
63	HHT HHT	10	0.10	86	0.86
64	HHT 111	8	0.08	94	0.94
65	11	2	0.02	96	0.96
66	111	3	0.03	99	0.99
67		0	0.00	99	0.99
68	1	1	0.01	100	1.00

Tabular and Graphical Representation of Samples

Consider an investigation intended to examine the tensile strength of structural concrete. The experiment consists of preparing test concrete cylinders, 0.2 m in diameter and 0.3 m in length, and performing tests of tensile strength 30 days later. Table 2.7 gives the results of a sample of 100 observations of tensile strength (in kilonewtons/m²) from such an investigation.

To develop a frequency histogram of the values shown in Table 2.7, we have to examine each observation and place a tally next to the appropriate value in the frequency histogram, as Table 2.8 shows. The frequency histogram can have specific values of y, as in this example, or *intervals* of uniform width. In the latter case, we will typically use the mid-cell value in each interval to characterize the observations in that interval. After we have completed the frequency histogram, we divide the absolute frequency by the sample size, in this case 100, to obtain the relative frequency. Figure 2.12 shows the plot of the relative frequency. These follow exactly the same principle as the probability function and cumulative probability function seen earlier.

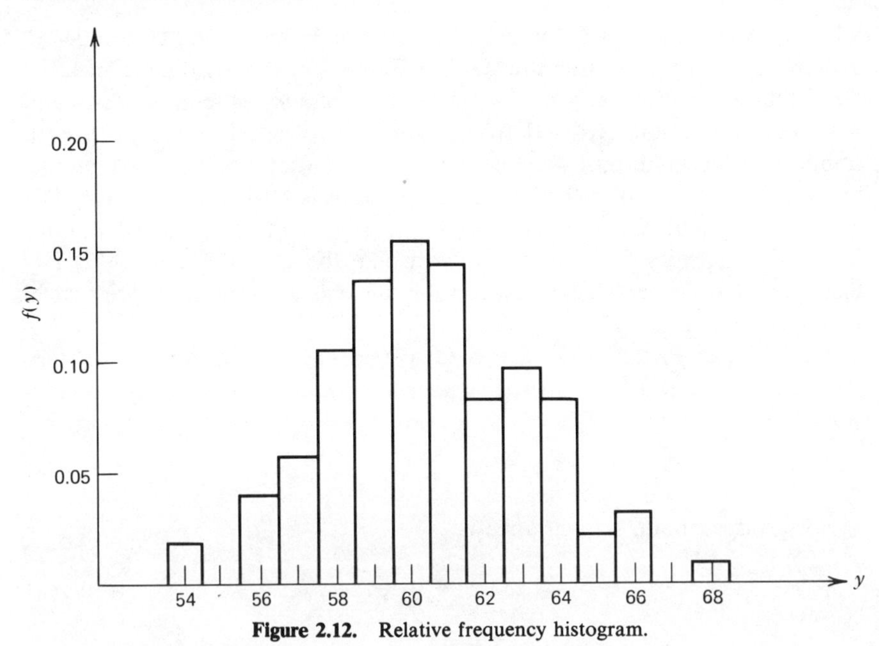

Figure 2.12. Relative frequency histogram.

Sample Statistics

After we have tabulated and graphically displayed the set of n observations, we compute various descriptive measures called *sample statistics*. These descriptive measures are classified according to the type of information they reveal about the sample, and hence about the population. Measures that reveal the sample values' tendency to occur toward the middle value of the group are called measures of *central tendency*. Those that show the tendency of the values to spread out over a range are called measures of *dispersion*.

The most important measures of central tendency are the arithmetic mean, the median, and the mode. The arithmetic or *sample mean* $\bar{y}$ of n observations $y_1, \ldots, y_n$ is

$$\bar{y} = \frac{1}{n} \sum_{i=1}^{n} y_i \tag{2.70}$$

For grouped data, like those in Table 2.8, a more convenient form of (2.70) is

$$\bar{y} = \frac{1}{n} \sum_{j=1}^{k} f_j y_j \tag{2.70a}$$

where f_j is the absolute frequency of the value y_j and k is the number of groups of values of y. For the data in Table 2.8, the arithmetic mean is 60.47 kilonewton/m^2, as you can verify. The *median* is the middle value in a set of observations ordered from highest to lowest. If there is an even number of observations, the median is the average of the two middle values. Table 2.8 shows that the median value is 60 kilonewton/m^2. The *mode* is the value that occurs most frequently in the set of observations. You can determine it from a histogram like the one in Figure 2.12. For example, the mode of the distribution of tensile strength values is 60 kilonewton/m^2.

The most important measures of dispersion are the variance, the standard deviation, and the range. The *sample variance* of a set of n observation is

$$s^2 = \frac{1}{n-1} \sum_{i=1}^{n} (y_i - \bar{y})^2 \tag{2.71}$$

For grouped data, the variance is

$$s^2 = \frac{1}{n-1} \sum_{j=1}^{k} f_j (y_j - \bar{y})^2 \tag{2.71a}$$

By expanding the right side of (2.71), we get

$$s^2 = \frac{1}{n-1} \sum_{i=1}^{n} y_i^2 - \frac{2\bar{y}}{n-1} \sum_{i=1}^{n} y_i + \frac{n\bar{y}^2}{n-1}$$

By multiplying each of the second and third terms in this expression by n/n and noting that $\bar{y} = \sum_{i=1}^{n} y_i / n$, we arrive at the more convenient computational form for the sample variance,

$$s^2 = \frac{1}{n-1} \sum_{i=1}^{n} y_i^2 - \frac{n\bar{y}^2}{n-1} \tag{2.72}$$

For grouped data, we have the equation

$$s^2 = \frac{1}{n-1} \sum_{j=1}^{k} f_j y_j^2 - \frac{n\bar{y}^2}{n-1} \tag{2.72a}$$

For the data in Table 2.8, we calculate a sample variance 7.22 (kN^2/m^4). The *standard deviation* is simply the square root of the variance, or

$$s = \sqrt{s^2} \tag{2.73}$$

For our example problem the standard deviation is 2.69 (kN/m^2). The *range* of a set of n observations is the difference between the highest and lowest values in the sample,

$$r = y_{\max} - y_{\min} \tag{2.74}$$

For the tensile strength data in Table 2.8, the range is $r=68-54=14\ \mathrm{kN/m^2}$.

An extremely important theorem relating to samples holds that, if a random sample of size n of a random variable Y is taken from a population having mean μ_Y and variance σ_Y^2, then the sample mean $\bar{y}$ is a value of a random variable whose distribution has mean μ_Y. If the population of Y is infinite, the variance of the population of sample means is

$$\sigma_{\bar{Y}}^2 = \frac{\sigma_Y^2}{n} \tag{2.75}$$

For a finite population of size N, the variance of sample means is

$$\sigma_{\bar{Y}}^2 = \frac{\sigma_Y^2}{n}\left[\frac{N-n}{N-1}\right] \tag{2.75a}$$

The factor $(N-n)/(N-1)$ is called the correction factor for finite populations, and is usually close to 1 unless the sample size is a significant portion of the population.

Sampling Distributions

In the previous sections we discussed how to compute the important sample statistics from a sample of n observations. If we are to be able to fully exploit the information that sample statistics provide, we must know the probability distribution of the statistic. We shall now discuss the sampling distributions for those sample statistics that are most important in the analysis of experimental data; namely, the normal, Student's t, Chi-Square, and F distributions. Each of these is a continuous distribution of a quantity called a sample statistic.

Normal Distribution. Earlier we described the normal distribution as a common probability distribution for a continuous random variable Y, implying that Y was a real-world, physical variable. We saw that the normal distribution was completely characterized by two parameters, its mean μ and variance σ^2. Let us now adopt a shorthand notation for referring to a normal distribution—we will say that a random variable Y is $N(\mu,\sigma^2)$. We also saw that the transformation $z=(y-\mu)/\sigma$ was useful, producing the standard normal distribution which we shall denote $N(0,1)$. A table of the standard normal distribution is given in Table I in Appendix A.

Many statistical procedures assume that the random variable of interest is normally distributed. One of the principal means of justifying such an assumption is the *central limit theorem.* This theorem states that if $y_1, y_2, \ldots, y_k$ is a sequence of k independent random variables with means μ_j, $j=1,\ldots,k$ and variances σ_j^2, $j=1,\ldots,k$ and we form another random variable $u=y_1+y_2+\cdots+y_k$, then the statistic

$$z=\frac{u-\sum_{j=1}^{k}\mu_j}{\sqrt{\sum_{j=1}^{k}\sigma_j^2}} \tag{2.76}$$

has an approximately $N(0,1)$ distribution.

An important result that derives from the central limit theorem is that, if $\bar{y}$ is the mean of a random sample of size k taken from a population having mean μ and variance σ^2, then

$$z=\frac{\bar{y}-\mu}{\sigma/\sqrt{k}} \tag{2.77}$$

is the value of a random variable whose probability density function approaches $N(0,1)$ as k approaches ∞. Note that we made no statements about the nature of the population of the random variable Y. Thus the sampling distribution of the mean of a population for which the variance σ^2 is known is $N(0,1)$.

The quantity z in (2.77) is known as a *statistic*. A statistic is a real-valued function of a random variable y. We shall make extensive use of various statistics in the analysis of experimental data. Their chief use is that they enable us to draw conclusions about *populations* based on information extracted from *samples*; that is, they are employed in what we call *statistical inference*. The *z-statistic* in (2.77) illustrates a point that is typically true about the statistics we employ in statistical inference—it contains both *sample information* (k and $\bar{y}$) and *population information* (μ and σ). Often the population information is not known, but is *hypothesized.* For instance, we might hypothesize that the mean μ has some specific value. Hence, the statistic might also contain *hypothesis information.* We shall enlarge on this idea in later sections when we discuss hypothesis testing.

The z-statistic is useful when the population variance σ^2 is known, but such is not always the case. If σ^2 is unknown but the sample size k is large, we can exploit a form of the central limit theorem and simply substitute the sample standard deviation s for σ in (2.77) to yield

$$z=\frac{\bar{y}-\mu}{s/\sqrt{k}} \tag{2.78}$$

Thus if k is large and σ is unknown, the z-statistic in (2.78) has approximately the $N(0,1)$ distribution.

Student's t Distribution. When σ is unknown but k is not large, little is known about the precise distribution of the statistic $(\bar{y}-\mu)/s/\sqrt{k}$. But if we can exact the additional assumption that the sample comes from a normal distribution, then the *t-statistic*

$$t=\frac{\bar{y}-\mu}{s/\sqrt{k}} \tag{2.79}$$

has the Student-t distribution with parameter $\nu=k-1$. As seen in Figure 2.13, this distribution is symmetrical about zero. Like the $N(0,1)$ distribution, Student's-t distribution has mean zero, but its variance depends on ν and hence on the sample size k. As we might expect, this distribution approaches the standard normal distribution as $\nu \rightarrow \infty$.

Table II (Appendix A) contains values of t_α as a function of ν, where t_α is such that the fraction α of the distribution lies to the right of t_α. Each row in Table II contains values of t that form right-hand tails of area α, as illustrated in Figure 2.14. The quantity α is actually a probability as we shall see later.

EXAMPLE. Eleven solid propellant igniters are test fired to measure ignition delay time. The sample mean is $\bar{y}=0.0734$ sec and the standard deviation is 0.0012 sec. What is the probability that the true mean ignition delay μ exceeds 0.074 sec? The population variance σ^2 is unknown, but ignition delay is known to be a normally distributed random variable.

$$t=\frac{\bar{y}-\mu}{s/\sqrt{k}}=\frac{0.0734-0.0740}{0.0012/\sqrt{11}}$$

$$t=-1.659$$

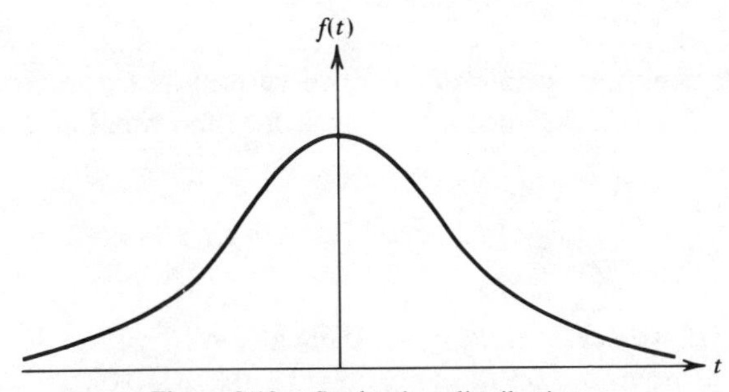

Figure 2.13. Student's-t distribution.

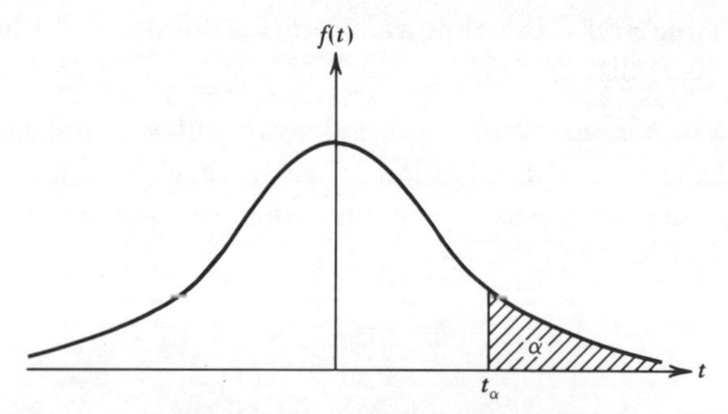

Figure 2.14. Tabulated values of t.

Now the probability that the mean ignition delay exceeds 0.0740 sec is found, by interpolating between the values of $t_{0.10}$ and $t_{0.05}$ at $\nu = 11-1=$ 10 degrees of freedom, to be 0.0674. Note that we used $|t|$ from the above calculation, because we must always deal with the right tail of the t distribution. Since the distribution is symmetric about zero, we encounter no difficulty in doing this.

Chi-Square Distribution. So far we have discussed only sampling distributions of the mean, but in experimental data analysis we are also concerned with the sample variance s^2. If the population from which the sample is taken is unknown, or if σ^2 is unknown, little is known about the population of s^2. If s^2 is the variance of a sample of k observations from a normal population having variance σ^2, then the statistic

$$\chi^2 = \frac{(k-1)s^2}{\sigma^2} \tag{2.80}$$

has the chi-square distribution with parameter ν, where χ^2_α is a value such that fraction α of the population lies to the right of χ^2_α. Figure 2.15 illustrates the chi-square distribution. Table III (Appendix A) contains values of χ^2_α as a function of $\nu = k-1$ degrees of freedom.

EXAMPLE. A random sample of 10 observations is taken from a normal population having the variance $\sigma^2 = 17.5$. Find the probability that s^2 is less than 5.0.

$$\chi^2 = \frac{(10-1)(5.0)}{17.5} = 2.571$$

From Table III we see that $P(\chi^2_\alpha < 2.571)$ at $\nu = 10-1=9$ degrees of freedom is, by interpolation, 0.978. Thus, $\alpha = 1-0.978 = 0.022$.

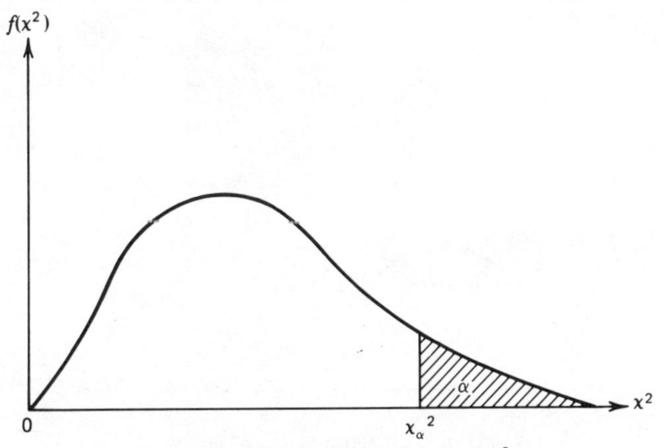

Figure 2.15. Tabulated values of χ_α^2.

Snedecor's F Distribution. Sometimes in analyzing experimental data we are faced with comparing the sample variances of two independent random samples, attempting to determine if the samples came from populations having *equal* variances.

If s_1^2 and s_2^2 are the variances of independent random samples of size k_1 and k_2, respectively, taken from two normal distributions having the same variance, then the statistic

$$F=\frac{s_1^2}{s_2^2} \tag{2.81}$$

has the Snedecor F distribution with the two parameters $\nu_1 = k_1 - 1$ and $\nu_2 = k_2 - 1$ degrees of freedom. Tables IV (Appendix A) contain values for $F_{0.10}$, $F_{0.05}$, $F_{0.025}$, and $F_{0.01}$ for various combinations of ν_1 and ν_2. Figure 2.16 shows the F distribution. Again we deal with right-tail probabilities, so that the shaded area represents the probability α of a value of the F distribution exceeding the value F_{ν_1,ν_2}.

EXAMPLE. In manufacturing solid propellant igniters, it is necessary to determine if using two different lots of one of the raw materials leads to different variances for the ignition delay variable. Eleven igniters made from lot 1 had a sample variance of 0.00000144 sec². Nine igniters made from lot 2 had a sample variance of 0.00000081 sec². Then

$$F=\frac{0.00000144}{0.00000081}=1.78$$

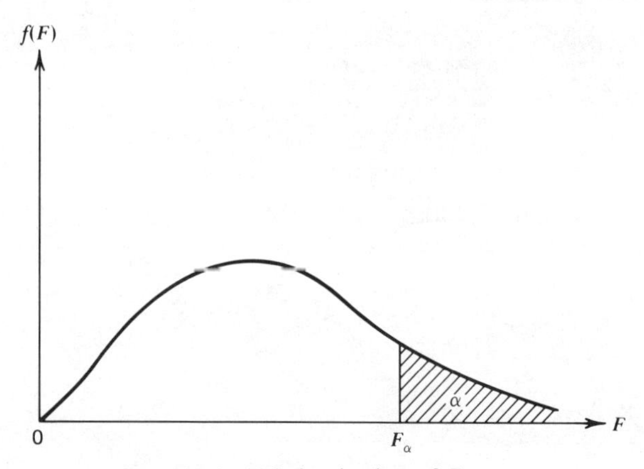

Figure 2.16. Tabulated values of F_α.

From the $F_{0.05}$ Table IV in Appendix A at $\nu_1=10$ and $\nu_2=8$, $F_{0.05,10,8}=3.35$. Since $F<3.35$, we would conclude that the two lots do not lead to difference variances for ignition delay. We shall consider this kind of statistical procedure further in a later section on hypothesis testing.

ESTIMATION OF POPULATION PARAMETERS

Statistical inference is concerned with inferring information about the population of a random variable y based on results obtained from a sample. Since the population of y is characterized by its parameters (for example, the mean μ and variance σ^2 are the important parameters of a normal distribution), we usually attempt to infer certain facts about these parameters. Indeed, in our discussion of sampling distributions, we concentrated on distributions of the sample mean and sample variance. Now the most useful approach to inferring information about population parameters is to *estimate* them using sample data. For example, the sample mean is an estimate of the population mean, and the sample variance is likewise an estimate of the population variance.

An estimate of an unknown population parameter, computed from sample data, is a numerical value of an *estimator*, which is simply a sample statistic. The process of dealing with estimators is called *estimation*. In experimental data analysis, two kinds of estimators are useful to us. A *point estimator* is a statistic that provides a single numerical value of the unknown parameter. An *interval estimator* gives a random interval within which we are willing to state the true parameter lies with some probability.

Point Estimation

Basically, point estimation is concerned with computing a *sample statistic* $\hat{\theta}$, which is regarded as an estimate of the population parameter $\hat{\theta}$. There are two characteristics that we desire the point estimator $\hat{\theta}$ to possess. These are *unbiasedness* and *minimum variance*. A point estimator $\hat{\theta}$ is said to be *unbiased* if the mean of its sampling distribution is the population parameter θ, whereas $\hat{\theta}$ is said to have minimum variance if the variance of its sampling distribution is less than that for any other estimator θ'. The sample statistics $\bar{y}$ and s^2 are unbiased estimators of μ and σ^2, respectively.

Interval Estimation

We cannot expect a point estimate to have precisely the same value as the population parameter. Consequently, it is sometimes preferable to identify some interval within which we are reasonably certain the true parameter lies. That is, we wish to make the probability statement

$$P(\hat{\theta}_l \leqslant \theta \leqslant \hat{\theta}_u) = 1 - \alpha \tag{2.82}$$

where $\hat{\theta}_l$ and $\hat{\theta}_u$ are the lower and upper ends of this interval, respectively. The procedures by which we compute such quantities as $\hat{\theta}_l$ and $\hat{\theta}_u$ are called *interval estimation*, and $\hat{\theta}_l$ and $\hat{\theta}_u$ are termed interval estimators. The interval

$$\hat{\theta}_l \leqslant \theta \leqslant \hat{\theta}_u \tag{2.83}$$

is usually called a $100(1-\alpha)\%$ *confidence interval* for the parameter θ.

To illustrate the construction of a confidence interval, suppose we have a large sample of size k from a population having the unknown mean μ and the known variance σ^2. We saw earlier that the z statistic applied in this case. Let us construct the $100(1-\alpha)\%$ confidence interval for the unknown population mean μ. Now we are willing to have μ lie *outside* our stated interval with probability α. Since the distribution of the z statistic, that is, $N(0,1)$, is symmetric about its mean zero, we can assign probability $\alpha/2$ that μ will lie to the left of our constructed interval and $\alpha/2$ that it lies to the right of it. Hence we have

$$-z_{\alpha/2} < \frac{y-\mu}{\sigma/\sqrt{k}} < z_{\alpha/2} \tag{2.84}$$

Rearranging this expression produces

$$\left(\bar{y} - z_{\alpha/2} \cdot \frac{\sigma}{\sqrt{k}}\right) < \mu < \left(\bar{y} + z_{\alpha/2} \cdot \frac{\sigma}{\sqrt{k}}\right) \tag{2.85}$$

Thus we can claim with probability $1-\alpha$ that the interval (2.85) contains the true mean μ. If σ^2 *is unknown but* k *is large*, say greater than 30, we have

$$\left(\bar{y}-z_{\alpha/2}\cdot\frac{s}{\sqrt{k}}\right)<\mu<\left(\bar{y}+z_{\alpha/2}\cdot\frac{s}{\sqrt{k}}\right) \tag{2.86}$$

simply by substituting the sample standard deviation s for σ in (2.85). *If we cannot assume that the sample came from a normal population*, we must use the t statistic to develop the interval estimator with $v=k-1$

$$\left(\bar{y}-t_{\alpha/2}\cdot\frac{s}{\sqrt{k}}\right)<\mu<\left(\bar{y}+t_{\alpha/2}\cdot\frac{s}{\sqrt{k}}\right) \tag{2.87}$$

For the variance σ^2, we have the $100(1-\alpha)\%$ confidence interval

$$\frac{(n-1)s^2}{\chi_2^2}<\sigma^2<\frac{(n-1)s^2}{\chi_1^2} \tag{2.88}$$

where χ_1^2 and χ_2^2 cut off left- and right-hand tails of area (probability) $\alpha/2$ of the chi-square distribution with $v=k-1$ degrees of freedom.

STATISTICAL TESTS OF HYPOTHESIS

A statistical *hypothesis* is an assumption about the distribution of a random variable y. For example, we may state the hypothesis that the mean of the distribution of the muzzle velocities y of a 7.62-mm rifle is 1285 m/sec. A statistical *test of hypothesis* is a procedure that allows us to objectively "reject" or "accept" ("not reject") the stated hypothesis based on information obtained from a sample. If we reject the stated hypothesis, usually framed in a negative way and called the *null hypothesis*, we then "accept" an alternative hypothesis which must then be true.

In accepting or rejecting a null hypothesis, which we shall denote H_0, we usually incur some probability of being wrong. That is, we typically approach a statistical test of hypothesis by acknowledging a probability α of *rejecting* the null hypothesis H_0 when it is in fact *true*. The quantity α is called the *significance level* of the test, and is usually a value such as 0.01, 0.05, or 0.10. The error incurred by rejecting a null hypothesis that is true is called *Type I* error. Similarly, we incur with probability β an error, called *Type II* error, when we *accept* a null hypothesis that is in fact *false*. Figure 2.17 summarizes the ways in which these errors can occur in statistical hypothesis testing.

The region containing those values for which we reject the null hypothesis H_0 is called the *rejection region*. The region containing the values for

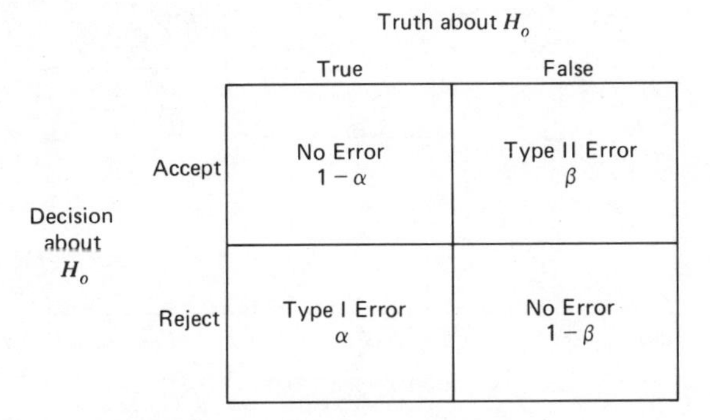

Figure 2.17. Errors in hypothesis testing.

which the null hypothesis is accepted is called the *acceptance region*. The value of the parameter θ, which separates the acceptance and rejection regions, is called the *critical value* and denoted as c. The relationship of c to the null hypothesis value $\theta=\theta_0$ depends on the nature of the alternative hypothesis. The three types of alternative hypotheses, denoted H_1, that can be constructed in hypothesis testing are as follows:

$$\theta>\theta_0 \tag{2.89}$$

$$\theta<\theta_0 \tag{2.90}$$

$$\theta\neq\theta_0 \tag{2.91}$$

Forms (2.89) and (2.90) are called *one-sided tests*, whereas (2.91) is called a *two-sided test*. When $\theta>\theta_0$, the test is called a *right-sided test*; and when $\theta<\theta_0$, it is referred to as a *left-sided test*. Figure 2.18 illustrates the relationship of the rejection region to the null hypothesis value θ_0 in each of these three cases.

Referring to Figure 2.18(a), we see that we reject the null hypothesis H_0: $\theta=\theta_0$ if the sample value $\hat{\theta}>c$. In Figure 2.19, we see how we can incur Type I error: that is, α is the probability that the sample value $\hat{\theta}$ actually belongs to the probability density function $f(\theta_0)$ but causes the rejection of H_0. Figure 2.19 reveals that, as we make α smaller, β becomes larger, and vice versa. We typically choose a value of α, determine the corresponding c, and compute the value of β that is incurred. If β is unacceptably large, we would choose a larger value of α and repeat the process until an acceptable value of β is obtained.

Statistical hypothesis testing is essentially an experimental procedure. It is important that a rigorous discipline be assumed when engaging in the statistical analysis of an experiment. We recommend the following eight-step procedure as an objective and disciplined approach to hypothesis

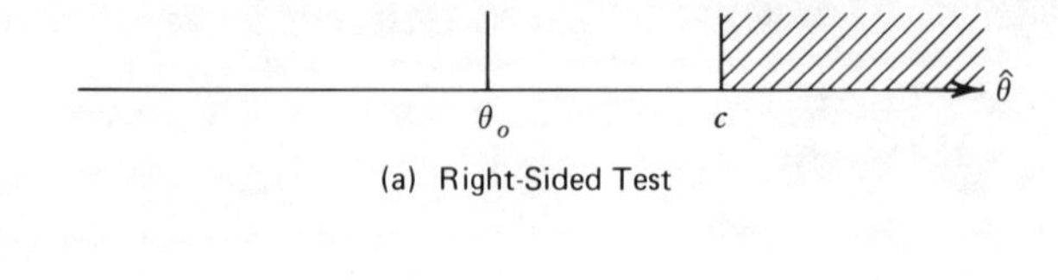

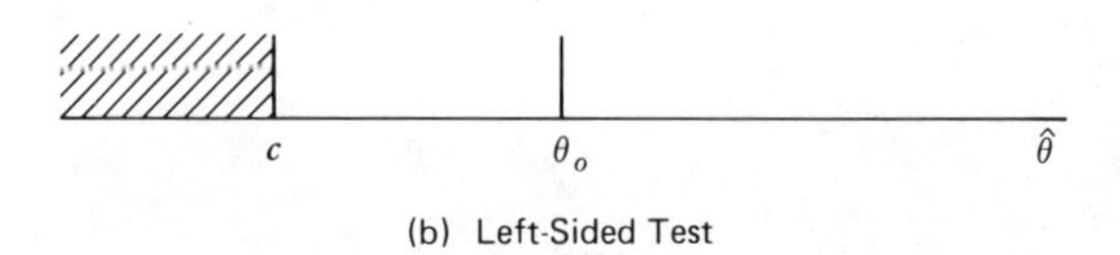

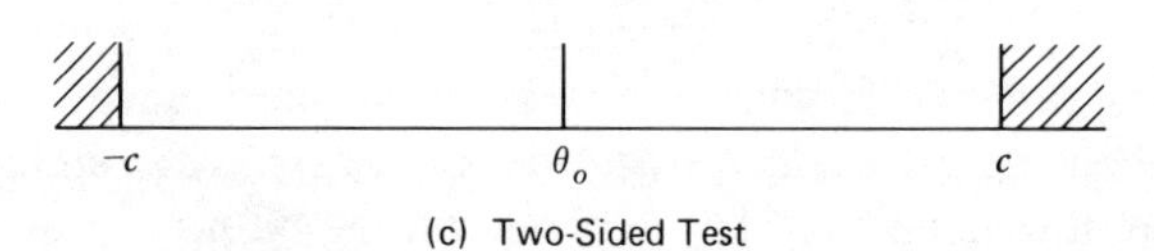

Figure 2.18. Rejection regions for hypothesis tests.

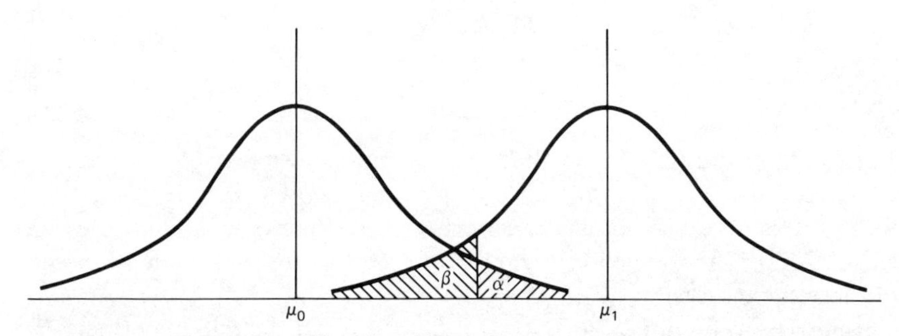

Figure 2.19. Relationship of Type I and Type II errors to the null hypothesis $H_0: \mu = \mu_0$.

testing. The procedure is illustrated using a test concerning the mean μ of a normal population having known variance. Suppose that the random variable y is the hourly production rate of V-8 cylinder blocks for automotive engines, and the variance is known to be $\sigma^2 = 64$.

STEP 1. State the null hypothesis

$$H_0: \mu = 200 \text{ blocks/hr}$$

State the alternative hypothesis

$$H_1: \mu < 200 \text{ blocks/hr}$$

STEP 2. Choose a significance level α.

$$\alpha = 0.05$$

STEP 3. Select the appropriate test statistic.

$$z_0 = \frac{\bar{y} - \mu_0}{\sigma/\sqrt{n}} = \frac{y - 200}{8/\sqrt{n}}$$

Thus the distribution of z_0 is $N(0,1)$

STEP 4. Compute the rejection value c.

$$-z_{1-\alpha} = \frac{c - 200}{8/\sqrt{n}}$$

$$\therefore c = -z_{1-\alpha}\left(\frac{8}{\sqrt{n}}\right) + 200$$

Hence if $\bar{y} < c$, reject H_0 and accept H_1.

STEP 5. Perform the experiment. Select the sample size n and collect the observations $y_1, \ldots, y_n$.

STEP 6. Perform all computations.

$$n = 10 \qquad \bar{y} = 194.5 \text{ blocks/hr}$$

$$c = (-1.645)\left(\frac{8}{\sqrt{10}}\right) + 200 = 195.8 \text{ blocks/hr}$$

STEP 7. Make the *statistical* decision. Reject H_0, since $\bar{y} < c$.

STEP 8. Make the corresponding *physical* decision (i.e., managerial, scientific, or engineering decision). We would conclude from this test of hypothesis that the hourly production rate is *less than* 200 blocks/hr.

This eight-step procedure applies whether we are testing hypotheses concerning means, variances, or, as we shall consider later, the *form* of the distribution.

Tests on Means

We shall consider several tests of hypotheses concerning the *mean* of a random variable y. There are three bases upon which these tests are distinguished:

1. Whether the variance σ^2 is known or unknown.
2. Whether we are comparing *one* mean μ to a specific value μ_0, or comparing *two* means μ_1 and μ_2.
3. Whether we frame the alternative hypothesis as $\mu < \mu_0$, $\mu > \mu_0$, or $\mu \neq \mu_0$.

Table 2.9. Tests of hypothesis on means with known variance σ^2

Hypotheses	Test statistic	Rejection region
H_0: $\mu = \mu_0$ H_1: $\mu \neq \mu_0$		$\lvert z_0 \rvert > z_{\alpha/2}$
H_0: $\mu = \mu_0$ H_1: $\mu < \mu_0$	$z_0 = \dfrac{\bar{y} - \mu_0}{\sigma/\sqrt{n}}$	$z_0 < -z_\alpha$
H_0: $\mu = \mu_0$ H_1: $\mu > \mu_0$		$z_0 > z_\alpha$
H_0: $\mu_1 - \mu_2 = \delta_0$ H_1: $\mu_1 - \mu_2 \neq \delta_0$		$\lvert z_0 \rvert > z_{\alpha/2}$
H_0: $\mu_1 - \mu_2 = \delta_0$ H_1: $\mu_1 - \mu_2 < \delta_0$	$z_0 = \dfrac{(\bar{y}_1 - \bar{y}_2) - \delta_0}{\sqrt{\dfrac{\sigma_1^2}{n_1} + \dfrac{\sigma_2^2}{n_2}}}$	$z_0 < -z_\alpha$
H_0: $\mu_1 - \mu_2 = \delta_0$ H_1: $\mu_1 - \mu_2 > \delta_0$		$z_0 > z_\alpha$

Tables 2.9 and 2.10 summarize the null and alternative hypotheses, test statistic, and rejection region for each of these conditions.

In Table 2.9 and Table 2.10, the quantity δ_0 is a *hypothesized difference* between means μ_1 and μ_2. If we test the hypothesis that the means μ_1 and μ_2 are *equal*, that is,

$$H_0: \mu_1 = \mu_2$$

with one of the alternative hypotheses

$$H_1: \mu_1 \neq \mu_2 \quad \text{or} \quad H_1: \mu_1 < \mu_2 \quad \text{or} \quad H_1: \mu_1 > \mu_2$$

then clearly the difference is $\delta_0 = 0$.

The quantity s_p^2 in the t statistic in Table 2.10 is called the *pooled variance* and is calculated using this expression

$$s_p^2 = \frac{(n_1 - 1)s_1^2 + (n_2 - 1)s_2^2}{n_1 + n_2 - 2} \tag{2.92}$$

Clearly s_p^2 is simply the weighted sample variance of the combined samples.

Table 2.10. Tests of hypothesis on means of normal distributions with variance unknown

Hypotheses	Test statistic	Rejection region
H_0: $\mu = \mu_0$ H_1: $\mu \neq \mu_0$		$\lvert t_0 \rvert > t_{\alpha/2, n-1}$
H_0: $\mu = \mu_0$ H_1: $\mu < \mu_0$	$t_0 = \dfrac{\bar{y} - \mu_0}{s/\sqrt{n}}$	$t_0 < -t_{\alpha, n-1}$
H_0: $\mu = \mu_0$ H_1: $\mu > \mu_0$		$t_0 > t_{\alpha, n-1}$
H_0: $\mu_1 - \mu_2 = \delta_0$ H_1: $\mu_1 - \mu_2 = \delta_0$	$t_0 = \dfrac{(\bar{y}_1 - \bar{y}_2) - \delta_0}{s_p \sqrt{\dfrac{1}{n_1} + \dfrac{1}{n_2}}}$ where $\nu = n_1 + n_2 - 2$ or	$\lvert t_0 \rvert > t_{\alpha/2, \nu}$
H_0: $\mu_1 - \mu_2 = \delta_0$ H_1: $\mu_1 - \mu_2 < \delta_0$	$t_0 = \dfrac{(\bar{y}_1 - \bar{y}_2) - \delta_0}{\sqrt{\dfrac{s_1^2}{n_1} + \dfrac{s_2^2}{n_2}}}$	$t_0 < -t_{\alpha, \nu}$
H_0: $\mu_1 - \mu_2 = \delta_0$ H_1: $\mu_1 - \mu_2 > \delta_0$	where $\nu = \dfrac{\left(\dfrac{s_1^2}{n_1} + \dfrac{s_2^2}{n_2}\right)^2}{\dfrac{(s_1^2/n_1)^2}{n_1+1} + \dfrac{(s_2^2/n_2)^2}{n_2+1}} - 2$	$t_0 > t_{\alpha, \nu}$

Tests on Variances

As we saw in earlier sections, there are two sets of hypotheses that are applied to variances of normal distributions:

1. That the population variance is σ_0^2.
2. That the variances σ_1^2 and σ_2^2 are equal.

The two important sampling distributions that arose from these tests were the Chi-Square distribution and Snedecor's F distribution. Table 2.11 summarizes the hypotheses, test statistics, and rejection regions for the test of hypotheses concerning variances.

Table 2.11. Tests on variances of normal distributions

Hypotheses	Test statistic	Rejection region
$H_0: \sigma^2 = \sigma_0^2$ $H_1: \sigma^2 \neq \sigma_0^2$		$\chi_0^2 > \chi_{\alpha/2,n-1}^2$ or $\chi_0^2 < \chi_{1-\alpha,n-1}^2$
$H_0: \sigma^2 = \sigma_0^2$ $H_1: \sigma^2 < \sigma_0^2$	$\chi_0^2 = \dfrac{(n-1)s^2}{\sigma_0^2}$	$\chi_0^2 < \chi_{1-\alpha,n-1}^2$
$H_0: \sigma^2 = \sigma_0^2$ $H_1: \sigma^2 > \sigma_0^2$		$\chi_0^2 > \chi_{\alpha,n-1}^2$
$H_0: \sigma_1^2 = \sigma_2^2$ $H_1: \sigma_1^2 \neq \sigma_2^2$	$F_0 = \dfrac{s_1^2}{s_2^2}$	$F_0 > F_{\alpha/2,n_1-1,n_2-1}$ *or* $F_0 < F_{1-\alpha/2,n_1-1,n_2-1}$
$H_0: \sigma_1^2 = \sigma_2^2$ $H_1: \sigma_1^2 < \sigma_2^2$	$F_0 = \dfrac{s_2^2}{s_1^2}$	$F_0 > F_{\alpha,n_2-1,n_1-1}$
$H_0: \sigma_1^2 = \sigma_2^2$ $H_1: \sigma_1^2 > \sigma_2^2$	$F_0 = \dfrac{s_1^2}{s_2^2}$	$F_0 > F_{\alpha,n_1-1,n_2-1}$

Goodness-of-Fit Tests

The tests of hypothesis discussed in previous sections dealt with *parameters* of samples and the populations from which they were drawn. Another very important area of hypothesis testing deals with determining the *type of distribution* from which a sample is taken. Using a sample $y_1, \ldots, y_k$, we often wish to test the hypothesis that the cumulative distribution function $F(y)$ of the sample is the same as that for a specified population $F_Y(y)$. Stated in the negative manner of a null *hypothesis* H_0, we would say "the cumulative distribution function $F(y)$ of the sample *is not different from* the population CDF $F_Y(y)$."

Chi-Square Test. There are several different techniques for testing a hypothesis such as that stated above. The two applied most frequently are the *Chi-Square test* and the *Kolmogorov-Smirnov test*. The Chi-Square test for goodness-of-fit compares the frequency f_i of sample observations having value y_i (or having values in the ith range in the case of continuous random variable y) to the *expected* frequency e_i. It uses as a test statistic

$$\chi^2 = \sum_{i=1}^{m} \frac{(f_i - e_i)^2}{e_i} \tag{2.93}$$

which has $\nu = m - p - 1$ degrees of freedom, where m is the number of intervals and p is the number of estimated parameters.

It should be obvious from the form of the Chi-Square test statistic that when the distribution of the sample is very close to that of the population, small values of χ^2 are obtained. Conversely, when the f_i and e_i differ widely, large values of χ^2 result. Consequently, we would expect to reject the null hypothesis H_0 for large values of χ^2 and accept H_0 for small values of χ^2.

We can also see from (2.93) that the test statistic is very sensitive to small values of the expected frequency e_i. Therefore, we usually require that $e_i \geqslant 5$. When this condition is not met, we combine intervals, usually working from the ends of the distribution toward the central values, until e_i for the combined intervals is at least 5. Since the Chi-Square test also loses sensitivity for very small values of ν, the degrees of freedom, we usually require that $m \geqslant 5$ intervals. Therefore, the minimum sample size for which the Chi-Square test for goodness-of-fit is appropriate is about 30 observations. Let us illustrate the Chi-Square goodness-of-fit test with an example, employing the eight-step procedure outlined earlier.

EXAMPLE. As we saw earlier in this chapter, the measurement of the radioactive disintegration of a radioisotope with a Geiger counter is hypothesized to be a Poisson process

$$f(y) = \frac{e^{-\lambda}\lambda^y}{y!} \qquad y = 0, 1, 2, \ldots$$

Suppose we apply this measurement to a sample of uranium. Let us test this hypothesis using the Chi-square test for goodness-of-fit.

STEP 1. H_0: the sample distribution is Poisson distributed. H_1: the sample distribution is *not* Poisson distributed.

STEP 2. $\alpha = 0.05$.

STEP 3. The test statistic is

$$\chi^2 = \sum_{i=1}^{m} \frac{(f_i - e_i)^2}{e_i}$$

with $\nu = m - p - 1$ degrees of freedom. Since we must use sample data to estimate the Poisson parameter λ, $p = 1$ and $\nu = m - 2$.

STEP 4. Reject H_0 if $\chi^2 > \chi^2_{\alpha, m-2}$

STEP 5. Perform the experiment. For 100 minutes of data, the results are shown in Table 2.12.

Table 2.12. Results of α-emission counting experiment

No. of α particles emitted	Observed intervals, f_i	Expected intervals e_i	$\frac{(f_i-e_i)^2}{e_i}$
0	1 } 5	1.5 } 7.9	1.06
1	4	6.4	
2	17	13.4	0.97
3	18	18.6	0.02
4	25	19.5	1.55
5	12	16.3	1.13
6	10	11.3	0.15
7	6	6.8	0.09
8	3 } 7	3.5 } 6.2	0.10
9	2	1.6	
10	1	0.7	
11	0	0.3	
12	1	0.1	
	100	100.0	5.07

From the data, our sample estimate $\bar{y}$ of the Poisson parameter λ is as follows:

$$\bar{y}=\frac{1}{100}\sum_{i=0}^{12} f_i y_i$$

$$=\frac{1}{100}\left[(0)(1)+(1)(4)+(2)(17)+\ldots+(11)(0)+(12)(1)\right]$$

$$=4.18 \text{ particles/min}$$

Substituting $\bar{y}=4.18$ for λ in the Poisson model produces the e_i values in Table 2.12.

STEP 6. Perform the calculations for the test of hypothesis. First we must combine intervals so that $e_i \geqslant 5$. This process, as shown in Table 2.12, produces 8 intervals. Then

$$\chi^2=\frac{(5-7.9)^2}{7.9}+\frac{(17-13.4)^2}{13.4}+\ldots+\frac{(6-6.8)^2}{6.8}+\frac{(7-6.2)^2}{6.2}$$

$$\chi^2=1.06+\ldots+0.97+0.09+0.10$$

$$\chi^2=5.07$$

Now $\nu=m-p-1=8-1-1=6$ degrees of freedom. From Table III in Appendix A, $\chi^2_{0.05,6}=12.59$.

STEP 7. Make the statistical decision. Since

$$\chi^2 < \chi^2_{0.05,6}$$

we do not reject the null hypothesis that the sample values are Poisson distributed.

STEP 8. Make the physical decision. The α-particle emission count is a Poisson-distributed random variable with mean $\lambda = 4.18$ counts/min.

Note that, although we have stated that the goodness-of-fit test compares the sample CDF with the population CDF, in the above example we dealt entirely with *frequencies* f_i and e_i in the Chi-Square test. Actually there is no discrepancy here when one recalls that the probability of a value b is $P(b) = F(b) - F(b-1)$ for a discrete random variable y, where $F(b)$ is the value of the CDF at $y = b$. Similarly, for a continuous random variable y, the Chi-Square test deals with the frequency f_i of values of y in an *interval*; that is, $P(a \leqslant y \leqslant b) = F(b) - F(a)$.

Kolmogorov-Smirnov Test. The Kolmogorov-Smirnov test for goodness-of-fit concerns the agreement between an *observed* cumulative distribution of sample values $F(y)$ and a specified *continuous* distribution function $F_Y(y)$. It is generally more efficient than the Chi-Square test for small samples ($n \leqslant 30$), and it can even be used for very small samples where the Chi-Square test cannot be applied at all ($n \leqslant 10$). Unlike the Chi-Square test, however, the Kolmogorov-Smirnov test cannot be employed with *discrete* distributions.

REGRESSION AND CORRELATION

Many experimental activities have as their objective the development of a model or formula by which a response y can be *predicted* given the values of one or more controllable variables x_i, $i = 1, \ldots, n$. Indeed, it has been stated many times over that such a capability is one of the major foci of this book. The approach by which such relationships are developed from experimental data is called the *method of least-squares*. The procedure used in least-squares is usually called regression.

The least-squares method rests on two main assumptions: (1) that a cause and effect relationship exists between x and y; and (2) that the x values can be observed without error. When one or both of these assumptions is violated, we usually forfeit our objective of developing a regression equation, in favor of establishing a measure of the extent to which x and y vary together. This tendency to "vary together" is called *correlation*. In this section we lay the foundation for much of our work in this book by considering the topics regression and correlation. We shall first focus on the relationship between a single response y and a single independent variable x.

Linear Regression

The problem we consider in *linear regression* is that of employing the method of least squares to estimate the unknown linear relation

$$\eta(x) = \alpha + \beta x \tag{2.94}$$

where α is the *intercept* and β the *slope* of a straight line, and $\eta(x)$ is the *true* value of the response at a specified x. In general, an observed value of the response, denoted y, will differ from the value of $\eta(x)$ in (2.94) by an error ε, so that we actually measure values

$$y = \alpha + \beta x + \varepsilon \tag{2.95}$$

Thus ε is a random variable, and we usually choose our estimating line so that the mean of ε is zero. That is, we use the k observations in a sample of size k to estimate the coefficients $\hat{\alpha}$ and $\hat{\beta}$ in the equation

$$\hat{y} = \hat{\alpha} + \hat{\beta} x \tag{2.96}$$

If $\hat{\alpha}$ and $\hat{\beta}$ are *constants*, then the error $\hat{e}_i$ incurred in predicting the value of $\hat{y}_i$ from (2.96), given x_i, is

$$\hat{e}_i = y_i - \hat{y}_i \tag{2.97}$$

where y_i is the observed value of $\eta(x)$ at x_i and $\hat{y}_i$ is the predicted value. We shall attempt to choose $\hat{\alpha}$ and $\hat{\beta}$ so as to minimize these errors $\hat{e}_i$. This concept is illustrated in Figure 2.20.

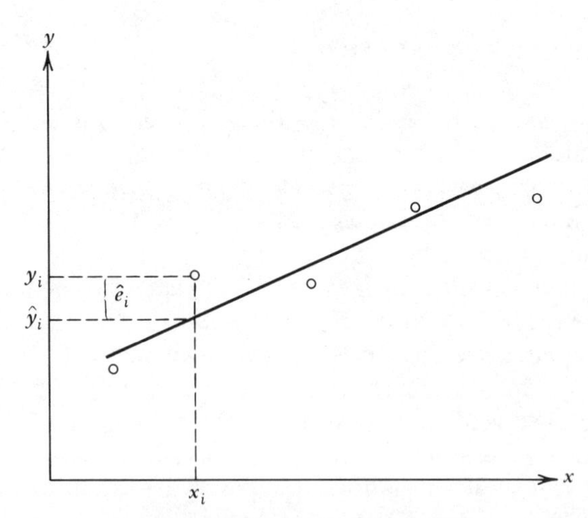

Figure 2.20. Method of least-squares.

Since we cannot minimize the $\hat{e}_i$ individually, we choose to minimize a criterion that aggregates these $\hat{e}_i$ into a single quantity. The criterion we choose to minimize is $\Sigma_{i=1}^{k}\hat{e}_i^2$. That is, we seek values of $\hat{\alpha}$ and $\hat{\beta}$ so as to minimize the sum of squares of the $\hat{e}_i$ quantities; hence, the expression "least-squares." The ensuing development shows how we can exploit this concept to compute $\hat{\alpha}$ and $\hat{\beta}$.

We seek to minimize the quantity

$$S=\sum_{i=1}^{k}\hat{e}_i^2=\sum_{i=1}^{k}(y_i-\hat{y}_i)^2$$

or

$$S=\sum_{i=1}^{k}\left[y_i-(\hat{\alpha}+\hat{\beta}x_i)\right]^2 \tag{2.98}$$

Recalling the *optimality criterion* from Chapter 1, we must take the first partial derivatives of S with respect to $\hat{\alpha}$ and $\hat{\beta}$, respectively, equate these partial derivatives of zero, and solve a system of two linear algebraic equations in two unknowns $\hat{\alpha}$ and $\hat{\beta}$.

We thus have

$$2\sum_{i=1}^{k}\left[y_i-(\hat{\alpha}+\hat{\beta}x_i)\right](-1)=0$$

$$2\sum_{i=1}^{k}\left[y_i-(\hat{\alpha}+\hat{\beta}x_i)\right](-x_i)=0$$

which upon rearrangement yields

$$\begin{aligned}\sum_{i=1}^{k}y_i&=\hat{\alpha}k+\hat{\beta}\sum_{i=1}^{k}x_i\\ \sum_{i=1}^{k}x_iy_i&=\hat{\alpha}\sum_{i=1}^{k}x_i+\hat{\beta}\sum_{i=1}^{k}x_i^2\end{aligned} \tag{2.99}$$

which are called the *least-squares normal equations*. Solution of these normal equations for the coefficients $\hat{\alpha}$ and $\hat{\beta}$ gives the equations

$$\begin{aligned}\hat{\beta}&=\frac{k\sum x_iy_i-\left(\sum x_i\right)\left(\sum y_i\right)}{k\sum x_i^2-\left(\sum x_i\right)^2}\\ \hat{\alpha}&=\frac{y_i-\hat{\beta}\sum x_i}{k}\end{aligned} \tag{2.100}$$

EXAMPLE. The electrical resistance of a certain metal is known to depend on the temperature of the metal. Measurements of resistance (y) were made at five different temperatures (x) with the following results:

resistance y (ohms)	3.5	7.2	12.6	16.4	20.2
temperature x (°K)	100	200	300	400	500

The intermediate calculations are

$$k=5 \qquad \sum x_i = 1{,}500 \qquad \sum y_i = 59.9$$

$$\sum x_i^2 = 550{,}000 \qquad \sum x_i y_i = 22{,}230$$

The regression coefficients are

$$\hat{\beta} = \frac{(5)(22{,}230)-(1500)(59.9)}{(5)(550{,}000)-(1500)^2} = \frac{21{,}300}{500{,}000} = 0.0426$$

$$\hat{\alpha} = \frac{(59.9)-(0.0426)(1500)}{5} = -0.8$$

Thus the regression equation is

$$\hat{y} = -0.8 + 0.0426x$$

Hypothesis Tests on the Slope and Intercept

Having computed sample estimates $\hat{\alpha}$ and $\hat{\beta}$ of the true but unknown regression coefficients α and β in (2.94), it is often necessary to test such null hypotheses as

$$H_0\colon \hat{\alpha} = \alpha_0$$

$$H_0\colon \hat{\beta} = \beta_0 \tag{2.101}$$

Although no assumptions on the distribution of the y_i values was necessary to compute the estimated regression (2.96), hypothesis testing with the estimates $\hat{\alpha}$ and $\hat{\beta}$ will require the assumptions that the true relation between y and x is linear, as expressed in (2.94) and that the y_i values are independently normally distributed with mean $\alpha + \beta x_i$ and the common variance σ^2. If as in (2.95),

$$y_i = \alpha + \beta x_i + \varepsilon_i \tag{2.102}$$

it follows that the ε_i are independently normally distributed with mean zero and common variance σ^2. This concept is illustrated in Figure 2.21.

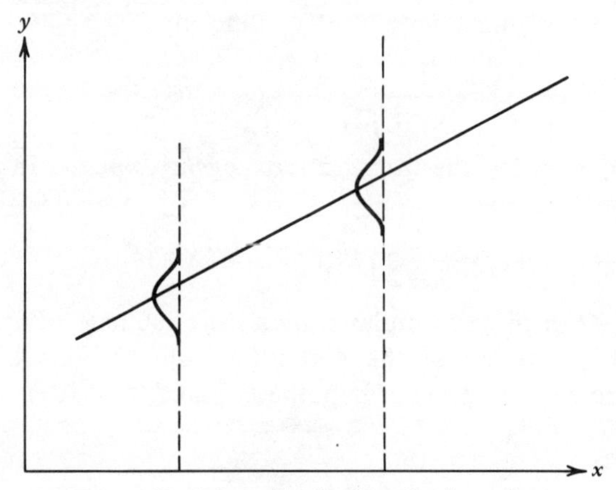

Figure 2.21. Normally distributed values of y_i.

To facilitate hypothesis testing with $\hat{\alpha}$ and $\hat{\beta}$, it is convenient to define certain "sums of squares" terms as follows:

$$S(X^2)=k\sum_{i=1}^{k}x_i^2-\left(\sum_{i=1}^{k}x_i\right)^2 \tag{2.103}$$

$$S(Y^2)=k\sum_{i=1}^{n}y_i^2-\left(\sum_{i=1}^{k}y_i\right)^2 \tag{2.104}$$

$$S(XY)=k\sum_{i=1}^{n}x_iy_i-\left(\sum_{i=1}^{k}x_i\right)\left(\sum_{i=1}^{k}y_i\right) \tag{2.105}$$

Using these expressions, we see that (2.100) can be rewritten as

$$\hat{\beta}=\frac{S(XY)}{S(X^2)}$$

$$\hat{\alpha}=\bar{y}-\hat{\beta}\bar{x} \tag{2.106}$$

Note that $\bar{x}$ and $\bar{y}$ are simply the sample means of x and y, respectively, and that the sample variances s_x^2 and s_y^2 can be obtained directly from the above sums of squares relations; that is

$$s_x^2=\frac{S(X^2)}{k(k-1)}$$

$$s_y^2=\frac{S(Y^2)}{k(k-1)} \tag{2.107}$$

The variance σ^2 is the variance of both the y_i values and the ε_i values, and is a function of the deviations $(y_i-\hat{y}_i)$ where $\hat{y}_i$ is the regression estimate at x_i. Thus an estimate of σ^2 is found by

$$s_e^2 = \frac{1}{k-2} \sum_{i=1}^{k} \left[y_i - (\hat{\alpha} + \hat{\beta} x_i) \right]^2 \tag{2.108}$$

The quantity s_e^2 is called the *standard error of the estimate*. In terms of the sums of squares relations

$$s_e^2 = \frac{1}{k-2} \left[S(Y^2) - \hat{\beta} S(XY) \right] \tag{2.109}$$

The divisor $k-2$ in these formulas *unbiases* the estimate of σ^2.

Recalling the structure of the t statistic from (2.79), we can form t statistics for use with the tests of hypothesis stated in (2.101). The statistics

$$t = \frac{(\hat{\alpha} - \alpha_0)}{s_e} \cdot \sqrt{\frac{kS(X^2)}{S(X^2) + (k\bar{x})^2}}$$

$$t = \frac{(\hat{\beta} - \beta_0)}{s_e} \cdot \sqrt{\frac{S(X^2)}{k}} \tag{2.110}$$

are each t-distributed with $v = k-2$ degrees of freedom. If we test the hypothesis H_0: $\alpha = 0$ or H_0: $\beta = 0$, which is often the case, the above statistics become simply

$$t = \frac{\hat{\alpha}}{s_e} \cdot \sqrt{\frac{kS(X^2)}{S(X^2) + (k\bar{x})^2}}$$

$$t = \frac{\hat{\beta}}{s_e} \cdot \sqrt{\frac{S(X^2)}{k}} \tag{2.111}$$

Confidence intervals for α and β can be obtained from

$$\hat{\alpha} \pm t_{\alpha/2, k-2} \cdot s_e \sqrt{\frac{S(X^2) + (k\bar{x})^2}{kS(X^2)}}$$

and

$$\hat{\beta} \pm t_{\alpha/2, k-2} \cdot s_e \sqrt{\frac{k}{S(X^2)}} \tag{2.112}$$

EXAMPLE. From our previous example,

$$S(X^2) = (5)(550{,}000) - (1500)^2 = 250{,}000$$

$$S(Y^2) = (5)(899.85) - (59.9)^2 = 911.2$$

$$S(XY) = (5)(22{,}230) - (1500)(59.9) = 21{,}300$$

$$s_e^2 = \tfrac{1}{3} \left[911.2 - (0.0426)(21{,}300) \right] = 1.273$$

The confidence intervals for $\hat{\alpha}$ and $\hat{\beta}$ at $\alpha = 0.05$ (do not confuse α = the probability of Type I error with $\hat{\alpha}$ = the estimated regression coefficient) are as follows:

$$(\hat{\alpha}) \quad -0.8 \pm (3.183)(1.273)^{1/2}\left[\frac{250{,}000 + (5)^2(300)^2}{5(250{,}000)}\right]^{1/2} = [-6.53, 4.93]$$

$$(\hat{\beta}) \quad 0.0426 \pm (3.183)(1.273)^{1/2}\left[\frac{5}{250{,}000}\right]^{1/2} = [0.027, 0.059]$$

Curvilinear Regression

In these cases where a simple linear relationship such as equation (2.94) fails to hold, we can often find a suitable nonlinear form. Some of the common nonlinear forms include

$$\eta(x) = \beta_0 + \beta_1 x + \beta_2 x^2 + \ldots + \beta_p x^p \tag{2.113}$$

$$\eta(x) = \alpha\beta^x \tag{2.114}$$

$$\eta(x) = \alpha x^\beta \tag{2.115}$$

$$\eta(x) = \alpha e^{\beta x} \tag{2.116}$$

The form in (2.113) is called *polynomial regression*. The exponential forms in (2.114)–(2.116) can be linearized through appropriate logarithmic transformations to yield

$$\log\eta(x) = \log\alpha + x\log\beta \tag{2.114a}$$

$$\log\eta(x) = \log\alpha + \beta\log x \tag{2.115a}$$

$$\ln\eta(x) = \ln\alpha + \beta x \tag{2.116a}$$

In (2.114a) and (2.115a), either the common logarithm (log) or the natural logarithm (ln) can be used to linearize (2.114) and (2.115), but the natural logarithm is more convenient for (2.116a) because it eliminates the ($\log e$) term. It is then a simple matter to apply the standard least-squares formulas (2.100) to compute the estimates of these linearized forms. Of course, we must be careful to substitute the logarithms of the x_i and y_i values where appropriate.

The polynomial regression in (2.113) is more involved, however. Here we must resort to the basic least-squares approach in (2.98), taking the $(p+1)$ partial derivatives, and forming $p+1$ *normal equations* to yield

$$\begin{aligned}
\sum y &= k\hat{\beta}_0 + \hat{\beta}_1\sum x + \ldots + \hat{\beta}_p\sum x^p \\
\sum xy &= \hat{\beta}_0\sum x + \hat{\beta}_1\sum x^2 + \ldots + \hat{\beta}_p\sum x^{p+1} \\
&\vdots \quad \vdots \\
\sum x^p y &= \hat{\beta}_0\sum x^p + \hat{\beta}_1\sum x^{p+1} + \ldots + \hat{\beta}_p\sum x^{2p}
\end{aligned} \tag{2.117}$$

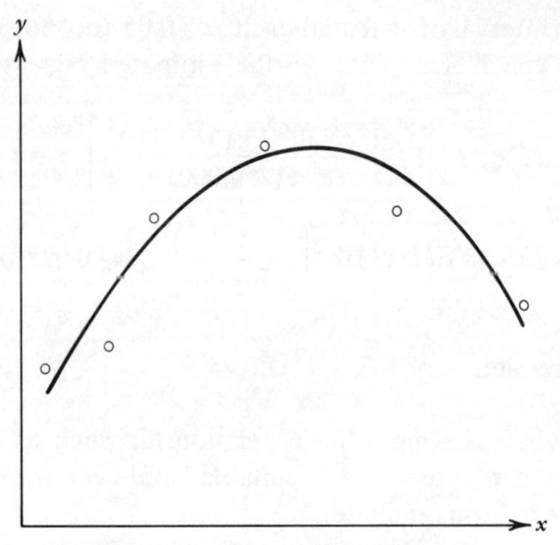

Figure 2.22. Polynomial least-squares.

These are $p+1$ linear algebraic equations in $p+1$ unknowns. Their solution yield the $p+1$ estimates $\hat{\beta}_0, \hat{\beta}_1, \ldots, \hat{\beta}_p$.

Figure 2.22 illustrates the least-squares concept for a polynomial.

EXAMPLE. To illustrate the fitting of a polynomial model by the method of least squares, consider the problem of finding the optimum age (hours) of a premixed binder so that the flexural strength (lb/in.^{2}) of a propellant grain manufactured from the binder is maximized. The following experimental data were obtained:

x Binder age, hr	y Flexural strength, lb/in.^{2}
0	154
2	181
4	194
6	208
8	199
10	182
30	1118

Inspection of the data plotted in Figure 2.23 suggests that a second degree polynomial should yield a good fit.

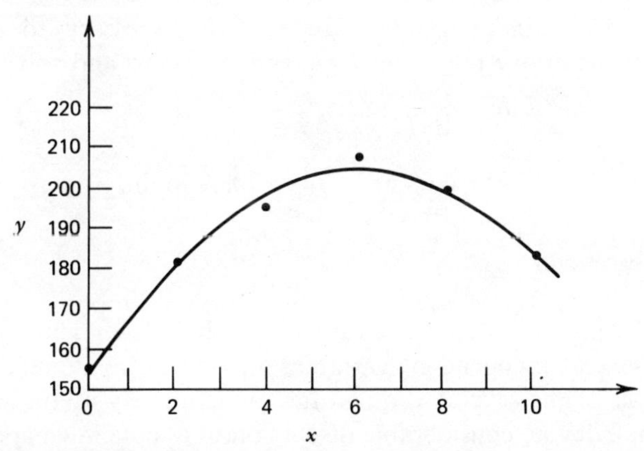

Figure 2.23. Second-degree polynomial regression relating flexural strength to binder age.

The polynomial regression normal equations needed for this analysis are

$$\sum y = k\hat{\beta}_0 + \hat{\beta}_1 \sum x + \hat{\beta}_2 \sum x^2$$

$$\sum xy = \hat{\beta}_0 \sum x + \hat{\beta}_1 \sum x^2 + \hat{\beta}_2 \sum x^3$$

$$\sum x^2y = \hat{\beta}_0 \sum x^2 + \hat{\beta}_1 \sum x^3 + \hat{\beta}_2 \sum x^4$$

First calculating the required sums from the above data, we obtain

$$k = 6 \qquad \sum x = 30 \qquad \sum x^2 = 220 \qquad \sum x^3 = 1800 \qquad \sum x^4 = 15{,}664$$

$$\sum xy = 5{,}798 \qquad \sum x^2y = 42{,}252 \qquad \sum y = 1118$$

thus giving us the following three equations in three unknowns:

$$1118 = 6\hat{\beta}_0 + 30\hat{\beta}_1 + 220\hat{\beta}_2$$

$$5798 = 30\hat{\beta}_0 + 220\hat{\beta}_1 + 1800\hat{\beta}_2$$

$$42{,}252 = 220\hat{\beta}_0 + 1800\hat{\beta}_1 + 15{,}664\hat{\beta}_2$$

The solution to these equations is

$$\hat{\beta}_0 = 153.1 \qquad \hat{\beta}_1 = 16.72 \qquad \hat{\beta}_2 = -1.375$$

giving us the second-degree polynomial equation

$$\hat{y} = 153.1 + 16.72x - 1.375x^2$$

The graph of this equation is also shown in Figure 2.23.

To find the optimum binder age x, it is necessary to equate the derivative of the above polynomial regression to zero and solve for x.

$$\frac{d\hat{y}}{dx} = 16.72 - 2.75x = 0$$

$$x^* = 6.08 \text{ hr} \qquad \hat{y}^* = 203.9 \text{ lb/in.}$$

Multiple Regression

We have already considered the case in which a dependent variable y is related to several independent variables x_i, $i = 1, \ldots, n$. Prediction models for y in terms of the several x_i quantities are extremely useful in optimization. We shall devote considerable discussion to a topic in Chapter 3 called *response surface methodology*, but it is necessary to introduce the fundamental method of response surface analysis, *multiple regression*, in this section.

As with linear and curvilinear regression, we must stipulate that the x-values can be observed without error. The y-values will be considered to relate to the x_i quantities in a manner.

$$\eta(x) = \beta_0 + \beta_1 x_1 + \beta_2 x_2 + \ldots + \beta_n x_n \tag{2.118}$$

where x is the n vector of independent variables. As we saw earlier, the observed values of y contain a random error, so that

$$y = \beta_0 + \beta_1 x_1 + \beta_2 x_2 + \ldots + \beta_n x_n + \varepsilon \tag{2.119}$$

We shall seek to estimate the multiple regression

$$\hat{y} = \hat{\beta}_0 + \hat{\beta}_1 x_1 + \hat{\beta}_2 x_2 + \ldots + \hat{\beta}_n x_n \tag{2.120}$$

Applying the least-squares criterion, we seek to minimize

$$S = \sum_{i=1}^{n} \left[y_i - \left(\hat{\beta}_0 + \hat{\beta}_1 x_1 + \ldots + \hat{\beta}_n x_n \right) \right]^2 \tag{2.121}$$

which, upon taking the $n+1$ partial derivatives with respect to $\hat{\beta}_0, \hat{\beta}_1, \ldots, \hat{\beta}_n$, equating these derivatives to zero, and rearranging, yields the $n+1$ *multiple regression normal equations.*

$$\begin{aligned}
\sum y &= k\hat{\beta}_0 + \hat{\beta}_1 \sum x_1 + \hat{\beta}_2 \sum x_2 + \ldots + \hat{\beta}_n \sum x_n \\
\sum x_1 y &= \hat{\beta}_0 \sum x_1 + \hat{\beta}_1 \sum x_1^2 + \hat{\beta}_2 \sum x_1 x_2 + \ldots + \hat{\beta}_n \sum x_1 x_n \qquad (2.122) \\
&\vdots \qquad\qquad \vdots \\
\sum x_n y &= \hat{\beta}_1 \sum x_n + \hat{\beta}_1 \sum x_1 x_n + \hat{\beta}_2 \sum x_2 x_n + \ldots + \hat{\beta}_n \sum x_n^2
\end{aligned}$$

which, upon solution produces the regression coefficients $\hat{\beta}_0, \hat{\beta}_1, \ldots, \hat{\beta}_n$.

EXAMPLE. A designed experiment was conducted to measure the yield $y(\%)$ from a chemical process as a function of reaction time x_1 (hours) and reactor temperature x_2 (°F). The following data were obtained:

x_1	x_2	y
3.0	297.5	62.7
3.0	322.5	76.2
3.0	347.5	80.8
6.0	297.5	80.8
6.0	322.5	89.2
6.0	347.5	78.6
9.0	297.5	90.1
9.0	322.5	88.0
9.0	247.5	76.1

Forming three equations in three unknowns as in (2.122) and solving for the coefficient $\hat{\beta}_0$, $\hat{\beta}_1$, and $\hat{\beta}_2$ yields the following results:

$$\hat{\beta}_0 = 64.69 \qquad \hat{\beta}_1 = 1.917 \qquad \hat{\beta}_2 = 0.0127$$

Thus the prediction equation for process yield as a function of reaction time and temperature is

$$\hat{y} = 64.7 + 1.917x_1 + 0.0127x_2$$

Correlation

So far in this section we have studied problems in which (1) a cause and effect relationship was assumed to exist between one or more variables x (or x_i, $i = 1,\ldots,n$) and a response y, and (2) the independent variable x (or x_i, $i = 1,\ldots,n$) could be observed without error. Although these assumptions apply in many experimental situations, we often encounter the case where one or both of these conditions are violated; that is, the cause and effect relationship may be unclear, and/or the x's as well as the y's may be random variables. For example, the tensile strength y of copper may well be related to its hardness x. But can it be stated that tensile strength is a *function* of hardness? Such a functional relationship could in fact exist, but in the process of obtaining experimental measurements of x and y we would very likely observe that each is a random variable.

In those cases where the assumptions underlying regression do not apply, or where we are not interested in a functional relationship but rather the degree of association between two variables x and y, we use a technique called *correlation analysis*. Correlation is a measure of the extent to which two variables x and y vary together. The statistic employed in correlation analysis is the *sample correlation* r, defined as

$$r=\frac{\operatorname{Cov}(x,y)}{s_x s_y} \tag{2.123}$$

The term $\operatorname{Cov}(x,y)$ is a shorthand expression for the *covariance of* x *and* y, and is computed from a sample of k observations (x_i,y_i) as

$$\operatorname{Cov}(x,y)=\frac{1}{k-1}\sum_{i=1}^{k}(x_i-\bar{x})(y_i-\bar{y}) \tag{2.124}$$

Hence

$$r=\frac{\left[\sum(x_i-\bar{x})(y_i-\bar{y})\right]/(k-1)}{\left[\frac{\sum(x_1-\bar{x}^2)}{(k-1)}\right]^{\frac{1}{2}}\left[\frac{\sum(y_i-\bar{y})^2}{(k-1)}\right]^{\frac{1}{2}}} \tag{2.125}$$

which can be simplified to

$$r=\frac{k\left(\sum x_i y_i\right)-\left(\sum x_i\right)\left(\sum y_i\right)}{\left\{\left[k\sum x_i^2-\left(\sum x_i\right)^2\right]\left[k\sum y_i^2-\left(\sum y_i\right)^2\right]\right\}^{1/2}} \tag{2.126}$$

EXAMPLE. The following measurements were made of the carbon content x (%) and the permeability index y of 11 sinter mixes:

x	4.5	5.1	4.2	4.6	5.5	4.9	4.4	5.0	4.7	4.5	5.1
y	20	14	12	16	15	21	13	20	18	23	14

The correlation coefficient r for the variables x and y is 0.0162.

Having computed the sample correlation coefficient r, we often wish to test the significance of the relationship represented by this quantity. That is, we can test whether the *true* correlation coefficient ρ has some specific value ρ_0 over its range $-1\leqslant\rho\leqslant+1$. The null hypothesis for this test is $H_0:\rho=\rho_0$. For large samples, say $k\geqslant 30$, the sample statistic is

$$Z=\frac{1}{2}\ln\frac{1+r}{1-r} \tag{2.127}$$

It can be shown that Z is approximately normally distributed with mean

$$\mu_z=\frac{1}{z}\ln\frac{1+\rho}{1-\rho} \tag{2.128}$$

and variance

$$\sigma_z^2=\frac{1}{k-3} \tag{2.129}$$

Employing the transformation to the standard normal random variable z, we have

$$z = \frac{Z - \mu_z}{\sigma_z}$$

or

$$z = \frac{\left(\frac{1}{2}\ln\frac{1+r}{1-r}\right) - \left(\frac{1}{2}\ln\frac{1+\rho}{1-\rho}\right)}{\frac{1}{\sqrt{k-3}}}$$

which upon rearrangement, and substitution of the null hypothesis value ρ_0, yields the test statistic

$$z = \frac{\sqrt{k-3}}{2}\ln\left[\frac{(1+r)(1-\rho_0)}{(1-r)(1+\rho_0)}\right] \tag{2.130}$$

This statistic is unreliable for small samples. If the sample size is smaller than 30 and a more restrictive null hypothesis $H_0: \rho = 0$ is tested, a t statistic

$$t = \frac{\sqrt{k-2}}{\sqrt{1-r^2}}|r| \tag{2.131}$$

can be used. This statistic has $\nu = k-2$ degrees of freedom.

EXAMPLE. In our previous example ($r = 0.0162$), if we test the null hypothesis $H_0: \rho = 0$ against the alternative hypothesis $H_1: \rho \neq 0$ at $\alpha = 0.05$, the value of the test statistic from (2.131) is

$$t = \frac{\sqrt{11-2}}{\sqrt{1-0.0162^2}}|0.0162| = 0.0486$$

Comparing this to $t_{1-\frac{\alpha}{2},k-2} = t_{0.975,9} = 2.262$. Since $t < 2.262$, we do not reject the null hypothesis. We conclude that x and y are uncorrelated.

ANALYSIS OF VARIANCE

In many experiments the main objective is to determine the effect of various *factors*, also called treatments, on some response variable y. For instance, in the evaluation of tire wear in a certain road test, it may be necessary to determine the effect of the tire brand. Or it may be important to determine the heat treatment temperature on the hardness of a fabricated metal part. In the case of the tire wear experiment, the tire brand represents a level of a *qualitative* factor, whereas the heat treatment temperature in the second experiment is a level of a *quantitative* factor. A quantitative factor is one that can be controlled over a continuum of values. We usually choose to perform the experiment at a few, equally spaced, levels of the factor.

Such carefully chosen conditions give rise to the term *designed experiment*. The principles of experimental design and the statistical procedures by which such experiments are analyzed are called *analysis of variance*. We shall examine the basic principles of analysis of variance in this section, but a much more extensive treatment of that subject is the focus of Chapter 3. In order to develop the conceptual framework of analysis of variance, we shall first examine experiments involving a single factor. The common term used in referring to such a procedure is *one-way analysis of variance*, and is abbreviated *one-way ANOVA*. Then we shall extend these principles to the analysis of experiments involving two factors; not surprisingly, these procedures are called *two-way analysis of variance* or two-way ANOVA.

One-Way Analysis of Variance

In an earlier section in this chapter we discussed the problem of testing hypotheses about the difference of means of two normal populations. The t statistic, incorporating the values of the two sample means and variances, was employed to test the null hypothesis H_0: $\mu_1 - \mu_2 = \delta$. Now consider the case in which we have samples from three or more populations. The t statistic is no longer suitable.

Suppose $x_1, x_2, \ldots, x_k$ are levels of some factor x, which we postulate as having some effect on a response variable of interest y. We set up an experiment in which n_1 observations are made of the response y at level x_1, n_2 observations at level x_2, and so on, with n_j observations at a given level x_j. The levels x_j are called *treatments*, there being k treatments in the experimental design. The observations $(y_{1j}, y_{2j}, \ldots y_{nj})$ constitute a *random sample* of size n_j at the treatment level x_j. The layout of a one-way ANOVA experiment is shown in Table 2.13.

We say that these y_{ij} values come from a normal distribution having mean μ_j and variance σ^2. Note that there may be a distinct mean μ_j associated with level x_j, but the variance σ^2 is the same over all levels of x.

Let $\bar{y}_{.j}$ be the mean of the j-th sample, and $\bar{y}_{..}$ the mean of all k samples.

Table 2.13. One-way experimental layout

x_1	x_2	$\cdots$	x_j	$\cdots$	x_k
y_{11}	y_{12}		y_{1j}		y_{1k}
y_{21}	y_{22}	$\cdots$	y_{2j}	$\cdots$	y_{2k}
$\vdots$	$\vdots$			$\vdots$	
$y_{n_1,1}$	$y_{n_2,2}$		$\vdots$		$y_{n_k,k}$
			$y_{n_j,j}$		
T_1	T_2	$\cdots$	T_j	$\cdots$	T_k

where $T_j = \sum_{i=1}^{n_j} y_{ij}$

That is,

$$\bar{y}_{.j} = \frac{1}{n_j} \sum_{i=1}^{n_j} y_{ij} = \frac{T_j}{n_j} \tag{2.132}$$

$$\bar{y}_{..} = \frac{1}{n} \sum_{j=1}^{k} \sum_{i=1}^{n_j} y_{ij} = \frac{T}{n} \tag{2.133}$$

where

$$n = \sum_{j=1}^{k} n_j \tag{2.134}$$

$$T = \sum_{j=1}^{k} T_j \tag{2.135}$$

Now in terms of the total variance,

$$S(Y^2) = \sum (y_{ij} - \bar{y}_{..})^2 = \sum y_{ij}^2 - n\bar{y}_{..}^2 \tag{2.136}$$

But $S(Y^2)$ can also be expressed as

$$\begin{aligned} S(Y^2) &= \sum (y_{ij} - \bar{y}_{..})^2 \\ &= \sum \left[(y_{ij} - \bar{y}_{.j}) + (\bar{y}_{.j} - \bar{y}_{..}) \right]^2 \\ &= \sum (y_{ij} - \bar{y}_{.j})^2 + \sum (\bar{y}_{.j} - \bar{y}_{..})^2 \end{aligned} \tag{2.137}$$

This is true since the cross-product term in the above expression, $(y_{ij} - \bar{y}_{.j})(\bar{y}_{.j} - \bar{y}_{..})$, is zero. We can observe that the first term on the right side of (2.137) represents the sum of squares of derivations of the y_{ij} from the sample mean $\bar{y}_{.j}$, whereas the second term is the sum of squares of deviations of the sample means $\bar{y}_{.j}$ from the overall mean $\bar{y}_{..}$. We can define these as

$$S_E = \sum (y_{ij} - \bar{y}_{.j})^2 \tag{2.138}$$

$$S_T = \sum (\bar{y}_{.j} - \bar{y}_{..})^2 \tag{2.139}$$

where S_E is called the "sum of squares for error" and S_T the "sum of squares for the treatment." That is, S_E reflects the variation *within* the samples, and S_T the variation *between* samples. Hence, the total variation is

$$S(Y^2) = S_E + S_T \tag{2.140}$$

A more convenient computational scheme for calculating these sums of squares terms is given by the following formulas:

$$S_T = \sum_{j=1}^{k} \frac{T_j^2}{n_j} - \frac{T^2}{n} \tag{2.141}$$

$$S_E = S(Y^2) - S_T \tag{2.142}$$

$$S(Y^2) = \sum_{j=1}^{k} \sum_{i=1}^{n_j} y_{ij}^2 - \frac{T^2}{n} \tag{2.143}$$

where $T = \sum_{j=1}^{k} T_j$ is stated in equation (2.135).

The appropriate test of hypothesis for one-way ANOVA is $H_0: \mu_1 = \mu_2 = \cdots = \mu_j = \mu_k$ against the alternative hypothesis H_1: the means μ_j are not all equal. Another way of stating the null hypothesis H_0 is that "the different treatment levels x_j of the factor X have no effect on the response y." Recall that the F statistic is the ratio of χ^2 statistics. Now under the null hypothesis H_0, the two quantities S_E/σ^2 and S_T/σ^2 are independent random variables having Chi-Square distributions with $n-k$ and $k-1$ degrees of freedom, respectively. Hence it follows that the test statistic

$$\frac{S_T/(k-1)\sigma^2}{S_E/(n-k)\sigma^2} = \frac{S_T/(k-1)}{S_E/(n-k)} \tag{2.144}$$

has the F distribution with $(k-1, n-k)$ degrees of freedom. We would reject H_0 at the α level of significance if

$$\frac{S_T/(k-1)}{S_E/(n-k)} > F_{k-1, n-1, \alpha}$$

and accept H_0 otherwise.

Table 2.14. ANOVA table for one-way analysis of variance

Source of variation	Sum of squares	Degrees of freedom	Mean square	F-test statistic
Between Samples	S_T	$k-1$	$\frac{S_T}{(k-1)}$	$\frac{S_T/(k-1)}{S_E/(n-k)}$
Within Samples	S_E	$n-k$	$\frac{S_E}{(n-k)}$	
Total	$S(Y^2)$	$n-1$		

The *ANOVA table*, shown in Table 2.14, is the typical approach to performing this test of hypothesis.
Another way of representing the F-test statistic is to denote the two mean square terms as MS_T and MS_E, respectively, so that $F = MS_T/MS_E$. The rationale for the F test is that we can conclude that the variation due to the treatment levels x_j is *significant* if it is relatively large as compared to the variation within the samples.

EXAMPLE. Four brands of tires, A, B, C, and D, are compared for thread loss after 25,000 miles of driving. Four tires of each brand were installed randomly on four automobiles. After 25,000 miles, the following tread losses (inches) were recorded.

A	B	C	D
0.72	0.68	0.66	0.70
0.76	0.70	0.64	0.68
0.73	0.69	0.68	0.72
0.75	0.67	0.66	0.72
T_A	T_B	T_C	T_D

The quantities needed for the analysis of variance are

$$T_A = 2.96 \qquad T_B = 2.74 \qquad T_C = 2.64 \qquad T_D = 2.82$$

$$T = T_A + T_B + T_C + T_D = 11.16$$

$$n = n_1 + n_2 + n_3 + n_4 = 4+4+4+4 = 16$$

Now the sums of square terms are

$$S(Y^2) = \left[(0.72)^2 + (0.76)^2 + \cdots + (0.72)^2 + (0.72)^2\right] - \frac{(11.16)^2}{16}$$

$$= 7.8012 - 7.7841$$

$$= 0.0171$$

$$S_T = \frac{\left[(2.96)^2+(2.74)^2+(2.64)^2+(2.82)^2\right]}{4} - \frac{(11.16)^2}{16}$$

$$= \frac{\left[8.766+7.5076+6.9696+7.9524\right]}{4} - 7.7841$$

$$= 7.7978 - 7.7841$$

$$= 0.0137$$

$$S_E = S(Y^2) - S_T = 0.0171 - 0.0137 = 0.0034$$

The ANOVA table for this analysis of variance problem is as follows:

Source of variation	Sum of squares	Degrees of freedom	Mean square	*F*-test statistic
Between brands	$S_T = 0.0137$	$k-1=3$	0.00457	16.15
Within brands	$S_E = 0.0034$	$n-k=12$	0.000283	
Total	$S(Y^2) = 0.0171$	$n-1=15$		

The tabulated value of F against which $F = 16.15$ is compared at $\alpha = 0.05$ is $F_{3,12,0.05} = 3.49$. Since $F > 3.49$, we reject the null hypothesis that there is no difference in tread wear among the four tire brands and conclude that the tire brand is a significant factor on tire wear.

Two-Way Analysis of Variance

Now let us consider the case in which we must design an experiment to study the effects of two variables A and B on a response variable y. Factor A has r levels $a_1, a_2, \ldots, a_r$, whereas factor B has s levels $b_1, b_2, \ldots, b_s$. For each combination of levels (a_i, b_j), we measure the response y_{ij}. Table 2.15 gives the data layout for a two-way ANOVA experiment. The row totals are denoted $T_{i.}$ and the column total $T_{.j}$. The row and column means are therefore

$$\bar{y}_{i.} = \frac{1}{s}\sum_{j=1}^{s} y_{ij} = \frac{T_{i.}}{s} \tag{2.145}$$

$$\bar{y}_{.j} = \frac{1}{r}\sum_{i=1}^{r} y_{ij} = \frac{T_{.j}}{r} \tag{2.146}$$

$$\bar{y}_{..} = \frac{1}{rs}\sum_{i=1}^{r}\sum_{j=1}^{s} y_{ij} = \frac{T}{rs} \tag{2.147}$$

$$T = \sum_{i=1}^{r} T_{i.} = \sum_{j=1}^{s} T_{.j} \tag{2.148}$$

Table 2.15. A two-way anova experimental layout

Factor A levels	Factor B levels b_1	b_2	$\cdots$	b_j	$\cdots$	b_s	Row levels
a_1	y_{11}	y_{12}	$\cdots$	y_{1j}	$\cdots$	y_{1s}	$T_{1.}$
a_2	y_{21}	y_{22}	$\cdots$	y_{2j}	$\cdots$	y_{2s}	$T_{2.}$
$\vdots$	$\vdots$			$\vdots$			$\vdots$
a_i	y_{i1}	y_{i2}	$\cdots$	y_{ij}	$\cdots$	y_{is}	$T_{i.}$
$\vdots$	$\vdots$			$\vdots$			$\vdots$
a_r	y_{r1}	y_{r2}	$\cdots$	y_{rj}	$\cdots$	y_{rs}	$T_{r}\cdot$
	$T_{.1}$	$T_{.2}$	$\cdots$	$T_{.j}$	$\cdots$	$T_{.s}$	T

As in one-way ANOVA, we seek to partition the total variation $S(Y^2)$ into its respective components. That is,

$$S(Y^2) = \sum_{i=1}^{r} \sum_{j=1}^{s} (y_{ij} - \bar{y}_{..})^2$$

$$= \sum \sum (y_{ij} - \bar{y}_{i.} - \bar{y}_{.j} + \bar{y}_{..})^2 + \sum \sum (\bar{y}_{i.} - \bar{y}_{..})^2 + \sum \sum (\bar{y}_{.j} - \bar{y}_{..})^2$$

$$= S_E + S_A + S_B \tag{2.149}$$

The ANOVA table is as shown in Table 2.16. As with the one-way ANOVA, more computationally tractable formulas are

$$S_A = \sum_{i=1}^{r} \frac{T_{i.}^2}{s} - \frac{T^2}{rs} \tag{2.150}$$

$$S_B = \sum_{j=1}^{s} \frac{T_{.j}^2}{r} - \frac{T^2}{rs} \tag{2.151}$$

Table 2.16. ANOVA table for two-way analysis of variance

Source of variation	Sum of squares	Degrees of freedom	Mean Square	F-test statistic
Due to factor A	S_A	$r-1$	$S_A/(r-1)$	MS_A/MS_E
Due to factor B	S_B	$s-1$	$S_B/(s-1)$	MS_B/MS_E
Due to error	S_E	$(r-1)(s-1)$	$S_E/(r-1)(s-1)$	
Total	$S(Y^2)$	$rs-1$		

$$S(Y^2) = \sum_{i=1}^{r} \sum_{j=1}^{s} y_{ij}^2 - \frac{T^2}{rs} \tag{2.152}$$

$$S_E = S(Y^2) - (S_A + S_B) \tag{2.153}$$

EXAMPLE. An experiment was conducted to evaluate the effects of three different machinists (a_1, a_2, a_3) and three different machines (b_1, b_2, b_3) on the percent defective parts produced. The results of this equipment were as follows:

	b_1	b_2	b_3	$T_{i.}$
a_1	1.2	1.1	0.9	3.2
a_2	1.6	1.7	1.4	4.7
a_3	1.7	1.6	1.5	4.8
$T_{.j}$	4.5	4.4	3.8	$T = 12.7$

The pertinent sums of squares were

$$S_A = \frac{[(3.2)^2 + (4.7)^2 + (4.8)^2]}{3} - \frac{(12.7)^2}{9}$$

$$= \frac{[10.24 + 22.09 + 23.04]}{3} - \frac{161.29}{9}$$

$$= 18.457 \qquad - 17.921$$

$$= 0.536$$

$$S_B = \frac{[(4.5)^2 + (4.4)^2 + (3.8)^2]}{3} - 17.921$$

$$= \frac{[20.25 + 19.36 + 14.44]}{3} - 17.921$$

$$= 18.017 - 17.921$$

$$= 0.096$$

$$S(Y^2) = [(1.2)^2 + (1.6)^2 + \ldots + (1.4)^2 + (1.5)^2] - 17.921$$

$$= 18.570 - 17.921$$

$$= 0.649$$

$$S_E = 0.649 - (0.536 + 0.096)$$

$$= 0.017$$

The two-way ANOVA table is therefore as follows:

Source of variation	Sum of squares	Degrees of freedom	Mean square	F-test statistic
Due to Machinists	$S_A = 0.536$	2	0.268	63.8
Due to Machines	$S_B = 0.096$	2	0.048	11.4
Error	$S_E = 0.017$	4	0.0042	
Total	$S(Y)^2 = 0.649$	8		

For each test of hypothesis, the tabulated value of F is $F_{2,8,0.05} = 4.46$. Since $F_{\text{test}} > 4.46$ for both Factor A and Factor B, we would reject the null hypotheses that (1) the different machinists have no effect on percent defective product, and (2) the machines have no effect on percent defective product.

SUMMARY

We have endeavored in this chapter to present the fundamentals of statistical analysis in sufficient detail to enable the reader to (1) understand the basic principles of experimentation, and (2) be able to apply the more important statistical procedures used to analyze experiments. The reader is urged to have at hand texts that present a more detailed view of statistics.

But the groundwork has been laid for a more extensive presentation of *designed experiments* and *response surface methodology* in Chapter 3. The reader will find frequent use for the basic statistical concepts discussed in this chapter when applying the experimental procedures outlined in the following chapters.

REFERENCES

1. Box, G. E. P., and N. R. Draper, *Evolutionary Operation*, Wiley, New York, 1969.
2. Cochran, W. G., and G. M. Cox, *Experimental Designs*, Second Edition, Wiley, New York, 1957.
3. Davies, O. L., *Design and Analysis of Industrial Experiments*, Second Edition, Hafner Publishing Company, New York, 1956.
4. Draper, N. R., and H. Smith, *Applied Regression Analysis*, Wiley, New York, 1966.
5. Fedorov, V. V., *Theory of Optimal Experiments*, Academic Press, New York, 1972.
6. Graybill, F. A., *An Introduction to Linear Statistical Models*, vol. 1, McGraw-Hill, New York, 1961.

7. Guttman, I., S. S. Wilks, and J. S. Hunter, *Introductory Engineering Statistics*, Second Edition, Wiley, New York, 1971.
8. Hicks, C. R., *Fundamental Concepts in the Design of Experiments*, Second Edition, Holt, Rinehart and Winston, New York, 1973.
9. Lipson, C., and N. J. Sheth, *Statistical Design and Analysis of Engineering Experiments*, McGraw-Hill, New York, 1973.
10. Miller, I., and J. E. Freund, *Probability and Statistics for Engineers*, Second Edition, Prentice-Hall, Englewood Cliffs, NJ, 1977.
11. Montgomery, D. C., *Design and Analysis of Experiments*, Wiley, New York, 1976.
12. Myers, R. H., *Response Surface Methodology*, Allyn and Bacon, Boston, MA, 1971.
13. Natrella, M. G., *Engineering Design Handbook: Experimental Statistics*, AMC Pamphlet AMCP 706-110, Headquarters, U.S. Army Material Command, July, 1963.

Chapter 3

Experimental Design Fundamentals

Chapter 2 presented a brief and rather straightforward treatment of the fundamental concepts of probability and statistics. These basic principles are essential to an appropriate analysis of experimental data, such as one would obtain in experimentally seeking the optimum conditions for a real-world physical process or system.

The present chapter explores the fundamental concepts and principles of *experimental design*, which are the procedures by which we plan, conduct, and analyze experiments. This chapter enlarges on the concepts of *regression* and *analysis of variance* presented in Chapter 2. We shall examine procedures for planning *designed experiments* based on the principles of analysis of variance, and we shall consider the basic approaches of *response surface methodology* that follow from our brief look at regression. But the pace quickens in this chapter, as we adopt a more rigorous mathematical notation based on matrix algebra and employ fewer physical examples to illustrate the concepts presented. Realistic examples are examined in later chapters.

REGRESSION ANALYSIS

Matrix Representation

In the last chapter, the least-squares regression line for approximating a function of a single variable was introduced. In this section the multiple-dimension regression generalization is considered, and the properties of these estimators further examined. This process of data analysis and approximation is called *regression analysis* and will lead directly to the response surface methodologies.

The general regression problem is quite similar in appearance to that of the single variate case. A scalar response y is given by an unknown function $\eta(\mathbf{x})$, which is now a scalar valued function of the input vector $\mathbf{x}$. This is written as

$$y=\eta(\mathbf{x}) \tag{3.1}$$

When η is assumed linear, (2.94) becomes

$$\eta=\beta_0+\beta_1 x_1+\beta_2 x_2+\cdots+\beta_n x_n \tag{3.2}$$

We intend to estimate this function based upon observations of a response variable y which is subject to errors ε as before,

$$y=\beta_0+\beta_1 x_1+\beta_2 x_2+\cdots+\beta_n x_n+\varepsilon \tag{3.3}$$

using the least squares technique of minimizing the squared error. The result will be estimates $\hat{\beta}_0, \hat{\beta}_1, \ldots, \hat{\beta}_n$ of the functional parameters $\beta_0, \beta_1, \ldots, \beta_n$, so that we have the *predicted* response $\hat{y}$, where

$$\hat{y}=\hat{\beta}_0+\sum_{i=1}^{n} \hat{\beta}_i x_i \tag{3.4}$$

It is still possible, but more laborious, to determine the general solution to the normal equations in the multidimensional case without recourse to matrix algebra, but matrix notation will provide a concise description of the regression process for any dimension. A summary of matrix properties is given in Appendix B.

First, consider (3.3). Let the values of the estimated β_i be given as the n vector $\boldsymbol{\beta}$, usually expressed as a *column vector* which is transposed to

$$\boldsymbol{\beta}^T=(\beta_1, \beta_2, \ldots, \beta_n) \tag{3.5}$$

The values of the x_i for each observation are given by $\mathbf{x}$, the transpose of which is

$$\mathbf{x}^T=(x_1, x_2, \ldots, x_n) \tag{3.6}$$

Then (3.3) can be rewritten as

$$y=\beta_0+\mathbf{x}^T\boldsymbol{\beta}+\varepsilon \tag{3.7}$$

To estimate β_0 and $\boldsymbol{\beta}$ it is necessary to make observations of y at several combinations of the x_i. This can be represented by a *design matrix*, which consists of m rows of the experimental input values $\mathbf{x}_i^T$, where $\mathbf{x}_i^T$ corresponds to the x-values for the ith observation, so that

$$D=\begin{bmatrix} \mathbf{x}_1^T \\ \mathbf{x}_2^T \\ \vdots \\ \mathbf{x}_m^T \end{bmatrix}=\begin{bmatrix} x_{11} & x_{12} & x_{13} & \cdots & x_{1n} \\ x_{21} & x_{22} & x_{23} & \cdots & x_{2n} \\ \vdots & & & & \\ x_{m1} & x_{m2} & x_{m3} & \cdots & x_{mn} \end{bmatrix} \tag{3.8}$$

The design matrix D is m by n. If we augment this matrix with a column of 1's, we have the *experimental matrix* X,

$$X = \begin{bmatrix} & x_1^T \\ \mathbf{1} & x_2^T \\ & \cdots \\ & x_m^T \end{bmatrix} = \begin{bmatrix} 1\ x_{11} & x_{12} & x_{13} & \cdots & x_{1n} \\ 1\ x_{21} & x_{22} & x_{23} & \cdots & x_{2n} \\ \cdots & \cdots & & & \\ 1\ x_{m1} & x_{m2} & x_{m3} & \cdots & x_{mn} \end{bmatrix} \tag{3.9}$$

Equation (3.7) can be generalized to all m observations on $\mathbf{y}$ to

$$\mathbf{y} = X\boldsymbol{\beta} + \boldsymbol{\varepsilon} \tag{3.10}$$

This is the linear regression model for any number of variables and observations. To derive the least-squares estimators $\hat{\boldsymbol{\beta}}$ for $\boldsymbol{\beta}$ in the linear case, $\hat{\mathbf{y}}$ is given by

$$\hat{\mathbf{y}} = X\hat{\boldsymbol{\beta}} \tag{3.11}$$

It is desired to minimize the errors

$$\hat{\boldsymbol{\varepsilon}} = \mathbf{y} - \hat{\mathbf{y}} \tag{3.12}$$

in the least squares sense,

$$\min L = \hat{\boldsymbol{\varepsilon}}^T\hat{\boldsymbol{\varepsilon}} = (\mathbf{y} - \hat{\mathbf{y}})^T(\mathbf{y} - \hat{\mathbf{y}}) \tag{3.13a}$$

$$= (\mathbf{y} - X\hat{\boldsymbol{\beta}})^T(\mathbf{y} - X\hat{\boldsymbol{\beta}}) \tag{3.13b}$$

$$= \mathbf{y}^T\mathbf{y} - 2\hat{\boldsymbol{\beta}}^T X^T\mathbf{y} + \hat{\boldsymbol{\beta}}^T X^T X\hat{\boldsymbol{\beta}} \tag{3.13c}$$

since $\boldsymbol{\beta}^T X^T\mathbf{y} = \mathbf{y}^T X\boldsymbol{\beta}$. The solution to this problem requires the use of matrix calculus to obtain the minimizing point given where the vector derivative equals the zero vector,

$$\frac{\partial L}{\partial \hat{\boldsymbol{\beta}}} = -2X^T\mathbf{y} + 2X^T X\hat{\boldsymbol{\beta}} = \mathbf{0} \tag{3.14}$$

whose solution can be obtained by rearrangement to yield the general normal equations,

$$(X^T X)\hat{\boldsymbol{\beta}} = X^T\mathbf{y} \tag{3.14a}$$

or, whenever $X^T X$ is nonsingular,

$$\hat{\boldsymbol{\beta}} = (X^T X)^{-1} X^T\mathbf{y} \tag{3.14b}$$

Example. In the case of a single variable, $n = 1$, $\hat{\boldsymbol{\beta}}^T = (\hat{\beta}_0, \hat{\beta}_1)$, $\mathbf{y}^T = (y_1, y_2, \ldots, y_m)$ and

$$X = \begin{bmatrix} 1 & x_1 \\ 1 & x_2 \\ 1 & x_3 \\ \cdots & \\ 1 & x_m \end{bmatrix} \qquad X^T = \begin{bmatrix} 1 & 1 & 1 & \cdots & 1 \\ x_1 & x_2 & x_3 & \cdots & x_m \end{bmatrix}$$

Then

$$X^TX = \begin{bmatrix} \left(\sum 1 = m\right) & \left(\sum^m x_i\right) \\ \left(\sum^m x_i\right) & \left(\sum^m x_i^2\right) \end{bmatrix} \qquad X^Ty = \begin{bmatrix} \sum^m y_i \\ \sum^m x_i y_i \end{bmatrix}$$

Therefore

$$\hat{\beta} = (X^TX)^{-1}X^T\mathbf{y}$$

$$= \frac{1}{\left[m\left(\sum x_i^2\right) - \left(\sum x_i\right)^2\right]} \begin{bmatrix} \left(\sum x_i^2\right) & \left(-\sum x_i\right) \\ \left(-\sum x_i\right) & (m) \end{bmatrix} \begin{bmatrix} \left(\sum y_i\right) \\ \left(\sum x_i y_i\right) \end{bmatrix}$$

$$= \frac{1}{\left[m\left(\sum x_i^2\right) - \left(\sum x_i\right)^2\right]} \begin{bmatrix} \left(\sum x_i^2\right)\left(\sum y_i\right) - \left(\sum x_i\right)\left(\sum x_i y_i\right) \\ m\left(\sum x_i y_i\right) - \left(\sum x_i\right)\left(\sum y_i\right) \end{bmatrix}$$

When this last matrix operation is carried out, the result is the same as (2.100) with minor notation changes.

Statistical Properties of the Least-Squares Estimates

The least-squares estimates $\hat{\boldsymbol{\beta}}$ have been derived without reference to the statistical properties of the observed errors ε_i. Such assumptions are actually not necessary in the derivation of these estimates, since the normal equations (3.14) provide estimates of $\boldsymbol{\beta}$ that minimize the squared errors $\hat{\boldsymbol{\varepsilon}}^T\hat{\boldsymbol{\varepsilon}}$ regardless of the distribution of $\boldsymbol{\varepsilon}$. However, to determine further properties of these estimates, some assumptions need to be made.

A very mild set of assumptions on the errors ε_i are that the expected value of the errors is zero and that they are independently distributed with constant variance σ^2. Notationally,

$$E(\boldsymbol{\varepsilon}) = \mathbf{0} \tag{3.15}$$

$$\text{Cov}(\boldsymbol{\varepsilon}) = \sigma^2 I_n \tag{3.16}$$

I_n is the identity matrix of order n. The zero off-diagonal elements indicate the lack of correlation (independence) between observations. A later section deals with other variance structures in the observations.

If the hypothesized model given by (3.14b) is correct, the assumptions given in (3.15) and (3.16) allow us to calculate the expected value of $\hat{\boldsymbol{\beta}}$

$$\begin{aligned} E(\hat{\boldsymbol{\beta}}) &= E\left[(X^TX)^{-1}X^T\mathbf{y}\right] \\ &= E\left[(X^TX)^{-1}X^T(X\boldsymbol{\beta}+\boldsymbol{\varepsilon})\right] \\ &= E\left[\boldsymbol{\beta}\right] + E\left[\boldsymbol{\varepsilon}\right] \\ &= \boldsymbol{\beta} \end{aligned} \tag{3.17}$$

This says that under the mild assumptions given above the expected value of the least-squares estimates $\hat{\boldsymbol{\beta}}$ is the actual vector of parameters $\boldsymbol{\beta}$. Such an estimate is called an *unbiased estimator*. *Bias* in regression models generally occurs due to terms present in the actual process that are omitted from the fitted model. This topic will be covered later in this section.

In addition to showing that the estimators are unbiased, it is also possible to calculate the covariance structure of these estimates. It is easy to show, as in Myers [45], that

$$\begin{aligned} \operatorname{Cov}\hat{\boldsymbol{\beta}} &= E\left[(\hat{\boldsymbol{\beta}}-\boldsymbol{\beta})(\hat{\boldsymbol{\beta}}-\boldsymbol{\beta})^T\right] \\ &= (X^TX)^{-1}\sigma^2 \end{aligned} \tag{3.18}$$

The covariance between the estimates is thus determined by the transpose product of the experimental matrix and the level in the observation error σ^2. Since the covariance is determined by this matrix, the experimenter can alter the properties of his estimates by choice of experimental points. For instance, suppose he sets his experiments such that the columns within the experimental matrix are orthogonal to each other; that is, so that

$$\sum_{k=1}^{n} x_{i_k} x_{j_k} = 0 \tag{3.19}$$

If he also scales them so that

$$\sum_{k=1}^{n} x_{ik}^2 = N \tag{3.20}$$

then the transpose product X^TX is NI and

$$\operatorname{Cov}\hat{\boldsymbol{\beta}} = N^{-1}I\sigma^2 \tag{3.21}$$

Such a choice of points results in model estimates which are uncorrelated with each other. This is the basis of many of the experimental designs considered in the next section.

It is also possible to use (3.18) to determine the variance of predicted y values,

$$\operatorname{Cov}\hat{y} = \operatorname{Cov} X\hat{\boldsymbol{\beta}}$$

$$= X^T \operatorname{Cov}\hat{\beta} X \qquad (3.22)$$

$$= X^T(X^T X)^{-1} X\sigma^2$$

In particular the variance of any particular point y_k can be seen to be,

$$\operatorname{Var}\hat{y}_k = \mathbf{x}_k^T(\mathbf{x}_k^T\mathbf{x}_k)^{-1}\mathbf{x}_k\sigma^2 \qquad (3.23)$$

In both cases the variances can be seen to be determined in part by the choice of the experimental points given by the experimental matrix X.

Now, it is also possible to allocate the sum of squares of the measured responses y between the fitted model and the residuals and to use these to construct an analysis of variance table. The total sum of squares is given by

$$SS_y = \mathbf{y}^T\mathbf{y} \qquad (3.24)$$

and that due to the regression by

$$SS_R = \hat{\boldsymbol{\beta}}^T X^T X \hat{\boldsymbol{\beta}}$$

$$= \hat{\boldsymbol{\beta}}^T X^T \mathbf{y} \qquad (3.25)$$

The remainder is attributed to error, and can be obtained from the difference,

$$SS_E = SS_y - SS_R \qquad (3.26)$$

which is

$$SS_E = \mathbf{y}^T\mathbf{y} - \hat{\boldsymbol{\beta}}^T X^T \mathbf{y} \qquad (3.26a)$$

$$= (\mathbf{y} - \hat{\mathbf{y}})^T(\mathbf{y} - \hat{\mathbf{y}}) \qquad (3.26b)$$

The basic ANOVA table then becomes as shown in Table 3.1.

Table 3.1. ANOVA table for general regression

Source of variation	Sum of squares	Degrees of freedom	Mean square
Regression	$\hat{\boldsymbol{\beta}}^T X^T \mathbf{y}$	k	SS_R/k
Error	$\mathbf{y}^T\mathbf{y} - \hat{\boldsymbol{\beta}}^T X^T \mathbf{y}$	$n-k$	$SS_E/(n-k)$
Total	$\mathbf{y}^T\mathbf{y}$	n	

Table 3.2. ANOVA table for partitioned regression

Source of variation		Sum of squares	Degrees of freedom	Mean square
Regression	β_0	$n\bar{y}^2$	1	$SS(\beta_0)$
	$R\mid\beta_0$	$\boldsymbol{\beta}^T X^T \mathbf{y} - n\bar{y}^2$	$k-1$	$SS(R\mid\beta_0)/(k-1)$
Error		$\mathbf{y}^T\mathbf{y} - \boldsymbol{\beta}^T X^T \mathbf{y}$	$n-k$	$SS_E/(n-k)$
Total		$\mathbf{y}^T\mathbf{y}$	n	

The regression sum of squares SS_R is often further partitioned between that due "to the mean" or the constant term $\hat{\beta}_0$, and that due to the linear regression terms, $(\hat{\beta}_1, \hat{\beta}_2, \ldots, \hat{\beta}_k)$. These terms are given by

$$SS(\hat{\beta}_0) = n\bar{y}^2 = \frac{1}{n}\mathbf{y}^T\mathbf{y} \tag{3.27}$$

and

$$SS(R\mid\hat{\beta}_0) = \hat{\boldsymbol{\beta}}^T \mathbf{x}^T \mathbf{y} - n\bar{y}^2 \tag{3.28}$$

The latter term specifies the sum of squares due to the regression given that the constant term is already included in the model. This partition results in an ANOVA table of the form shown in Table 3.2.
The last column in the table is the "mean square," given by the sum of squares divided by the degrees of freedom for that sum. In the case of the error term it is possible to show that its mean square term is an unbiased estimate of the experimental variance σ^2.

The results presented so far have been made without any assumption about the distribution of the errors. It has been possible to calculate certain unbiased statistics, *point estimates*, without this distribution. However, to provide *interval estimates*, and to test hypotheses about the computed estimates, it is necessary to hypothesize an error distribution. The most common assumption is that the errors are independently distributed normal variables of zero mean and constant variance σ^2; that is,

$$\boldsymbol{\varepsilon} \sim N(\mathbf{0}, \sigma^2 I) \tag{3.29}$$

Variates of this type satisfy the assumptions made in equations (3.15) and (3.16). The assumption of a normal distribution is justified when experimental errors are likely to be from several additive sources, since, by the law of large numbers such errors will tend to be normally distributed.

When the errors are distributed normally it is then possible to show that the estimates $\hat{\boldsymbol{\beta}}$ are also distributed normally according to

$$\hat{\boldsymbol{\beta}} \sim N\left(\boldsymbol{\beta}, (X^T X)^{-1}\sigma^2\right) \tag{3.30}$$

and the individual confidence intervals for $\boldsymbol{\beta}$ given by

$$\beta_k = \hat{\beta}_k \pm Z_{1-\alpha/2}\sigma\left(\mathbf{x}_k^T\mathbf{x}_k\right)^{-1/2} \tag{3.31}$$

However, since σ is generally not known, an approximate confidence interval can be given using S as an estimate of σ, as follows:

$$\beta_k = \hat{\beta}_k \pm t_{k,1-\alpha/2}S\left(\mathbf{x}_k^T\mathbf{x}_k\right)^{-1/2} \tag{3.32}$$

In the case where the experimental matrix X is not orthogonal, (3.32) is somewhat misleading because the $\hat{\beta}_i$ are correlated as shown in (3.18). A more accurate statement is given by the joint confidence region for all the $\hat{\beta}_i$,

$$(\boldsymbol{\beta}-\hat{\boldsymbol{\beta}})^T X^T X(\boldsymbol{\beta}-\hat{\boldsymbol{\beta}}) \leqslant kS^2F(k,n-k,1-\alpha) \tag{3.33}$$

It is now possible to use the results of the ANOVA Tables 3.1 and 3.2 to test hypotheses about the significance of the estimated regression. Table 3.2 is typically used for this purpose. To test the hypothesis

$$H_0\colon \beta_0 = 0$$

$$H_1\colon \beta_0 \neq 0$$

we use the test statistic

$$F = \frac{MS(\beta_0)}{MS_E} \tag{3.34}$$

which is approximately distributed as $F_{1,n-k,1-\alpha}$.
To test the hypothesis

$$H_0\colon \beta_1 = \beta_2 = \ldots = \beta_k = 0$$

$$H_1\colon \beta_i \neq 0 \text{ for some } i$$

the test statistic

$$F = \frac{MS_{R|\beta_0}}{MS_E} \tag{3.35}$$

is compared to the tabulated value $F_{k-1,n-k,1-\alpha}$.

Bias and Lack-of-Fit

It was shown earlier that if the linear model hypothesized in (3.2) was correct, then the least-squares coefficients $\hat{\boldsymbol{\beta}}$ were *unbiased*. That meant that their expected value would be the actual value of the coefficient unbiased by any extraneous factors. This result will no longer be true when the hypothesized model is incorrect although the local approximation—which is our object—may still be adequate.

Consider the case where the model is hypothesized to contain n linear terms given by $\boldsymbol{\beta}_1$ and the actual model contains $n+k$ terms $\boldsymbol{\beta}^T=(\boldsymbol{\beta}_1^T,\boldsymbol{\beta}_2^T)$. Then the actual mechanism can be written

$$\eta(\mathbf{x})=X_1\boldsymbol{\beta}_1+X_2\boldsymbol{\beta}_2 \tag{3.36}$$

X_2 is the matrix containing the part of experimental matrix that could have been included for the terms given by $\boldsymbol{\beta}_2$ but was not. Equation (3.36) can be rewritten

$$\eta(\mathbf{x})=\left[\, X_1 \,\vdots\, X_2 \,\right]\begin{bmatrix}\boldsymbol{\beta}_1\\ \boldsymbol{\beta}_2\end{bmatrix} \tag{3.36a}$$

to show this clearer.

The least-squares estimators are then calculated using only X_1, by (3.14b), resulting in the estimates

$$\hat{\boldsymbol{\beta}}=(X_1^TX_1)^{-1}X_1^T\mathbf{y} \tag{3.37}$$

We can use this result to calculate the expected value of $\hat{\boldsymbol{\beta}}$.

$$\begin{aligned}E(\hat{\boldsymbol{\beta}})&=E\left[(X_1X_1)^{-1}X_1^T(X_1\boldsymbol{\beta}_1+X_2\boldsymbol{\beta}_2+\boldsymbol{\varepsilon})\right]\\&=\boldsymbol{\beta}_1+(X_1^TX_1)^{-1}X_1^TX_2\boldsymbol{\beta}_2\\&=\boldsymbol{\beta}_1+A\boldsymbol{\beta}_2\end{aligned} \tag{3.38}$$

The matrix A, given by

$$A=(X_1^TX_1)^{-1}X_1^TX_2$$

is called the *Alias Matrix*.

From (3.38) it can be seen that the extent of the bias is a function of both X_1 and X_2 as well as the magnitude of the missing terms given by $\boldsymbol{\beta}_2$. Even though the latter effect cannot be controlled, it is possible to control its expression into the $\hat{\boldsymbol{\beta}}$ estimate by careful choice of the X_1 points. In particular, it is obvious that any column in X_2 orthogonal to those in X_1 will not bias the estimates at all. This will be discussed in greater detail in the section on Response Surface Methods.

EXAMPLE. Consider the orthogonal design given by

$$X_1=\begin{bmatrix}1 & 1 & 1\\ 1 & 1 & -1\\ 1 & -1 & 1\\ 1 & -1 & -1\end{bmatrix}$$

to estimate the linear terms $\boldsymbol{\beta}_1^T=(\beta_0,\beta_1,\beta_2)$.

The effect of the missing quadratic terms $\boldsymbol{\beta}_2^T=(\beta_{11},\beta_{22},\beta_{12})$ is that of the pure quadratic terms $x_1^2\beta_{11}$ and $x_2^2\beta_{22}$ and the mixed quadratic term

$x_1x_2\beta_{12}$. The missing design columns would therefore be

$$X_2 = \begin{bmatrix} 1 & 1 & 1 \\ 1 & 1 & -1 \\ 1 & 1 & -1 \\ 1 & 1 & 1 \end{bmatrix}$$

and the full model $y = X_1\boldsymbol{\beta}_1 + X_2\boldsymbol{\beta}_2 + \varepsilon$. The alias matrix then is

$$A = \tfrac{1}{4}X_1^TX_2 = \tfrac{1}{4}\begin{bmatrix} 4 & 4 & 0 \\ 0 & 0 & 0 \\ 0 & 0 & 0 \end{bmatrix}$$

which indicates a bias structure of the form,

$$E(\beta_0) = \beta_0 + \beta_{11} + \beta_{22}$$
$$E(\beta_1) = \beta_1$$
$$E(\beta_2) = \beta_2$$

No bias is introduced by the β_{12} term since it is orthogonal to all the other columns. The pure quadratic terms β_{11} and β_{22} bias the constant term. A check of the first column of X_1 and the first two of X_2 show that these terms are estimated using the same column: β_{11} and β_{22} are in this case said to be *aliased* to β_0 since they are indistinguishable from each other. This topic will be considered in greater detail later.

The presence of bias terms also changes the expected mean squares of the ANOVA entries since they too are biased. ANOVA tables that have been considered so far implicitly assume that there is no bias. If bias does exist the results obtained are in error.

It is possible to partition the sum of squares attributed to error into lack-of-fit (LOF) and pure error sums whenever a pure error estimate can be made by replicating one or more points in a design. The SS_{LOF} can then be obtained from the difference between SS_E and SS_{PE}.

Designate

$y_{11}, y_{12}, \ldots, y_{1n_1}$ as n_1 repeat observations at $\mathbf{x}_1$

$y_{21}, y_{22}, \ldots, y_{2n_2}$ as n_2 repeat observations at $\mathbf{x}_2$

and

$y_{m1}, y_{m2}, \ldots, y_{mn_m}$ as n_m repeat observations at $\mathbf{x}_m$

The estimate of pure error is

$$S_e^2 = \frac{\sum_{i=1}^{m}\sum_{j=1}^{n_i}(y_{ij} - \bar{y}_i)}{\sum_{i=1}^{m} n_i - m} \tag{3.39}$$

whose expected value is σ^2.

This partitioning of the SS_E allows us to write a new ANOVA table and to test the significance of *lack-of-fit*. This ANOVA table is shown in Table 3.3.

Table 3.3. ANOVA table for lack-of-fit

Source of variation	Sum of squares	Degrees of freedom	Mean square
Regression	$\hat{\beta}_1 X_1 \mathbf{y}$	k	$\sigma^2 + \dfrac{(\beta_1 + A\beta_2)^T X^T X(\beta_1 + A\beta_2)}{k}$
Error	$\mathbf{y}^T\mathbf{y} - \hat{\beta}_1 X_1 \mathbf{y}$	$m-k$	
Lack-of-fit	by difference	$m-k-n_e$	$\sigma^2 + \dfrac{\beta_2^T (X_2 - X_1 A^T)(X_2 - X_1 A)\beta_2}{l}$
Pure error	$n_e S_e^2$	n_e	σ^2
Total	$\mathbf{y}^T\mathbf{y}$	m	

whence $l = m - k - n_e$ and $n_e = \Sigma_{i=1}^m n_i$. The SS_{LOF} is given by the difference

$$SS_{\text{LOF}} = \mathbf{y}^T\mathbf{y} - \hat{\beta}_1^T X_1^T \mathbf{y} - n_e S_e^2 \tag{3.40}$$

Residual Analysis

At this point in the analysis a fitted model, either a regression model or an analysis of variance model, has been completed and various components of variation can be compared. These partitionings of the total variation, sums of squares that are attributed to various sources, come about from calculations based upon our hypothetical models and assumed distributions of random error. Given these hypotheses, the cross sums of variation are compared to either reject or not reject certain hypotheses. Sometimes, in the case of replication, further partitioning (i.e., lack-of-fit and pure error within the residual variation) is possible, but in essence, we have exhausted the analyses based upon total variation at this point.

If we are to obtain further information from the data, it is now necessary to turn from the grouped sources of variation to the individual variations. Whether our object is confirmation of the analysis procedure already employed (to reject or to not reject, since direct confirmation is not possible) or to suggest ways in which it might be altered, the most likely place to look is at the individual residuals.

When our model is correct and the measurement error is NID $(0, \sigma^2)$, the residuals should approximate a NID $(0, \sigma^2)$ distribution. However, when the model is incomplete, where other representations are more likely, or other distributions of error occur [other distributions or perhaps nonhomogeniety of variance, NID $(\mathbf{0}, V)$], this may be apparent from examination of the *residuals*. Finally, the analysis of residuals can reveal the presence of wild values called *outliers* which can significantly affect the total error and the estimate of model terms.

Distribution of the Residuals. It will be shown under the discussion of the properties of residuals that the estimated residuals are correlated, due to the method of least squares. In spite of this the residuals should approximate a normal distribution of mean zero and constant variance σ^2 (estimated from S^2).

For small samples (e.g., $n<20$), this is primarily a matter of judgement. It is suggested in general that the residuals be plotted against the order of experimentation, against the fitted values, against the independent variables, and against any other factor that could possibly affect the variance. These methods will be discussed in turn, and the object in each case is a relatively constant band of values about the X axis with no discernible patterns of increase (or decrease) in the band itself.

It is sometimes useful to make a cumulative plot upon normal probability paper of the standardized residuals ε_i/S, or upon half normal paper of $|\varepsilon_i|/S$. The slope in either case should be relatively straight if the residuals are approximately normally distributed.

The standardized residuals should also be roughly distributed with 90% falling within $\pm 1.65S$ of the mean and 95% falling within $1.96S$ of the mean. Values falling beyond should be considered possible outliers.

Several factors may influence residuals not to behave as normally distributed variables. In particular there may be a systematic effect at work. However, the errors may arise from other distributions as well. Since the residuals arise from the least-squares procedure, which implicitly assumes a normal distribution, this may be masked. Nevertheless, it may be possible to detect two other main types of distribution from residuals: the uniform and double exponential:

$$f_u(t)=\begin{cases} \dfrac{1}{2b} & -b\leqslant t\leqslant b \\ 0 & \text{otherwise} \end{cases} \tag{3.41}$$

and

$$f_{DE}(t)=\frac{1}{2\sigma}e^{-|t|/\sigma} \tag{3.42}$$

These two distributions form two types of departure from the normal distribution, the uniform having less spread and the exponential more spread than the normal for given standard deviation. In particular, 100% of the uniform residuals should fall within $\pm S\sqrt{3}$ of the mean, whereas only about 86% of the exponential variables will fall within $\pm 1.96S$ of the mean. The uniform distribution represents a strict limit upon the size of the residual as well as an equally likely distribution within that space. This strict limit would make the detection of outliers much easier whenever a uniform distribution can be posited. Extreme values would be more likely with the exponential distribution and outliers consequently more difficult to analyze. Maximum likelihood estimation procedures for both distributions are treated in a later section.

Outliers. Residuals of apparently inordinate size (beyond about $2S$ for normal distributions) can arise from either unlikely events in the tail of the error distribution or from systematic errors such as misquoted data, invalid run conditions, and the like. In the least-squares method the square of the residual is used, so that a single outlier can contribute the most to error estimate, making significant effects more difficult to detect.

Because of this inflation of the error the tendency is to dismiss suspected outliers altogether, and repeat the analysis without them. However, to do this is to increase the arbitrariness of the procedure.* At the very least it should be ascertained whether any explanation is provided by experimental variation, improper transcription or other special conditions.† In addition it is strongly suggested that action on such values be reported along with any such results.

Outliers, by their spontaneous behavior, elude a strict treatment. Anscombe [2] outlines historical treatments of outliers and warns in particular against the unthinking use of specified rejection rules. He proceeds to derive an analysis of rejection rules based upon an insurance analogy:

> The premium payable may then be taken to be the protection increase in the variance of estimation errors due to using the rejection rule, when in fact all the observations come from a homogeneous normal source; the protection given is the reduction in variance (or mean squared error) when spurious readings are present. [2, p. 128]

Based upon the analogy he is able to calculate impartial rejection rules which balance the premium with the protection.

Other approaches to outliers include their retention by reducing their "weight" (weighted least squares will be introduced later) or by altering their value to that of the nearest nonsuspect value [3, "Winsorizing," p. 149] or some distance toward the grand mean or that of the expected value

*Anscombe [2] quotes the German astronomer Bessel in a work in 1838 to the effect that he did not reject observations because of large residuals: "We have believed that only through strict observance of this rule could we remove arbitrariness from our results."

†As noted above, the presence of several such values may suggest either systematic variations or an exponential error distribution.

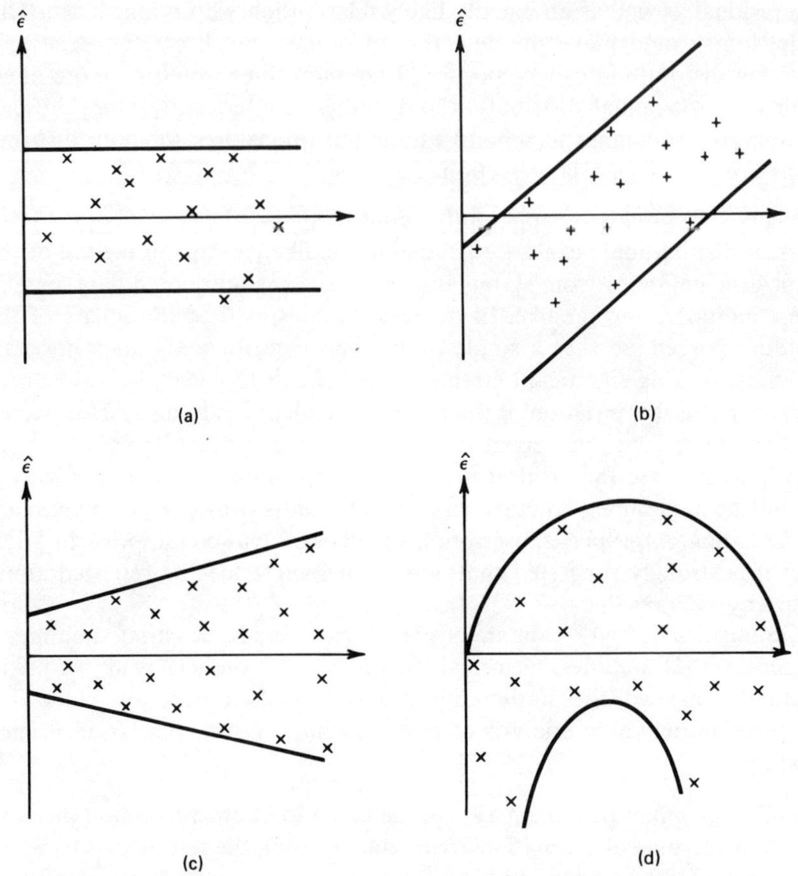

Figure 3.1. Various forms of variation behavior for the experimental error $\hat{\varepsilon}$.

at that point.* These later techniques would be preferable whenever the suspect value cannot be dismissed for a known cause.

Graphical Procedures. Typical scatter plots are shown in Figure 3.1 of residuals versus some ordinate. The first (3.1a) depicts the general situation, with residuals apparently evenly distributed with constant variance.

The second plot (3.1b) depicts a linear relation whose interpretation will vary with what the ordinate is. If the ordinate is an independent factor an error of calculation has occurred, since a linear factor has supposedly been estimated already. If the ordinate is the time order of experimentation, a time factor is present and should be explained, and if necessary included

*The empirical motivation for such a rule is also given in [25, 26].

within the model or subtracted from it. If the ordinate is the fitted y, there is an apparent relation between the size of ε and y. This can be caused by omitting the β_0 from the fitted model or may come from some missing interaction factor.

Plots of the type given in (3.1c) indicate a need for weighted least squares or a preliminary transformation of the y_i observations.

Plots of the final type, Figure (3.1d), indicate a missing quadratic factor, either in time or from one or more of the independent variables. If the ordinates are the y_i, a preliminary transformation of the data may be advisable.

Properties of Residuals. The residuals of regression models have certain properties which stem from the least squares procedure and which differ slightly from the actual errors they should approximate. First, whenever a constant term is included in the model

$$\mathbf{1}^T\hat{\boldsymbol{\varepsilon}} = \sum_{i=1}^{n} \hat{\varepsilon}_i = 0 \tag{3.43}$$

This can be observed from (3.14); that is,

$$-2(X^T\mathbf{y} - X^TX\hat{\boldsymbol{\beta}}) = \mathbf{0}$$

Since the first component, representing the derivative with respect to $\hat{\beta}_0$, is

$$-2\sum_{i=1}^{n}(y_i - x_{i0}\hat{\beta}_0 - x_{i1}\hat{\beta}_1 - \cdots - x_{im}\hat{\beta}_m) = 0 \tag{3.44}$$

which is

$$-2\sum_{i=1}^{n}(y_i - \hat{y}_i) = -2\sum_{i=1}^{n}\hat{\varepsilon}_i = 0 \tag{3.45}$$

This is consistent with the assumption of the $\boldsymbol{\varepsilon}$ being distributed according to NID $(0, I\sigma^2)$ but will always be satisfied by the least squares calculation so that it provides no additional information.

It can also be shown that

$$\sum_{i=1}^{n} \varepsilon_i \hat{y}_1 = 0 \tag{3.46}$$

since

$$\begin{aligned}
\sum \varepsilon_i \hat{y}_i &= \boldsymbol{\varepsilon}^T\hat{\mathbf{y}} = (\mathbf{y} - X\hat{\boldsymbol{\beta}})^T X\hat{\boldsymbol{\beta}} \\
&= \mathbf{y}^T X\hat{\boldsymbol{\beta}} - \hat{\boldsymbol{\beta}}^T X^T X\hat{\boldsymbol{\beta}} \\
&= \mathbf{y}^T X(X^TX)^{-1}X^T\mathbf{y} - \mathbf{y}^T X(X^TX)^{-1}X^TX(X^TX)^{-1}X^T\mathbf{y} \\
&= \mathbf{y}^T R\mathbf{y} - \mathbf{y}^T R\mathbf{y} \\
&= 0
\end{aligned}$$

where $R = X(X^TX)^{-1}X^T$. This result is also an outcome of least squares. Its practical significance is that no relation should exist between the residuals and the estimated response function y_i: a plot of ε_i versus y_i should show zero mean and no trend. These calculations provide no new information; however, they do serve as a check upon calculations.

Unlike the measurement errors, the residuals are correlated. This is another outcome of using least squares. Consider that there are n degrees of freedom associated with the n observations, from which r model coefficients and n residuals are to be made. Only $n-r$ degrees of freedom are available to the residuals so they must be correlated in some way. It is quite easy to calculate the variance of the residuals. First,

$$\begin{aligned} \hat{\boldsymbol{\varepsilon}} &= \mathbf{y} - \hat{\mathbf{y}} = \mathbf{y} - X\hat{\boldsymbol{\beta}} \\ &= \mathbf{y} - X(X^TX)^{-1}X^T\mathbf{y} \\ &= (I-R)y \end{aligned} \tag{3.47}$$

where R is defined as above. Using (3.47) and assuming that the postulated model is $E(\hat{y}) = X\boldsymbol{\beta}$, it follows that

$$\hat{\varepsilon} - E = E(\hat{\varepsilon}) = (I-R)(\mathbf{y} - X\boldsymbol{\beta}) = (I-R)\hat{\varepsilon} \tag{3.48}$$

Therefore

$$\begin{aligned} \operatorname{Var}(\hat{\varepsilon}) &= E\left\{\left[\hat{\boldsymbol{\varepsilon}} - E(\hat{\varepsilon})\right]\left[\hat{\boldsymbol{\varepsilon}} - E(\hat{\boldsymbol{\varepsilon}})\right]^T\right\} \\ &= (I-R)E(\hat{\boldsymbol{\varepsilon}}\hat{\boldsymbol{\varepsilon}}^T)(I-R)^T \\ &= (I-R)I\sigma^2(I-R)^T \\ &= (I-R)\sigma^2 \end{aligned} \tag{3.49}$$

since $E(\boldsymbol{\varepsilon}) = 0$, $E(\boldsymbol{\varepsilon}\boldsymbol{\varepsilon}^T) = I\sigma^2$ and $R^2 = R$. For models in which r is not close to n, these correlations do not significantly affect the graphical procedures outlined earlier. This point is discussed in Anscombe [2].

Some Computational Considerations of Least-Squares Methods

The pervasive use of computers and the widespread availability of quality software (in software libraries in particular, such as IMSL, Argonne, EISPACK, Harwell) has altered the methods, ranges and applicability of least squares techniques. Introductory texts no longer need contain simplified computation techniques for calculator use, since presumably everyone has access to either direct use of a computer or a time-sharing network. Nevertheless, a few comments about the practical side of numerical techniques are in order, even (or especially) for those who never intend to study an actual algorithm or create one of their own.

The most significant feature of digital computation—and the forms of most numerical analysis—is the representation of real variables by finite word length floating point numbers and the implication of data truncation, roundoff, and loss of precision during the course of computation. Floating point numbers have the form

$$\mathrm{fl}(a) = \pm 10^{q}(0.d_1, d_2 \ldots d_n) \tag{3.50}$$

where n gives the finite number of digits which can be stored. Data with more than n digits are truncated or rounded to n digits. In certain operations, such as addition and subtraction between numbers of divergent magnitudes, most of the significant digits will be lost during computation, and in general some roundoff occurs in every computation.

Analysis of such problems is the focus of numerical or algorithmic analysis. Fundamental approaches involve determining the sensitive area of calculations, streamlining algorithms, and analyzing algorithmic stability. Sensitive areas, such as the accumulation of sums and differences, are now often done in extended precision; for example, with numbers of $2n$ digits. The amount of roundoff in an algorithm is approximately proportional to the number of computations, so there is a premium on efficient algorithms. Finally, some algorithms are by their nature ill-fitted to finite precision computation and are numerically unstable so that alternative procedures must be developed.

Given the availability of quality software, many of these considerations are provided to the user already. Yet not completely! Consider the least squares quadratic polynomial problem given by

$$\eta = \beta_0 + \beta_1 t + \beta_2 t^2 \tag{3.51}$$

whose design matrix is typically given by

$$X = \begin{bmatrix} 1 & t_1 & t_1^2 \\ 1 & t_2 & t_2^2 \\ & \cdots & \\ 1 & t_n & t_n^2 \end{bmatrix}$$

The transpose matrix product is thus

$$X^T X = \begin{bmatrix} n & \sum t_i & \sum t_i^2 \\ & \sum t_i^2 & \sum t_i^3 \\ \text{sym} & & \sum t_i^4 \end{bmatrix}$$

Depending upon the t_i chosen, the magnitude of the sums n, Σt_i, Σt_i^2, Σt_i^3, and Σt_i^4 can vary considerably. When the t_i represent dates such as 1925, 1930, ..., 1975, these sums will be of the order 10, 10^3, 10^6, 10^9, and 10^{12} with a total range in magnitudes of 10^{11}. It is advisable in such cases to code the data so as to reduce spread in magnitudes. For instance, fit the model using

$$X_{ij} = \left(\frac{t_i - 1950}{25} \right)^j$$

The completed model,

$$\hat{y}_i = \hat{\beta}_0 + \hat{\beta}_1 X_{i1} + \hat{\beta}_2 X_{i2}$$
$$= \hat{\beta}_0 + \hat{\beta}_1 \left(\frac{t_i - 1950}{25} \right) + \hat{\beta}_2 \left(\frac{t_i - 1950}{25} \right)^2$$

can be returned to the original variables with coefficients that are estimated more precisely than otherwise would be the case.

An alternative would be to use orthogonal polynomials for equally spaced data. Since the matrix is diagonal, calculations are simplified and accumulation is avoided. It would still be a good idea, in general, to code the variables.

Coding. Coding will be used extensively in the next section to obtain certain properties in the estimators. Even when data is unplanned it is a good idea to code, as pointed out above.

The standard coding convention is

$$X_{ij} = \frac{\xi_{ij} - \bar{\xi}_j}{S_j} \tag{3.52}$$

where

$$S_j = \frac{1}{N} \sum_{i=1}^{n} \left(\xi_{ij} - \bar{\xi}_j \right)^2 \tag{3.53}$$

Another convention is

$$X_{ij} = \left[\frac{\xi_{ij} - \bar{\xi}_j}{d_j} \right] \tag{3.54}$$

where

$$d_j = \frac{\max_i(\xi_{ij}) - \min_i(\xi_{ij})}{2} \tag{3.55}$$

Solution of the Normal Equations. The least-squares estimator $\hat{\boldsymbol{\beta}}$ for $\boldsymbol{\beta}$ is given by the normal equations (3.14b)

$$\hat{\boldsymbol{\beta}}=(X^TX)^{-1}X^T\mathbf{y}$$

implying that the matrix product X^TX is inverted in solving for $\hat{\boldsymbol{\beta}}$. In practice the equivalent linear equation (3.14a)

$$X^TX\hat{\boldsymbol{\beta}}=X^T\mathbf{y}$$

is solved instead. This is because elimination procedures require far fewer operations to provide a solution than does the calculation of the inverse. In fact, the disparity is so great that regardless how many right sides had to be estimated with the same data matrices, it would always be cheaper to factorize (3.14a) each time than to solve the inverse once and simply multiply it on the left as in (3.14b). Actually, even standard factorization is not generally used in (3.14a) since as a symmetric positive definite even more efficient routines can be used.

Because of the gain in efficiency and precision, most packages will solve the normal equations by factorization even when the inverse is to be calculated for the variance matrix.

Linear Dependence and Sensitivity. There are times when lack of precision obscures not only answers but the lack of an answer. A unique estimate $\hat{\boldsymbol{\beta}}$ exists when X^TX is of full rank, when it is not singular. Now consider the problem posed in [15]. There are two ways noted in that paper for introducing dependencies within the data and thus to reduce the rank of the system. The first way is to estimate one of the X_{ij} from some linear combination of the remaining data values. Or, to obtain some theoretically expected result by altering the measured values to fit the theoretical distribution. Because of roundoff, it is possible that the singularity thus introduced can be overlooked and a plausible looking solution calculated.

The solutions obtained are the result of singular systems becoming nearly singular because of roundoff. The solutions to nearly singular systems have the unattractive, but hardly surprising, property of being extremely sensitive to minute changes in the observations. As the singularity is approached, the solution is increasingly unstable and arbitrary.

A measure of this relative sensitivity is given by the condition number of a matrix, Cond (A). If $\mathbf{x}$ is a solution to the equation

$$A\mathbf{x}=\mathbf{b}$$

and $\mathbf{x}+A\mathbf{x}$ is the solution to

$$A(\mathbf{x}+\Delta\mathbf{x})=\mathbf{b}+\Delta\mathbf{b}$$

where $\Delta\mathbf{b}$ is a small change in $\mathbf{b}$, then the relative sensitivity of solutions can be related to the condition number of A, denoted Cond (A), or

$$\frac{\|\Delta\mathbf{x}\|}{\|\mathbf{x}\|} \leqslant \text{Cond}(A)\frac{\|\Delta\mathbf{b}\|}{\|\mathbf{b}\|}$$

Exact expressions for Cond (A) are given in [28] and [40], but a good approximation is given by the ratio $\lambda_{max}/\lambda_{min}$ of the largest to smallest eigenvalues of A. As a matrix nears singularity, at least one of the eigenvalues approaches zero and the condition number becomes unbounded. Large condition numbers also occur for poorly scaled matrices, hence our earlier interest for scaling even for full rank systems.

Several techniques exist for dealing with poor conditioning. Ridge Analysis [35, 48] is a procedure for augmenting the poorly scaled matrix with a diagonal matrix to produce less sensitivity in the solution. The paper by Box et al. [15] cited above proposes an eigenvalue analysis to determine the type and nature of dependencies to remove them from the system. The result is a full-scale problem that will achieve realistic solutions. Finally, Singular Value Decomposition (SVD) [29, 41] is a numerically exact method of decomposing the system to full rank according to the error level of the data and the precision of the calculations. The output of this analysis includes both the estimates obtained from the full-rank portion of the system plus a null vector that can be added to the first estimate without appreciably increasing the squared error terms. The null vector is associated with the singular values; in the case that the singular value were exactly zero, the null vector belongs to the kernal of A and *any* multiple would not contribute to the squared error at all.* Several of these techniques will be discussed later.

Curve Fitting. Many times instead of a response surface of several variables we want to approximate a curve in one dimension. This curve fitting problem is also encountered in problems of more than one dimension that can be parameterized in a single variable. For instance, many optimization techniques require the estimation of an ascent direction followed by a line search along it. If the search begins at a point $\mathbf{x}_0$ and the search direction is given by $\mathbf{S}$, any point along the line can be specified by a step parameter λ,

$$\mathbf{x} = \mathbf{x}_0 + \lambda\mathbf{S} \tag{3.56}$$

The search direction may be a local estimate of first or second order. Typically the search will proceed too far for a simple linear estimate and a low-order polynomial must be used instead.

*The kernal of A, $K(A)$ contain the vectors which A maps to $\mathbf{0}$, so $\mathbf{k} \in K(A)$ implies $A\mathbf{k} = \mathbf{0}$ and any multiple α of $\mathbf{k}$ will not add to the squared error, SS. $SS = \|A(\mathbf{x}+\mathbf{k}) - b\| = \|A\mathbf{x} + A\mathbf{k} - \mathbf{b}\| = \|A\mathbf{x} - \mathbf{b}\|$.

Estimating the polynomial coefficients is a straightforward extension of the regression techniques mentioned so far. Here each column represents a power of the independent variable X. The X matrix takes the form

$$X = \left[\mathbf{X}^{(0)} \quad \mathbf{X}^{(1)} \quad \dots \quad \mathbf{X}^{(m)} \right] \tag{3.57}$$

$$= \begin{bmatrix} 1 & x_1 & x_1^2 & \dots & x_1^m \\ 1 & x_2 & x_2^2 & \dots & x_2^m \\ 1 & x_3 & x_3^2 & \dots & x_3^m \\ \dots & \dots & \dots & & \dots \\ 1 & x_n & x_n^2 & \dots & x_n^m \end{bmatrix}$$

Then the coefficients to be estimated are the terms of the model

$$y_i = \beta_0 + \beta_1 x_i^1 + \beta_2 x_i^2 + \dots + \beta_m x_i^m + \varepsilon_i \tag{3.58}$$

Whenever the range in orders of the column sums becomes larger than a factor of 100, it is a good idea to transform the variables in each column or at least to increase the precision of the calculations.

Terms can be added to the model up to the $m = n - 1$ **st** order. Such an estimate will have zero residuals but will very likely have poor interpolating qualities between the data points and particularly at each end of the interval. In estimating an optimum it is frequently preferable to estimate a smooth, low-order polynomial since the estimated optimum will be investigated in latter experiments.

Stepwise statistical procedures exist for testing the addition of each order of the regression model given that the lower order terms are already in the model. As this adds a degree of freedom to the model and subtracts one from the residual, each new term must add enough to the total model sum of squares to offset the increase in the error mean squares. Standard texts in regression analysis can be consulted for details, and many stepwise regression packages contain this feature.

Regression polynomial models can be made upon randomly or evenly spaced data. However, evenly spaced data provides a uniformly better model and should be preferred. If evenly spaced data is available, a set of orthogonal estimators are available for fitting polynomial models that provide particularly simple calculations for evaluating the relative contribution of each term.

Consider the following set of orthogonal polynomials of order i defined upon n equally spaced data points, $x_j, j=1,2,\ldots,n$.

$$\sum_{j=1}^{n} P_i(x_j)=0 \tag{3.59}$$

$$\sum_{j=1}^{n} P_i(x_j)P_k(x_j)=\begin{cases} 0 & i\neq k \\ C_i & i=k \end{cases}$$

The model for these orthogonal polynomials would be

$$y_j=\alpha_0+\sum_{i=1}^{m}\alpha_i P_i(x_j) \tag{3.60}$$

and the design matrix

$$X=\begin{bmatrix} 1 & P_1(x_1) & P_2(x_1) & \cdots & P_m(x_1) \\ 1 & P_1(x_2) & P_2(x_2) & \cdots & P_m(x_2) \\ \cdots & \cdots & \cdots & & \cdots \\ 1 & P_1(x_n) & P_2(x_n) & \cdots & P_m(x_n) \end{bmatrix} \tag{3.61}$$

Since the polynomials are orthogonal, the matrix transpose product is diagonal

$$X^TX=\mathrm{Diag}\left[N, \sum_{j=1}^{n} P_1^2(x_j), \sum_{j=1}^{n} P_2^2(x_j),\ldots, \sum_{j=1}^{n} P_m^2(x_j)\right] \tag{3.62}$$

Similarly, the α_i are easily calculated to be

$$\alpha_0=\frac{\sum_{j=1}^{n} y_j}{N} \tag{3.63}$$

and

$$\alpha_i=\frac{\sum_{j=1}^{n} y_i P_i(x_j)}{\sum_{j=1}^{n} P_i^2(x_j)} \qquad i=1,\ldots,m \tag{3.64}$$

Because the estimates are orthogonal, the sum of squares attributed to each coefficient are independent and can be easily tested against the residual mean square error.

$$SS_i = \alpha_i^2 \sum_{j=1}^{n} P_i^2(x_j) \tag{3.65}$$

$$= \frac{\left[\sum_{j=1}^{n} y_i P_i(x_j)\right]^2}{\left[\sum_{j=1}^{n} P_i^2(x_j)\right]^2} \tag{3.65a}$$

Generating the Polynomials. The orthogonal polynomials can be defined in terms of the average of the data $\bar{x}$, the spacing d between the data points, and the number n of data points. The first polynomials are given by

$$P_0(x) = 1$$

$$P_1(x) = \frac{x-\bar{x}}{d}$$

$$P_2(x) = \left(\frac{x-\bar{x}}{d}\right)^2 - \left(\frac{n^2-1}{12}\right)$$

$$P_3(x) = \left(\frac{x-\bar{x}}{d}\right)^3 - \left(\frac{3n^2-7}{20}\right)\left(\frac{x-\bar{x}}{d}\right)$$

and higher terms can be generated from the recursion formula

$$P_{i+1}(x) = P_i(x)P_1(x) - \frac{i^2(n^2-i^2)}{4(4i^2-1)} P_{i-1}(x) \qquad i = 1, 2 \ldots \tag{3.66}$$

EXPERIMENTAL DESIGNS FOR LINEAR AND QUADRATIC RESPONSE SURFACES

With regression methods we have a tool for estimating the response surface of a function for use, ultimately, in an optimizing process. This tool is suitable whenever data observations equal or exceed the number of variables to be estimated. It has been shown in (3.14), that the properties of these estimators are dependent upon the data values (and any scaling or transformation) through the experimental matrix X and its transpose product X^TX in particular. This section addresses the case in which the experimenter not only can observe the independent variables exactly, but can control them as well. In this case the experimenter is in a position to *design* experiments to achieve a suitable precision of results in an economical manner.

Before considering the various designs, let us consider the goals of our analysis. The overall goal is approximating a response to predict something about that function, generally its local behavior. Often the function itself is unknown and any reasonable approximation to the function in a local area is sufficient. The simplest approximations are no more than generalizations of the first terms in the Taylor's series approximation,

$$f(\mathbf{x}_0+\boldsymbol{\Delta}\mathbf{x}) \doteq f(\mathbf{x}_0)+\nabla f(\mathbf{x}_0)\boldsymbol{\Delta}\mathbf{x} \tag{3.67}$$

and

$$f(\mathbf{x}_0+\boldsymbol{\Delta}\mathbf{x}) \doteq f(\mathbf{x}_0)+\nabla f(\mathbf{x}_0)x+\boldsymbol{\Delta}\mathbf{x}^T\nabla^2 f(\mathbf{x}_0)\boldsymbol{\Delta}\mathbf{x} \tag{3.68}$$

which look much like the linear and quadratic regression models introduced in the last section

$$\hat{y}=\hat{\beta}_0+\sum_{i=1}^{k}\hat{\beta}_i(x_i-\bar{x}_i) \tag{3.69}$$

and

$$\hat{y}=\hat{\beta}_0+\sum_{i=1}^{k}\hat{\beta}_i(x_i-\bar{x}_i)+\sum_{i=1}^{k}\hat{\beta}_{ii}(x_i-\bar{x}_i)^2 + \sum_{i>j}^{k}\sum_{j=1}^{k}\hat{\beta}_{ij}(x_i-\bar{x}_i)(x_j-\bar{x}_j) \tag{3.70}$$

These are estimates about the point $\mathbf{x}_0$. In the case of the first model (3.69), it is not possible to optimize the response directly since, if not constant, it increases without bound. However, in Chapter 4 it will be shown that many constrained and unconstrained techniques depend upon the *gradient* of the response function, and this can easily be obtained from the linear model since

$$\nabla y=(\hat{\beta}_1,\hat{\beta}_2,\ldots,\hat{\beta}_k) \tag{3.71}$$

The quadratic model can be utilized directly by classical techniques for estimates of the optimum. Such models require at least $\frac{1}{2}(k+1)(k+2)$ experiments and in many cases the lesser order approximation, with its modest requirement of at least $(k+1)$ experiments, is preferred. The disparity in the number of points and the addition of quadratic terms leads to two sets of designs, the first- and second-order designs of response surfaces.

Coding Conventions

In the designs to follow it is convenient and useful to code or transform natural variables ξ_i according to one of the schemes given in the previous section (3.52) or (3.54). Either coding scheme results in coded variables that satisfy

$$\sum_{u=1}^{n} x_{iu}=0 \qquad i=1,2,\ldots,k \tag{3.72}$$

and implies that the estimates $\hat{\beta}_i$, $i=1,2,\ldots,k$ will be orthogonal to the estimate $\hat{\beta}_0$.

The variance of a given estimate is generally inversely proportional to $\Sigma_{u=1}^{n}x_{iu}^2$, the "spread" of the design, and it is common to adopt the convention that

$$\sum_{u=1}^{n} x_{iu}^2=N \qquad i=1,2,\ldots,k \tag{3.73}$$

for all designs. This convention is accomplished with the coding in (3.52). Designs of equal spread can be compared directly for their efficiency, hence the convention.

Most of the designs we investigate, and all of the linear designs in particular, will be *orthogonal*; that is, the matrix transpose product X^TX will be diagonal. Orthogonality is achieved with the following additional restrictions on the coded variables

$$\sum_{u=1}^{N} x_{iu}x_{ju}=0 \qquad i\neq j \qquad i,j=1,2,\ldots,N \tag{3.74}$$

The restrictions in (3.72)–(3.74) limit the design columns to being orthogonal to one another. Application of the three sets of restrictions given by (3.72) through (3.74) results in a product matrix of the form

$$X^TX=\begin{bmatrix} N & & 0 \\ & N & \\ & & \cdots & \\ 0 & & N \end{bmatrix}_{(k+1)\text{ square}} \tag{3.75}$$

It was shown by Myers [45] that these orthogonal designs are minimum variance; they produce model coefficient estimates of the least variance. This can easily be shown [45, p. 109]. The variance of $\hat{\beta}_i$ is given by $c_{ii}\sigma^2$, where c_{ii} is the ith diagonal element of $(X^TX)^{-1}$.

Let $a_{i,j}$ be the (i,j) element of X^TX and A_{pp} be the cofactor of a_{pp}. Then, if $A_{ij:pp}$ is the cofactor of a_{ij} in $A_{pp}[(i,j)\neq(p,p)]$, the determinant $|X^TX|$ can be evaluated using the Cauchy Expansion:

$$|X^TX| = a_{pp}A_{pp} - \sum_i \sum_j a_{pi}a_{pi}A_{ij:pp} \qquad \begin{matrix} p=0,1,\ldots,k \\ i,j\neq p \end{matrix}$$

$$= a_{pp}A_{pp} - Q \tag{3.76}$$

Since X^TX is positive definite, it can be seen that Q is a positive definite form in the a_{pi}. If the equation is multiplied by $\sigma^2/|X^TX|$

$$\sigma^2 = \frac{a_{pp}A_{pp}^2}{|X^TX|} - \frac{Q^2}{|X^TX|} \tag{3.77}$$

If the conventions (3.72)–(3.74) are adopted, then

$$\sigma^2 = \frac{N \operatorname{Var}\hat{\beta}_p - Q\sigma^2}{|X^TX|} \tag{3.78}$$

and therefore

$$\operatorname{Var}\hat{\beta}_p = \frac{\sigma^2}{N}\,\frac{1+Q}{|X^TX|} \tag{3.79}$$

Since X^TX is full rank and positive definite, $|X^TX|$ is positive, and $Q/|X^TX|$ is greater or equal to zero. Minimum variance occurs when it is zero, which requires all the a_{pi} $(i\neq p)$ to be zero: X^TX must be diagonal.

First Order Designs

Factorial Designs. A very convenient orthogonal design is the 2^k or full factorial design . Examples of factorial designs are shown in Figure 3.2. As in the case of ANOVA models, this design allows the estimation of both main effects ($\hat{\beta}_i$) and two factor interactions ($\hat{\beta}_{ij}$). For the case $k=3$, the design is given by

$$D_0 = \begin{bmatrix} 1 & 1 & 1 \\ 1 & 1 & -1 \\ 1 & -1 & 1 \\ 1 & -1 & -1 \\ -1 & 1 & 1 \\ -1 & 1 & -1 \\ -1 & -1 & 1 \\ -1 & -1 & -1 \end{bmatrix} \quad \begin{matrix} (a) \\ (b) \\ (c) \\ (d) \\ (e) \\ (f) \\ (g) \\ (h) \end{matrix}$$

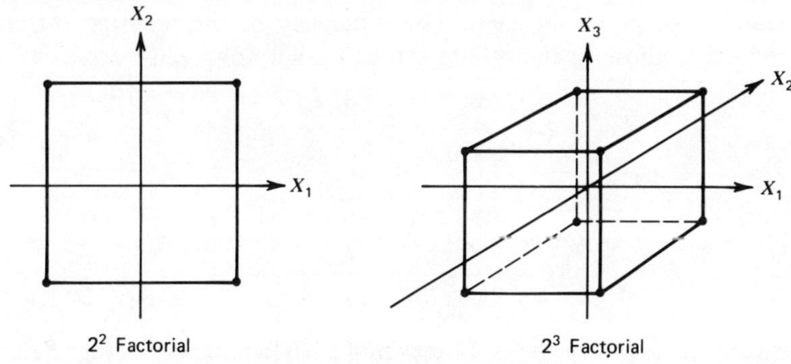

Figure 3.2. 2^k factorial designs.

This design meets all our scaling criteria and results in a diagonal moment matrix

$$X^T X = \begin{bmatrix} 8 & 0 & 0 & 0 \\ & 8 & 0 & 0 \\ & & 8 & 0 \\ \text{sym} & & & 8 \end{bmatrix}$$

The variance matrix $(X^T X)^{-1}$ is therefore given by $\frac{1}{8} I_4$. Consequently, this is a design of minimum variance, with the individual variances given by

$$\text{Var}(\hat{\beta}_i) = \tfrac{1}{8}\sigma^2 \qquad \text{Cov}(\hat{\beta}_i, \hat{\beta}_j) = 0$$

We can demonstrate the implications of minimum variance in another way. Consider the 4 point " one at a time" design given by

$$D_1 = \begin{bmatrix} 0 & 0 & 0 \\ 2 & 0 & 0 \\ 0 & 2 & 0 \\ 0 & 0 & 2 \end{bmatrix} \quad \begin{matrix} (0) \\ (1) \\ (2) \\ (3) \end{matrix}$$

where the design is scaled according to (3.73) and (3.74), but not (3.72). Here the coefficient estimates are given by

$$\hat{\beta}_1' = (1) - (0) \tag{3.80}$$

where (1) and (0) are the first two responses measured in this experiment. The variance of this estimate is $2\sigma^2$.

We can compare the *efficiency* of these two designs according to

$$E = \frac{(\text{Var}_1)(N_1)}{(\text{Var}_0)(N_0)} \tag{3.81}$$

This is the efficiency of the design D_0 with respect to D_1. N_0 and N_1 are the total number of points in each. The efficiency of the factorial estimates with respect to those of the one at a time design are given by

$$E=\frac{4\,\mathrm{Var}\,\hat{\beta}_1'}{8\,\mathrm{Var}\,\hat{\beta}_1}=\frac{8\sigma^2}{8\sigma^2/8}=8$$

and

$$E=\frac{4\,\mathrm{Var}\,\hat{\beta}_0'}{8\,\mathrm{Var}\,\hat{\beta}_0}=\frac{4\sigma^2}{8\sigma^2/8}=4$$

In this case the full factorial is 4 times more efficient in estimating $\hat{\beta}_0$ and 8 times more efficient in estimating $\hat{\beta}_i$ on a per observations basis. These efficiencies would hold for any minimum variance design with respect to the one at a time design.

As the example showed, the number of degrees of freedom for error is 2^k-k-1. For $k=3$ this means half of the available degrees of freedom are being used to estimate error. As k increases, this proportion increases as well. This has the advantage of reducing the correlation among residuals, but other uses for these degrees of freedom are possible, especially to estimate interaction coefficients or to test for lack of fit. In either case error estimates could be obtained from either adding experiments at the center of the design* or by replication of data points. It is this flexibility of uses that makes the full factorials so useful a design. In fact, the factorial is used in several quadratic designs so that if lack of fit is detected, the design is easily upgraded to estimate second-order terms.

Bias. The bias structure of the full factorials by second-order terms is quite simple.

$$E(\hat{\beta}_0)=\beta_0+\sum_{i=1}^{k}\beta_{ii} \tag{3.82a}$$

$$E(\hat{\beta}_i)=\beta_i \qquad i=1,2,\ldots,k \tag{3.82b}$$

We see that the main effects are unbiased by second-order terms, and that the constant term is biased by the sum of the pure quadratic terms. Bias by the interaction terms does not occur as their estimating columns are orthogonal to those of $\hat{\beta}_0$ and the $\hat{\beta}_i$.

The bias of the constant term can be used to obtain a single degree of freedom partition of the error sum of squares when center points are included in the experiment. The average at the center points, $\bar{y}_0$, is

*This will affect the scaling of equation (3.74) by reducing the spread somewhat unless the x_{iu} are adjusted upward to compensate.

compared to the average of all other points $\bar{y}_1 = \hat{\beta}_0$. The expected difference is

$$E(\bar{y}_1 - \bar{y}_0) = \sum_{i=1}^{k} \beta_{ii} \tag{3.83}$$

If n_0 is the number of center points and n_1 is the number (2^k) of design points, the portion of lack of fit allocated to curvature is

$$\frac{n_0 n_1 (\bar{y}_1 - \bar{y}_0)^2}{n_0 + n_1} \tag{3.84}$$

The remainder of lack of fit is attributed to interaction. Pure error is calculated using the corrected sum of squares of observations at the center point

$$SS_E = \sum_{i \in n_0} y_i^2 - \frac{\left(\sum y\right)^2}{n_0} \tag{3.85}$$

EXAMPLE. Consider a 2^3 factorial design with four center points, so that there are 12 design points in total. The degrees of freedom structure is as follows:

Source of Variation	Degrees of Freedom
Regression	4
Lack-of-fit	5
Curvature	1
Interaction	4
Pure Error	3
Total	12

When center points are used and curvature is not thought significant, the mixed interaction terms can be calculated in place of lack of fit. The columns for each term are added to the design matrix and the analysis proceeds as before. It is clear that because of the coding conventions, the new columns will be orthogonal both to the $\hat{\beta}_0$ estimating column and to the columns of the main effects.

The ultimate in factorial designs would include center points, replication, and estimation of lack of fit from third-order terms and curvature, pure error, and all first-order and mixed interaction terms. Although feasible, this is rarely done in practice, for interaction terms, pure quadratics and lack of fit would be obtained at less expense with a second-order model based upon the full factorial.

We can define *redundancy* in a design as the ratio of points in a design (or available degrees of freedom) to the number of model parameters estimated (degrees of freedom associated with the model). These extra points improve our error estimate, and our chances for detecting lack of fit, but are generally regarded (after some point) as luxuries rather than necessities. High redundancies imply inefficient experimentation when all that is desired is a reasonable linear response approximation.*

However, factorial experiments can be divided into orthogonal portions to reduce the redundancy while maintaining the desirable feature of orthogonality and minimum variance estimation. Consider the 2^3 design given below and divided into two four-point half fractions:

	x_0	x_1	x_2	x_3	x_1x_2	x_1x_3	x_2x_3	$x_1x_2x_3$
	1	1	1	1	1	1	1	1
D^+	1	1	−1	−1	−1	−1	1	1
	1	−1	−1	1	1	−1	−1	1
	1	−1	1	−1	−1	1	−1	1
	1	1	1	−1	1	−1	−1	−1
$D-$	1	1	−1	1	−1	1	−1	−1
	1	−1	−1	−1	1	1	1	−1
	1	−1	1	1	−1	−1	1	−1

The three columns of both fractions satisfy our coding conventions, so each represents a four point orthogonal design of redundancy 1.33 compared to 2.67 for the full factorial. A look at the remaining columns shows that we pay for discarding the extra points. For instance, in D^+ the column for x_1 is identical to that of x_2x_3. Similarly, x_2 and x_1x_3, x_3 and x_1x_2, and x_0 (for the constant term) and $x_1x_2x_3$ are the same. The same is true for D^- except that the sign of the column changes so that x_1 and $-x_2x_3$ are the same, and so on. Obviously this gives rise to a bias structure for each fraction of D.

for D^+:	for D^-:
$E(\beta_0)=\beta_0+\beta_{123}$	$E(\hat{\beta}_0)=\beta_0-\beta_{123}$
$E(\hat{\beta}_1)=\beta_1+\beta_{23}$	$E(\hat{\beta}_1)=\beta_1-\beta_{23}$
$E(\hat{\beta}_2)=\beta_2+\beta_{13}$	$E(\hat{\beta}_2)=\beta_2-\beta_{13}$
$E(\hat{\beta}_3)=\beta_3+\beta_{12}$	$E(\hat{\beta}_3)=\beta_3-\beta_{12}$

*This will be discussed further under simplex designs, the minimal linear designs.

Aliasing. In each fraction it is not merely that estimates are biased, it is that they are indistinguishable from each other. Such factors are said to be *aliased* together. Short of running the entire design, it is impossible to distinguish such aliases apart.

The column aliased to the constant column, the column for $x_1x_2x_3$ in this case, is called the *defining contrast*. In the half or 2^{k-1} fractional factorial there will be two defining contrasts. In this case it is given notationally as $I = ABC$ for the D^+ design and $0 = ABC$ for the D^- fraction. The defining contrast is the contrast which is completely lost by being aliased to the constant term; it is lost in the sense that its value is unavailable for contribution to the sum of squares attributed to the regression. Instead it is included as part of the sum of squares for the mean. In addition, but less importantly, it will not contribute to any gradient estimates.

The defining contrast specifies the design and implicitly the alias structure of the design. Referring again to a 2^{3-1} fractional factorial, let us observe what results when x_1x_3, $I = AC$, is chosen as the defining contrast.

	x_1	x_2	x_1x_2	x_3	x_2x_3	$x_1x_2x_3$	x_1x_3
D^+							
	1	1	1	1	1	1	1
	−1	1	−1	−1	−1	1	1
	1	−1	−1	1	−1	−1	1
	−1	−1	1	−1	1	−1	1
D^-							
	1	1	−1	−1	1	1	−1
	−1	−1	1	1	−1	1	−1
	1	1	1	−1	−1	−1	−1
	−1	1	−1	1	1	−1	−1

The most noticeable change is that x_1 and $-x_3$ have been aliased together by this choice, as well as x_2 and $x_1x_2x_3$, and x_1x_2 and x_2x_3. The aliasing of two first-order terms is an unattractive feature of this defining contrast: since all contributions due to x_3, direct or indirect, are aliased to other factors, this defining contrast is equivalent to saying that x_3 is not a factor at all in the function being estimated. If that is the case, a 2^2 design and analysis is the proper way to proceed.

Half factorials can be constructed once the defining contrast is known. Let the defining contrast be specified by $A^{\gamma_1}B^{\gamma_2}C^{\gamma_3}\ldots$ where the γ_i takes on the values of 1 or 0 depending upon whether the factor is included ($\gamma_i = 1$) or not in the contrast. Then form the equation

$$L_u = \gamma_1 Z_{1u} + \gamma_2 Z_{2u} + \ldots + \gamma_k Z_{ku} \qquad (3.86)$$

where $Z_{iu}=1$ for a "high" level (e.g., $x_{iu}=+1$) and $Z_{iu}=0$ for the "low" level (e.g., $x_{iu}=-1$). Proceed through the entire factorial design assigning values to L_u for each row. If S_u is now given as $S_u=L_u$ (modulo 2), the values of S divide the design into two parts which can be assigned as D^+ and D^-.

EXAMPLE. We can construct the 2^{4-1} for the defining contrast $I=ABCD$ using (3.86). The equation becomes

$$L=Z_1+Z_2+Z_3+Z_4$$

since

$$\gamma_1=\gamma_2=\gamma_3=\gamma_4=1$$

Then

x_1	x_2	x_3	x_4	L	S
1	1	1	1	4	0
1	1	1	−1	3	1
1	1	−1	1	3	1
1	1	−1	−1	2	0
1	−1	1	1	3	1
1	−1	1	−1	2	0
1	−1	−1	1	2	0
1	−1	−1	−1	1	1
−1	1	1	1	3	1
−1	1	1	−1	2	0
−1	1	−1	1	2	0
−1	1	−1	−1	1	1
−1	−1	1	1	2	0
−1	−1	1	−1	1	1
−1	−1	−1	1	1	1
−1	−1	−1	−1	0	0

L_u (modulo 2) assigns a value of 0 if L is even (no remainder) and a 1 if it is odd.

Either set can be used as the fractional factorial for $I=ABCD$.

It is also possible to evaluate the generalized interactions and aliases from the defining contrast before constructing the design. If we multiply two effects together we simply reduce their exponents modulo 2. For instance, for the 2^{3-1} with ABC as the contrast, A was aliased with

$$A(ABC) \equiv A^2BC \equiv BC$$

the BC interaction (using "$\equiv$" to mean "aliased to"). We can thus examine the alias structure of the fractional factorial constructed above with defining contrast $ABCD$.

$$\begin{array}{ll} A \equiv BCD & AB \equiv CD \\ B \equiv ACD & AC \equiv BD \\ C \equiv ABD & AD \equiv BC \\ D \equiv ABC & \end{array}$$

In this design first-order effects are aliased to third-order interactions only, and all second-order interations are aliased to other second-order interactions. This would be a suitable first-order design of redundancy $8/5$ or 1.60.

Smaller fractions can of course be made for larger designs. As with the smaller fractions, a price is extracted as the total number of points in the design is reduced and a greater number of effects are aliased together while others are sacrificed altogether.

For the quarter fraction two defining contrasts and two equations

$$L_1 = \gamma_1 Z_1 + \gamma_2 Z_1 + \dots + \gamma_k Z_k \tag{3.87a}$$

$$L_2 = w_1 Z_1 + w_2 Z_1 + \dots + w_k Z_k \tag{3.87b}$$

are necessary to define the point groupings. In addition, the generalized interaction of the two contrasts $(A^{\gamma_1}B^{\gamma_2}C^{\gamma_3}\dots)(A^{w_1}B^{w_2}C^{w_3}\dots)$, exponents reduced modulo 2, defines a third defining contrast. The four quarter factorials are then given by the four sets of $\{S_1, S_2\}$:

$$\{S_1=0,\ S_2=0\}, \{S_1=0,\ S_2=1\}, \{S_1=1,\ S_2=0\}, \{S_1=1,\ S_2=1\}$$

The existence of three defining contrasts implies three aliases for each effect. In general, the three defining contrasts are picked to avoid aliasing first-order terms, and it is of course desirable to alias them to higher order interactions (assumed not significant) as much as possible.

For higher 2^{k-p} fractions, a total of p defining contrasts must be chosen, and these will induce a further $2^p - p - 1$ generalized interactions to be sacrificed. Each effect will have a total of $2^p - 1$ aliases.

Other Factorial Designs. The logical extension of 2^k and 2^{k-p} factorials are the 3^k and 3^{k-p} factorials, factorial designs with three factor levels,

(1,0, −1).* For example, the nine points of the 3^2 are

$$D=\begin{bmatrix} 1 & 1 \\ 1 & 0 \\ 1 & -1 \\ 0 & 1 \\ 0 & 0 \\ 0 & -1 \\ -1 & 1 \\ -1 & 0 \\ -1 & -1 \end{bmatrix}$$

Designs of this type can provide orthogonal estimates for all terms including quadratics and interactions containing quadratic terms ($\beta_0, \beta_1, \beta_2, \beta_3, \beta_{11}, \beta_{22}, \beta_{12}, \beta_{112}, \beta_{122}, \beta_{1122}$). Because of this they are commonly treated as quadratic or second-order designs. Relative to 2^k designs they provide even higher redundancy and are rarely used as first-order designs except where they are to form the basis of Box-Benken type second order designs.

As with the 2^k factorials, 3^{-p} fractions of the 3^k designs can be constructed in a manner similar to the 2^{k-p} factorials. In this case, the computations are done modulo 3 since there are now three factor levels to be considered.

Blocking. Blocking arises whenever restrictions upon experimental randomization occurs, particularly restrictions on the number of experiments that can be run with some homogeneous factor such as equipment, test material, and the like. For factorial experiments the logical blocking is by fractional replicates. This allows the block effects to be kept orthogonal to the model estimates. Estimation proceeds as before except that a block effect sum of squares can be partitioned out of the general error sum of squares.

An analysis of blocking requirements is considered in Box and Hunter [13] and is particularly simple for first-order designs. If the designs can be broken into orthogonal subsets (such as fractional replicates), it is only additionally necessary to allocate center points, if any, to each subset equally.

Simplex Designs. Since it is impossible to get something for nothing, the lower bound for experimental design redundancy must be 1. In a k factor problem there are $k+1$ factors to estimate, and at least $k+1$ experiments are necessary for their estimation. If these points form an orthogonal, and

*Some expansion would be necessary to meet our spread convention (3.52).

thus minimum variance design, we would expect that the pointwise efficiency would be equal to that of larger designs such as the 2^k and 2^{k-p} factorials. Such a class of designs exists and they are called *simplexes* (or simplicies) or *regular simplexes*.

Consider the figures shown in Figure 3.3. In two dimensions the simplex is an equilateral triangle centered on the origin. In three dimensions, the simplex is a tetrahedron. In higher dimensions, these figures form hypertriangles. Note that the orientation of the simplex is not unique and any rotation of the figures pictured forms a simplex design.

For $k=2$ one design is given by (with one vertex along the positive y axis)

$$D=\begin{bmatrix} 0 & 2/\sqrt{2} \\ \sqrt{3/2} & -1/\sqrt{2} \\ -\sqrt{3/2} & -1/\sqrt{2} \end{bmatrix}$$

and an upright tetrahedron is given by

$$D=\begin{bmatrix} 0 & 0 & 3/\sqrt{3} \\ 0 & 2\sqrt{2/3} & -1/\sqrt{3} \\ \sqrt{2} & -\sqrt{2/3} & -1/\sqrt{3} \\ -\sqrt{2} & -\sqrt{2/3} & -1/\sqrt{3} \end{bmatrix}$$

Regardless of orientation a regular simplex is orthogonal and when scaled by our usual convention, the moment matrix will be diagonal with $k+1$ for its diagonal values, and each variance will be $(k+1)^{-1/2}$. We can

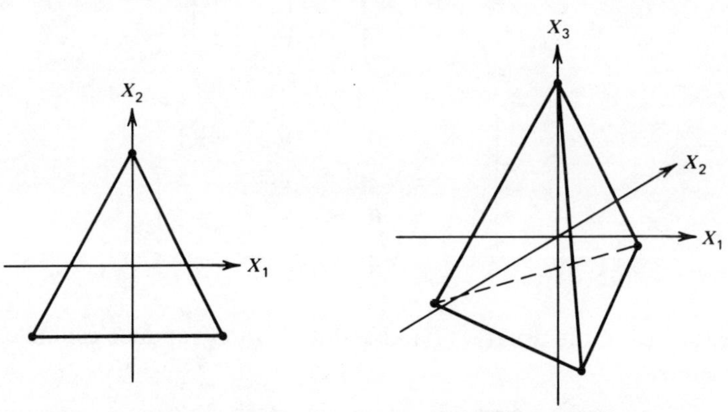

Figure 3.3. *k*-simplex designs.

compare this to the 2^k designs for $k=3$, for example

$$E=\frac{0.25\sigma^2\times 4}{0.125\sigma^2\times 8}=1$$

Both are minimum variance designs so this result is not unexpected.

Of course, something must be sacrificed in a design of this type. Since the redundancy is exactly 1 in a simplex design, there are no degrees of freedom to make either an error estimate or to estimate the fit of the model. In fact, the model will fit the data exactly and there are no residuals. The simplex provides a cheap estimate of major effects without any guarantee that the estimate adequately represents the data or not. This lack of a check on our results should not be taken lightly. However, there are situations in which the exact shape of the response surface is not needed, only its general tendencies. In that case, a linear model is certainly applicable and second-order effects may be discounted.

A second distinctive feature of simplex designs is their bias structure, which is dependent upon their orientation. For example, consider the following orthogonal simplex designs.

$$X_1=\begin{bmatrix} 1 & \sqrt{3} & 0 & 0 \\ 1 & -\sqrt{3}/3 & 2\sqrt{2/3} & 0 \\ 1 & -\sqrt{3}/3 & -\sqrt{2/3} & \sqrt{2} \\ 1 & -\sqrt{3}/3 & -\sqrt{2/3} & -\sqrt{2} \end{bmatrix}$$

$$X_2=\begin{bmatrix} 1 & 1 & -1 & -1 \\ 1 & -1 & 1 & -1 \\ 1 & -1 & -1 & 1 \\ 1 & 1 & 1 & 1 \end{bmatrix}$$

and

$$X_3=\begin{bmatrix} 1 & 0 & \sqrt{2} & -1 \\ 1 & -\sqrt{2} & 0 & 1 \\ 1 & 0 & -\sqrt{2} & -1 \\ 1 & \sqrt{2} & 0 & 1 \end{bmatrix}$$

All are scaled in the usual way and result in equal variance estimates of the parameters,

$$\text{Var}(\hat{\beta}_0)=0.25\sigma^2 \qquad \text{Var}(\hat{\beta}_i)=0.25\sigma^2 \qquad i=1,2,3$$

Even though all three are biased by the pure quadratics in the constant term in the same way the 2^k designs are

$$E(\hat{\beta}_0) = \beta_0 + \sum_{i=1}^{3} \beta_{ii}$$

the linear terms are biased in quite different ways:

Expected Value	Design 1	Design 2	Design 3
$\hat{\beta}_1$	$\beta_1 + \frac{2\sqrt{3}}{3}\beta_{11} - \frac{\sqrt{3}}{3}(\beta_{22}+\beta_{33})$	$\beta_1 + \beta_{23}$	β_1
$\hat{\beta}_2$	$\beta_2 + \sqrt{\frac{2}{3}}\,(\beta_{11} - \beta_{22}) - \frac{1}{3}\sqrt{3}\,(\beta_{12}+\beta_{13})$	β_2	$\beta_2 - \beta_{13}$
$\hat{\beta}_3$	$\beta_3 - \sqrt{\frac{2}{3}}\,\beta_{23}$	$\beta_3 + \beta_{12} + \beta_{13}$	$\beta_3 + \beta_{11} - \beta_{22}$

In view of the biases on the constant term, it is of course possible to add center points to the design and make an error estimation with a single degree of freedom to test for curvature. These addition points provide some security, but at the expense of the strongest feature of the design, its overall economy. If curvature is detected, the simplex can be augmented by placing points between each vertex to form a second-order design of minimum redundancy. This is sometimes done in practice for models of order two and even higher for systems of mixtures [30]. In general, however, the simplex is primarily useful for quick estimates of low-order functions.

Second-Order Response Surface Models

The adequacy of first-order models breaks down as the curvature of the response surface increases. The surface that results is a hyperplane of minimum squared error, which will obscure the true nature of the surface by filling the valleys and leveling the hills. The individual model coefficients may be heavily biased by the higher order terms and a gradient based upon them may lead to the waste of experiments by suggesting a nonascent direction.

In the models we have seen so far it is usually possible to test for the presence of curvature, since most estimates of the constant term are biased by a combination of the pure quadratic terms. When there are center points and sufficient degrees of freedom, the average of the center points, an unbiased estimate of $\hat{\beta}_0$, and the average of the design points ($\hat{\beta}_0$+bias terms) are compared. The sum of squares with a single degree of freedom so calculated is compared to the error estimate in order to test its significance. A low F ratio suggests that there is no reason to believe significant curvature exists.

Because the individual quadratic terms are all aliased to the constant term, there is no way of estimating the individual terms. To remove these aliased terms, it will be necessary to add a third level to each independent factor (not counting the center point). This section will deal with several approaches to estimating second-order models.

Form of the Model. The full second order is given by

$$\eta = \beta_0 + \sum_{i=1}^{k} \beta_i x_i + \sum_{i=1}^{k} \beta_{ii} x_i^2 + \sum_{i=1}^{k} \sum_{j>i}^{k} \beta_{ij} x_i x_j{}^* \tag{3.88}$$

whose common matrix form is given by, in the form of the Taylor's expansion of a function

$$\eta = \beta_0 + \boldsymbol{\beta}^T \mathbf{x} + \tfrac{1}{2}\mathbf{x}^T B \mathbf{x} \tag{3.88a}$$

The parameter vector $\boldsymbol{\beta}$ is given by

$$\boldsymbol{\beta}^T = (\beta_1, \beta_2, \ldots, \beta_k) \tag{3.89}$$

and the matrix B by

$$B = \begin{bmatrix} 2\beta_{11} & \beta_{12} & \cdots & \beta_{1k} \\ \beta_{12} & 2\beta_{22} & \cdots & \beta_{2k} \\ & \cdots & & \\ \beta_{1k} & \beta_{2k} & & 2\beta_{kk} \end{bmatrix} \tag{3.90}$$

Note that the B matrix is symmetric in the β_{ij}. This matrix is of particular importance when the response surface is being optimized since it is an estimate of the Hessian function which determines the nature of any stationary point.

In the linear model $k+1$ coefficient estimates were required. In the second-order model the total becomes $\frac{1}{2}(k+1)(k+2)$: $k+1$ first-order terms, k pure quadratic estimates β_{ii} and $\binom{k}{2} = \frac{1}{2}k(k-1)$ mixed quadratic

*Since order is unimportant for $x_i x_j$, $\beta_{ij} = \beta_{ji}$ and only the terms β_{ij}, $i = 1, 2, \ldots, k$ and $j = i+1, i+2, \ldots, k$ are estimated.

terms β_{ij}. This is the minimum number of experiments necessary to estimate the full second-order model.

A second-order design will differ from that of a first-order design in the number of independent rows in the design, since a greater number of independent experiments are now required, and by the columns representing the pure quadratic and mixed interaction terms. However, since experimentation generally represents an investment, we would also like to incorporate the first-order design within the second-order design to obtain the maximum use from that investment.

Now, designs of the 2^k full factorial type are capable of orthogonal estimates of the mixed quadratic terms already, so the new design must be constructed with the purpose of separating the biased quadratic terms from the constant term. The origin of this bias is easily observed. In the 2^k design the x_i columns contain the values ± 1 so that the x_i^2 columns contain only $(\pm 1)^2 = 1$ and is parallel to the column estimating the constant. To separate the two, at least one other level within the column x_i must be added. Two possibilities are 0 and $\pm\alpha$. The combination of these new levels would also have to be chosen in such a way as to preserve orthogonality among the various columns.

Central Composite Design. One method of augmenting factorial designs is given by the Central Composite Design. Added to the factorial design are the star or axial points given by

$$
\begin{array}{ccccc}
x_1 & x_2 & x_3 & \cdots & x_n
\end{array}
$$
$$
\begin{bmatrix}
+\alpha & 0 & 0 & & 0 \\
-\alpha & 0 & 0 & & 0 \\
0 & +\alpha & 0 & & 0 \\
0 & -\alpha & 0 & & 0 \\
0 & 0 & +\alpha & & 0 \\
0 & 0 & -\alpha & & 0 \\
 & & \cdots & & \\
0 & 0 & 0 & & +\alpha \\
0 & 0 & 0 & & -\alpha \\
0 & 0 & 0 & & 0
\end{bmatrix}
$$

The total design for $k=2$ and $k=3$ are illustrated in Figure 3.4. It can be seen that the axial points represent points directed along each of the coordinate points directed along each of the coordinate axes. For the choice $\alpha=1$, the points lie on the face of the hypercube defined by the factorial corner points. One or more center points are included in the design.

The total number of points added by the axial and center points is given by $T=2k+n_2$, where n_2 is the number of center points. If we call the 2^k factorial points F, then the total number of points in this design will be

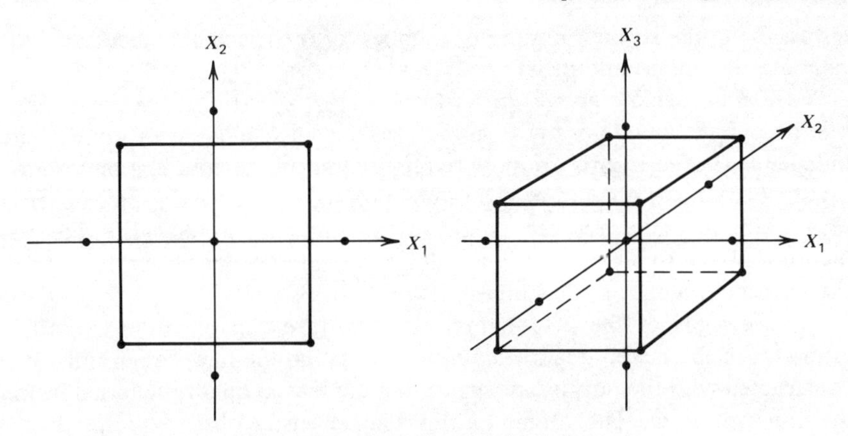

Figure 3.4. Central composite designs.

$N = F + T = 2^k + 2k + n_2$. As k increases, the factorial can be replaced with a suitable fraction of the full factorial so that the redundancy of the model does not grow too large.

This choice of augmenting points is suitable for our scaling conventions. Note that new columns are orthogonal with each other and with the quadratic columns. The new quadratic columns are no longer parallel to the constant column although they are not yet orthogonal, since

$$\sum_{i=1}^{N} x_i^2 = F + 2\alpha^2$$

However, if the quadratic model is transformed from (3.88) to

$$\eta = \beta_0 + \sum \beta_i x_i + \sum \beta_{ii}(x_i^2 - \bar{x}_i^2) + \sum \beta_{ii}\bar{x}_i^2 + \sum\sum \beta_{ij} x_i x_j \quad (3.91)$$

$$= \beta_0' + \sum \beta_i x_i + \sum \beta_{ii}(x_i^2 - \bar{x}_i^2) + \sum\sum \beta_{ij} x_i x_j \quad (3.91a)$$

then for suitable choice of α the design can be made orthogonal.

Consider the general central composite design with the quadratic estimates corrected for the mean of x_i^2. Then X is given by (for $k=2$)

$$\begin{array}{cccccc} & x_1 & x_2 & x_1^2 & x_2^2 & x_1x_2 \end{array}$$

$$\begin{bmatrix} 1 & 1 & 1 & 1-c & 1-c & 1 \\ 1 & 1 & -1 & 1-c & 1-c & -1 \\ 1 & -1 & 1 & 1-c & 1-c & -1 \\ 1 & -1 & -1 & 1-c & 1-c & 1 \\ 1 & \alpha & 0 & \alpha^2-c & -c & 0 \\ 1 & -\alpha & 0 & \alpha^2-c & -c & 0 \\ 1 & 0 & \alpha & -c & \alpha^2-c & 0 \\ 1 & 0 & -\alpha & -c & \alpha^2-c & 0 \\ 1 & 0 & 0 & -c & -c & 0 \end{bmatrix}$$

where

$$c=\frac{1}{N}\sum x_i^2=\frac{F+2\alpha^2}{F+T}$$

Then X^TX is given by

$$\begin{array}{c} x_0 \\ x_1 \\ x_2 \\ x_1^2 \\ x_2^2 \\ x_1x_2 \end{array}\overset{\begin{array}{cccccc} x_0 & x_1 & x_2 & x_1^2 & x_2^2 & x_1x_2 \end{array}}{\begin{bmatrix} 9 & & & & & \\ & 4+2\alpha^2 & & & & \\ & & 4+2\alpha^2 & & & \\ & & & p & q & \\ & & & q & p & \\ & & & & & 4 \end{bmatrix}}$$

where the missing terms are zero and

$$p=\frac{FT-4F\alpha^2-4\alpha^4+2(F+T)\alpha^4}{F+T}$$

$$q=\frac{FT-4F\alpha^2-4\alpha^4}{F+T}$$

p and q are the product of the quadratic estimators. X^TX will be orthogonal when

$$FT-4F\alpha^2-4\alpha^4=0$$

where

$$\alpha=\left(\frac{QF}{4}\right)^{1/4}$$

and

$$Q=\left[(F+T)^{1/2}-F^{1/2}\right]^2$$

A logical extension of the 2^k factorial designs with the necessary three factor levels for second-order models is the 3^k and 3^{k-p} factorial designs. When the quadratic model (3.91) is corrected for the mean $\overline{x_i^2}$, the 3^k will

Table 3.4. Values of α for central composite designs with 1, 2, and 3 center points

k	$n_2=1$	$n_2=2$	$n_2=3$
2	1.000	1.078	1.147
3	1.216	1.287	1.353
4	1.414	1.483	1.547
5	1.596	1.662	1.724
6	1.761	1.824	1.885
7	1.910	1.970	2.029
8	2.045	2.103	2.160

always be orthogonal. The 3^k design consists of the three coded factor levels 0 and ± 1 taken at every combination. For instance, the experimental matrix for the complete 3^2 design is given by

$$
\begin{array}{cccccc}
x_0 & x_1 & x_2 & x_1-\overline{x_1^2} & x_2-\overline{x_2^2} & x_1x_2
\end{array}
$$

$$
X=\begin{bmatrix}
1 & 1 & 1 & 1/3 & 1/3 & 1\\
1 & 1 & 0 & 1/3 & -2/3 & 0\\
1 & 1 & -1 & 1/3 & 1/3 & -1\\
1 & 0 & 1 & -2/3 & 1/3 & 0\\
1 & 0 & 0 & -2/3 & -2/3 & 0\\
1 & 0 & -1 & -2/3 & 1/3 & 0\\
1 & -1 & 1 & 1/3 & 1/3 & -1\\
1 & -1 & 0 & 1/3 & -2/3 & 0\\
1 & -1 & -1 & 1/3 & 1/3 & 1
\end{bmatrix}
$$

which is clearly orthogonal

$$
X^TX=\begin{bmatrix}
9 & & & & & 0\\
 & 6 & & & & \\
 & & 6 & & & \\
 & & & 2 & & \\
 & & & & 2 & \\
0 & & & & & 4
\end{bmatrix}
$$

As k increases, fractions of the 3^k can be constructed, although the arithmetic (3.91) is now mod 3 and three fractions will be constructed for each defining contrast.

It has been shown that orthogonal designs possess certain attractive properties in the first-order case, for instance minimum variance and uncorrelated estimates. It can also be shown that for first-order designs, orthogonality also implies *rotatability*. A rotatable design is one in which the variance of the response surface is a function only of distance from the design center and therefore not a function of the relative orientation of the design to the actual canonical orientation of the surface. This can easily be demonstrated. For our typical scaling,

$$
\begin{aligned}
\operatorname{Var}(\hat{y}) &= \operatorname{Var}(\hat{\beta}_0+\hat{\beta}_1x_1+\ldots+\hat{\beta}_kx_k)\\
&= \operatorname{Var}\hat{\beta}_0 + x_1^2\operatorname{Var}\hat{\beta}_1+\ldots+x_k^2\operatorname{Var}\hat{\beta}_k\\
&= \frac{\sigma^2}{N}(1+x_1^2+\ldots+x_k^2)\\
&= \frac{\sigma^2}{N}(1+\rho^2)
\end{aligned}
\tag{3.92}
$$

and ρ is the Euclidian distance from the design center. The same cannot be said for orthogonal second-order designs. The 3^2 design illustrated above is only orthogonal in certain orientations. Box and Hunter illustrate this with plots of $N\,\mathrm{Var}(\beta_{ii})/\sigma^2$ and $N\,\mathrm{Cov}(\beta_{ij},\beta_{kl})/\sigma^2$ for various rotations of the 3^2 design with the standard scaling, and it is only for certain orientations that the design is orthogonal. Similarly, we can show that the variance function of the estimate y is given by

$$\begin{aligned}\mathrm{Var}\,\hat{y} &= \mathrm{Var}\,\hat{\beta}^T\mathbf{x}_0 \\ &= \mathbf{x}_0^T\,\mathrm{Var}\,\hat{\beta}^T\mathbf{x}_0 \\ &= \mathbf{x}_0^T(X^TX)^{-1}\mathbf{x}_0\sigma^2\end{aligned} \tag{3.93}$$

for any point $\mathbf{x}_0^T=(1,x_1,x_2,\ldots,x_k)$. For the 3^2 design scaled in the usual way, this becomes

$$\mathrm{Var}\,\hat{y} = \frac{\sigma^2}{9}(5-3x_1^2-3x_2^2+2x_1^4+2x_2^4+x_1^2x_2^2)$$

or

$$V(\mathbf{x}_0) = \frac{N}{\sigma^2}\mathrm{Var}(\hat{y}) = 5-3x_1^2-3x_2^2+2x_1^4+2x_2^4+x_1^2x_2^2$$

where $V(\mathbf{x}_0)$ is the *variance function* of y on a per unit basis.

The variance function plays a central role in the evaluation of second-order designs. The four plots shown in Figure 3.5 are rotatable designs: a rotatable pentagonal design with five center points, a rotatable hexagonal design with a single center point, and two rotatable central composite designs with one and five center points. In all four, variance is only a function of radial distance from the design center, whereas the last two designs indicate how change in the variance may be altered with center points to obtain a nearly constant variance within the design region. This *Uniform Precision* estimate may be preferable to the uncorrelated estimates of an orthogonal design, although either property can be obtained in addition to rotatability.

The analysis of second-order designs is ably covered in both Box and Hunter [14] and Myers [45] at length and only a summary will be presented here. The overall object of the analysis is to discover what conditions are necessary to obtain a spherical (or rotatable) error variance, and what design choices can be made once this property is obtained. Attention is focused upon the matrix product X^TX, which can be considered the matrix of design moments.

$$[\underbrace{ii\text{---}i}_{s_i}\,\underbrace{j\text{---}j}_{s_j}\,\underbrace{k\text{---}k}_{s_k\cdots}] = \frac{1}{N}\sum_{u=1}^{N} x_{iu}^{s_i}x_{ju}^{s_j}x_{ku}^{s_k}\cdots$$

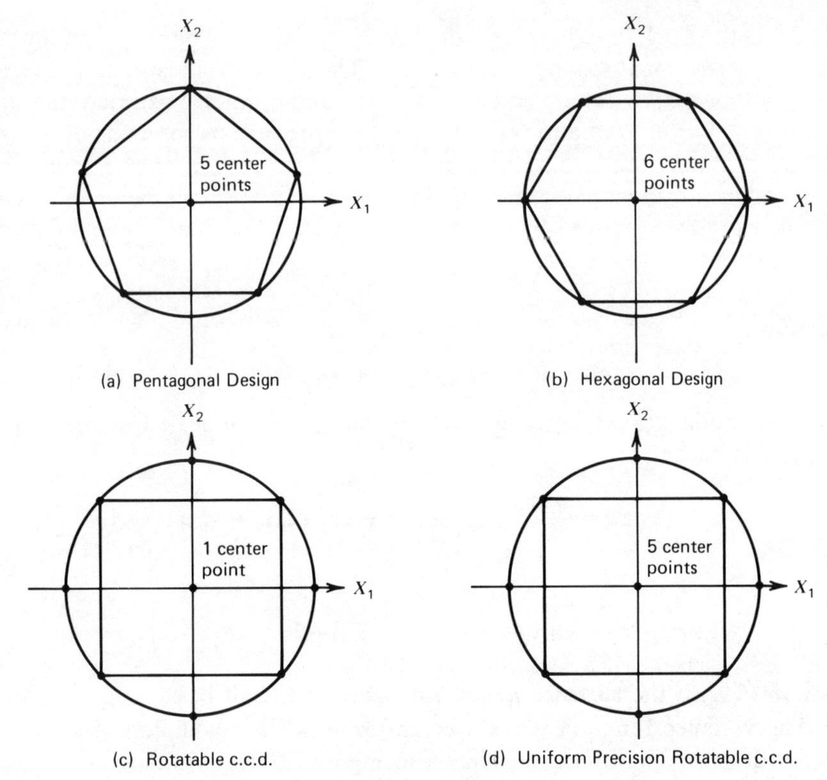

Figure 3.5. Various second-order response surface designs.

or, for $k=2$,

$$N^{-1}X^TX=\begin{bmatrix} [0] & [01] & [02] & [11] & [22] & [12] \\ [01] & [11] & [12] & [111] & [122] & [112] \\ [02] & [12] & [22] & [112] & [222] & [122] \\ [11] & [111] & [112] & [1111] & [1122] & [1112] \\ [22] & [122] & [222] & [1122] & [2222] & [1222] \\ [12] & [112] & [122] & [1112] & [1222] & [1122] \end{bmatrix}$$

For our usual scaling it is obvious that

$$[0]=N^{-1}\sum^{N} x_0=1$$

$$[11]=[22]=1$$

$$[01]=[02]=[12]=0$$

To obtain rotatability, it is necessary that the variance function $V(\mathbf{x})$ be only a function of distance from the design center. However, another way to approach the problem is to observe that the variance function must remain invariant under any rotation of the moment matrix. Comparing terms between the moment matrix and any rotation, it is possible to show that all odd moments (moments with an odd number of any component indice) must be zero and the pure fourth moments $[iiii]$ must be equal to three times the mixed fourth moment $[iijj]$. Therefore,

$$[111]=[222]=[122]=[112]=[1222]=[1112]=0$$

and if

$$\lambda_2=[ii]$$
$$\lambda_4=[iijj]$$

then

$$3\lambda_4=[iiii]$$

and the moment matrix (again for $k=2$) becomes

$$N^{-1}X^TX=\begin{bmatrix} 1 & 0 & 0 & \lambda_2 & \lambda_2 & 0 \\ 0 & \lambda_2 & 0 & 0 & 0 & 0 \\ 0 & 0 & \lambda_2 & 0 & 0 & 0 \\ \lambda_2 & 0 & 0 & 3\lambda_4 & \lambda_4 & 0 \\ \lambda_2 & 0 & 0 & \lambda_4 & 3\lambda_4 & 0 \\ 0 & 0 & 0 & 0 & 0 & \lambda_4 \end{bmatrix} \tag{3.94}$$

By our scaling convention, of course $\lambda_2=[ii]=1$.

These results are general and apply to designs of any number of variables k and order d of design: all odd moments are identically zero (in any orientation) and certain relations hold among certain of the even moments $\lambda_0, \lambda_2, \ldots, \lambda_{2d}$.

For the center composite introduced earlier, rotatability can be obtained by specification of the axial lengths α. The mixed fourth moment (unscaled) is given by

$$\lambda_4=[iijj]=F$$

and the pure fourth moment (unscaled) by

$$3\lambda_4=[iiii]=F+2\alpha^4$$

so therefore

$$\alpha=F^{1/4}$$

To obtain the proper scaling the linear columns are multiplied by $(N/F)^{1/2}$ and the quadratic columns by (N/F).

The moment matrix given in (3.94) can be inverted to obtain the precision matrix $N(X^TX)^{-1}$ for any value of k to obtain

$$N(X^TX)^{-1}=\left\{\begin{array}{c|ccc|cccc|ccc} a & & & & d & d & \cdots & d & & & \\ \hline & 1 & & & & & & & & & \\ & & 1 & & & & & & & & \\ & & & \ddots & & & & & & & \\ & & & 1 & & & & & & & \\ \hline d & & & & b & c & \cdots & c & & & \\ d & & & & c & b & & & & & \\ \vdots & & & & \vdots & & \ddots & & & & \\ d & & & & c & & & b & & & \\ \hline & & & & & & & & \lambda_4^{-1} & & \\ & & & & & & & & & \ddots & \\ & & & & & & & & & & \lambda_4^{-1} \end{array}\right\} \tag{3.95}$$

where blank cells indicate zero submatrices,

$$a=2\lambda_4^2(k+2)A \tag{3.95a}$$

$$b=[(k+1)\lambda_4-(k-1)]A \tag{3.95b}$$

$$c=(1-\lambda_4)A \tag{3.95c}$$

$$d=-2\lambda_4 A$$

and

$$A=[2\lambda_4\{(k+2)\lambda_4-k\}]^{-1} \tag{3.95d}$$

The terms of this matrix represent the individual precisions $N\text{Var}(b_{ij})/\sigma^2$ of the terms in the regression model. For the case in which $\lambda_4=1$, the estimates are uncorrelated and the design is orthogonal. In general we can show from the above that

$$\begin{aligned} V(\rho)=A\{&2(k+2)\lambda_4^2+2\lambda_4(\lambda_4-1)(k+2)\rho^2 \\ &+[(k+1)\lambda_4-(k-1)]\rho^4\} \end{aligned} \tag{3.96}$$

when $\lambda_4 = 1$ this reduces to

$$V(\rho) = \tfrac{1}{2}(k + 2 + \rho^4) \tag{3.96a}$$

The precision of the individual terms and the variance function for $\hat{y}$ for rotatable designs is thus a function of λ_4 and ρ. In general, we will be able to satisfy the scaling and rotatability conventions with the proper choice of some factor such as the axial length α and the proper scale factor, and then modify the exact value of λ_4 by adding center points to obtain a certain variance profile $V(\rho)$ or orthogonality.

We can solve (3.96) for the value of λ_4 which is necessary for $V(0) = V(1)$, the Uniform Precision Estimate. These are given in Table 3.5.

In practice it is not always possible to obtain Uniform Precision exactly. For a $k = 2$ CCD, five center points yields a value of $\lambda_4 = 0.81$ for a design which is nearly Uniform Precision. Additional center points can be added to obtain a λ_4 of one for an orthogonal design, assuming that the standard scaling is used. These values are summarized in Table 3.6.

General expressions for bias of any rotatable second-order design can be given in terms of factors already calculated and higher order moments due to the missing terms from the alias matrix

$$E(\hat{\beta}_0) = \beta_0 - 2\lambda_4 A \sum_{f=1}^{k} \sum_{g=f}^{k} \sum_{h=g}^{k} \sum_{i=1}^{k} [fghii]\, \beta_{fgh} \tag{3.97}$$

$$E(\hat{\beta}_i) = \beta_i + 3\lambda_4 \beta_{iii} + \lambda_4 \sum_{h \neq i}^{k} \beta_{hhi} \tag{3.98a}$$

$$E(\hat{\beta}_{ii}) = \beta_{ii} + \{(k+2)\lambda_4 - k\} A [iiiii]$$

$$+ (1 - \lambda_4) A \sum_{f=1}^{k} \sum_{g=f}^{k} \sum_{h=g}^{k} \sum_{i=1}^{k} [fghii]\, \beta_{fgh} \tag{3.98b}$$

$$E(\hat{\beta}_{ij}) = \beta_{ij} + \lambda_4^{-1} \sum_{f=1}^{k} \sum_{g=f}^{k} \sum_{h=g}^{k} [fghii]\, \beta_{fgh} \tag{3.98c}$$

Table 3.5. Values of λ_4 required for uniform precision estimates

k	2	3	4	5	6	7	8
λ_4	0.7844	0.8385	0.8704	0.8918	0.9070	0.9184	0.9274

Box and Hunter [14].

Table 3.6. Constants for some orthogonal (ORTH) and uniform precision (UP) central composite designs

k	2	3	4	5	5(1/2)*	6(1/2)*	7(1/2)*	8(1/4)†
F	4	8	16	32	16	32	64	64
α	1.414	1.682	2.000	2.378	2.000	2.378	2.828	2.828
UP: n_2	5	6	7	10	6	9	14	13
T	9	12	15	22	16	21	28	29
N	13	20	31	52	32	53	92	93
λ_4	0.81	0.86	0.86	0.89	0.89	0.90	0.92	0.93
ORTH: n_2	8	9	12	17	10	15	22	20
T	12	15	20	27	20	27	36	36
N	16	23	36	59	36	59	100	100
λ_4	1.00	.99	1.00	1.01	1.00	1.01	1.00	1.00

Box and Hunter [14].
*Indicates CCD with 2^{k-1} fractional factorial.
†Indicates CCD with 2^{k-2} fractional factorial.
$T=2k+n_2$
$N=F+T$

For designs in which the fifth moments are zero, most of these biases disappear and the expected values reduce to

$$E(\hat{\beta}_0)=\beta_0 \tag{3.99a}$$

$$E(\hat{\beta}_i)=\beta_i+3\lambda_4\beta_{iii} \tag{3.99b}$$

$$E(\hat{\beta}_{ii})=\beta_{ii} \tag{3.99c}$$

$$E(\hat{\beta}_{ij})=\beta_{ij} \tag{3.99d}$$

and only the first-order coefficients are biased by the pure cubic terms.

So far we have restricted our illustrations of second-order experimental designs to two common designs, the 3^k factorial and the central composite designs, although our analysis is applicable to all second-order designs. Of the two the central composite is the more versatile and certainly preferable given its rotatability. In addition, the central composite can be built upon an initial factorial design and it will also be shown later that the composite designs can readily be taken in convenient blocks when necessary. However, there are a good many other candidate designs that are useful for particular purposes, or are built upon first-order designs other than the 2^k and 2^{k-p} factorials.

Other Designs. The 3^k factorial designs are not rotatable, although some of their fractions are and can form the basis of a class of second-order designs suggested by Box and Behnken [10] and sometimes referred to as Box-Behnken designs. The designs consist of blocks in an orthogonal arrangement which are each orthogonal, consisting of a 2^2 factorial plan. Consider the example given by Box and Behnken for $k=4$ in which the *'s denote half of a 2×2 submatrix containing a 2^2 design

$$D=\begin{array}{c} \begin{matrix} x_1 & x_2 & x_3 & x_4 \end{matrix} \\ \left[\begin{array}{cccc} * & * & & \\ & & * & * \\ \hline * & & * & \\ & * & & * \\ \hline & * & & * \\ * & & * & \end{array}\right] \end{array} \tag{3.100}$$

Since the submatrices are orthogonal and the combinations of submatrices will be as well the design satisfies the linear rotatability criteria and promises all odd moments to be zero. This particular fraction of the 3^k also satisfies the rotatability criteria on the fourth moments. This design can be given explicitly by

$$D = \begin{bmatrix} 1 & 1 & 0 & 0 \\ 1 & -1 & 0 & 0 \\ -1 & 1 & 0 & 0 \\ -1 & -1 & 0 & 0 \\ 0 & 0 & 1 & 1 \\ 0 & 0 & 1 & -1 \\ 0 & 0 & -1 & 1 \\ 0 & 0 & -1 & -1 \\ 0 & 0 & 0 & 0 \\ \hdashline 1 & 0 & 0 & 1 \\ 1 & 0 & 0 & -1 \\ -1 & 0 & 0 & 1 \\ -1 & 0 & 0 & -1 \\ 0 & 1 & 1 & 0 \\ 0 & 1 & -1 & 0 \\ 0 & -1 & 1 & 0 \\ 0 & -1 & -1 & 0 \\ 0 & 0 & 0 & 0 \\ \hdashline 0 & 1 & 0 & 1 \\ 0 & 1 & 0 & -1 \\ 0 & -1 & 0 & 1 \\ 0 & -1 & 0 & -1 \\ 1 & 0 & 1 & 0 \\ 1 & 0 & -1 & 0 \\ -1 & 0 & 1 & 0 \\ -1 & 0 & -1 & 0 \\ 0 & 0 & 0 & 0 \end{bmatrix} \tag{3.100a}$$

A total of three center points are included. This design will readily block into three segments as indicated in (3.100), provided that one center point is allocated to each of the blocks. It is noted in the paper by Box and Behnken that this particular design is none other than a rotation of the $k=4$ central composite, although this is not true in general. This paper can be consulted for further designs and calculations for using them.

Another class of rotatable designs are regular figures such as pentagons, hexagons, and various polygons inscribed in hyperspheres, as well as combinations of regular figures inscribed within each other. Care must be used in equiradial figures, however, since no single equiradial set can estimate more than a first-order model. This is owing to the fact that

$$\sum_{i=1}^{k} x_{iu}^2 = \rho^2 x_{ou} \tag{3.101}$$

and the design will be singular for the second-order case. In addition, $\hat{\beta}_0$ and the $\hat{\beta}_{ii}$ cannot be separately estimated by such a design. Either problem may be overcome with the addition of a second equiradial set or center points. λ_4 is given by

$$\lambda_4 = \frac{Nk \sum_{w=1}^{s} n_w \rho_w^4}{(k+2)\left(\sum_{w=1}^{s} n_w \rho_w^2\right)^2} \tag{3.102}$$

for s rotatable sets of n_w points apiece and radius ρ_w.

In two dimensions, pentagonal and hexagonal designs are the minimal possible rotatable sets for use with only center points. Using (3.102), we can obtain

$$\lambda_4 = \frac{(n_1+n_2)k}{(k+2)n_1} \tag{3.102a}$$

so for the pentagonal design ($n_1 = 5$)

n_2	1	3	5
λ_4	0.6	0.8	1.0

so $n_2 = 3$ is approximately Uniform Precision and $n_2 = 5$ is orthogonal.

According to Box and Hunter [14], the hexagonal design has all moments of order five to zero so the bias equations of (3.99) apply and only the linear terms are biased. For the hexagonal design, close to uniform precision ($\lambda_4 = 0.75$) is obtained with three center points and orthogonality ($\lambda_4 = 1.0$) for six center points.

Concentric circles can also be combined to obtain rotatable $k = 2$ designs. Rotatability is achieved with the proper ratio of radii, with the ratio a function of whether uniform precision or orthogonality is desired.

Table 3.7. Radii for equispaced points on concentric circles

n_1	5	5	5	6	6	7
n_2	6	7	8	7	8	8
ρ_2/ρ_1 ($\lambda_4 = 0.7844$)	0.414	0.438	0.454	0.407	0.430	0.404
ρ_2/ρ_1 ($\lambda_4 = 1.00$)	0.204	0.267	0.304	0.189	0.250	0.176

Box and Hunter [14].

In three dimensions regular figures such as tetrahedrons ($n=4$), octahedrons ($n=6$), and cubes ($n=8$) inscribed on a sphere can be augmented with center points to form nonsingular rotatable designs of order one or of order two in combination, whereas icosahedrons ($n=12$) and dodecahedrons ($n=20$) with center points form nonsingular second-order rotatable designs.

In five or more dimensions, only three regular figures exist. The regular simplex of $k+1$ points and the 2^k hypercube (factorial points) are already known. There is also a generalization of the octahedron of 2^k points called the cross-polytope. The combination of the factorial and cross-polytope is the central composite design since the points of the cross polytope are the 2^k axial points mentioned earlier. It has already been shown that designs of this type can be more rotatable.

First-order orthogonal simplices can also form the basis of second-order experiments. A second-order rotatable set of $\frac{1}{2}k(k+1)$ points can be obtained from the simplex as the set of bisecting points between any two simplex vertices. The simplex itself is not second-order rotatable, but can be made so with the addition of a second simplex transformed from the first by the transformation $-I_{k+1}$. The total number of points is $\frac{1}{2}(k+1)(k+4)$, $k+1$ more than is required to estimate the model. As k increases, this redundancy forms a decreasing portion of the total number of experiments to be performed, and the total number of points used is half or less of those used in central composite or Box-Behnken designs, based upon factorial experiments.

To illustrate an augmented Simplex design, consider the first-order orthogonal design for $k=3$ (an arbitrary scaling constant)

$$D_1=\alpha\begin{bmatrix} 1 & 0 & 0 \\ -1/3 & 2\sqrt{2/9} & 0 \\ -1/3 & -\sqrt{2/9} & \sqrt{2/3} \\ -1/3 & -\sqrt{2/9} & -\sqrt{2/3} \end{bmatrix}$$

to which is added a reflection of itself

$$D_2=-D_1=\alpha\begin{bmatrix} -1 & 0 & 0 \\ 1/3 & -2\sqrt{2/9} & 0 \\ 1/3 & \sqrt{2/9} & -\sqrt{2/3} \\ 1/3 & \sqrt{2/9} & \sqrt{2/3} \end{bmatrix}$$

to form the first rotatable set. The second set is obtained from the $\binom{k+1}{2}$ pairs of vertices combined to form

$$x_{iu}=\frac{x_{ju}+x_{ku}}{2}\begin{cases} j=1,2,\ldots,k+1 \\ k=j+1,\ldots,k+1 \\ i=1,2,\ldots,1/2k(k+1) \end{cases}$$

$$D_2=\begin{bmatrix} 1/3 & \sqrt{2/9} & 0 \\ 1/3 & -\frac{1}{2}\sqrt{2/9} & \frac{1}{2}\sqrt{2/3} \\ 1/3 & -\frac{1}{2}\sqrt{2/9} & -\frac{1}{2}\sqrt{2/3} \\ -1/3 & \frac{1}{2}\sqrt{2/9} & \frac{1}{2}\sqrt{2/3} \\ -1/3 & \frac{1}{2}\sqrt{2/9} & -\frac{1}{2}\sqrt{2/3} \\ -1/3 & -\sqrt{2/9} & 0 \end{bmatrix}$$

The entire design consists of 14 points. α and β are set to satisfy the scaling convention

$$\lambda_2=1=[ii]=\frac{1}{14}\left[\alpha^2\sum_{u=1}^{8}x_{iu}^2+\beta^2\sum_{u=9}^{14}x_{iu}^2\right]$$

and to obtain a specified level for λ_4, for instance, using $\lambda_4=0.7844$ and (3.102)

$$0.7844=\frac{42}{5}\cdot\frac{8\alpha^4+6\left(\frac{\beta}{\sqrt{3}}\right)^4}{\left[8\alpha^2+6\left(\frac{\beta}{\sqrt{3}}\right)^2\right]^2}$$

whence $\alpha=2.10721$ and $\beta=1.79963$ or $\alpha=1.24887$ and $\beta=3.84205$.

The completed, scaled design is rotatable, second order, and given by

$$D = \begin{vmatrix} 1.98357 & 0 & 0 \\ -0.66119 & 1.87013 & 0 \\ -0.66119 & -0.93506 & 1.61958 \\ -0.66119 & -0.93506 & -1.61958 \\ -1.98357 & 0 & 0 \\ 0.66119 & -1.87013 & 0 \\ 0.66119 & 0.93506 & -1.61958 \\ 0.66119 & 0.93506 & 1.61958 \\ 0.76463 & 1.08135 & 0 \\ 0.76463 & -0.54067 & 0.93647 \\ 0.76463 & -0.54067 & -0.93647 \\ -0.76463 & 0.54067 & 0.93647 \\ -0.76463 & 0.54067 & -0.93647 \\ -0.76463 & -1.08135 & 0 \end{vmatrix}$$

Unlike the Simplex design, this design is not saturated; the second simplex allows us 4 degrees of freedom for an estimate of the residual error with which to test the adequacy of the model. Center points could be added to make an error estimate so that the residual could be further partitioned, although for this number of degrees of freedom, no test would be very sensitive. Estimates from this model are less precise than larger models, although presumably of approximately the same efficiency. The number of design levels used here limits its usefulness, although this presents no difficulties, for example, in simulation modeling where experimental levels can be arbitrarily set.

Design Blocking. Blocking of experiments was introduced earlier as a restriction upon experimentation caused by some external factor. For instance, experiments must be run on several pieces of equipment or with different batches of material. In many cases, these restrictions limit randomization to less than complete designs. Our object in choosing a blocking is to prevent this block effect from influencing the estimation of model coefficients.

In the first-order case, this blocking was especially easy for factorial experiments, since fractions of the complete factorial could be used to keep block effects orthogonal to polynomial estimates. This is the basic strategy for second-order experiments as well, although the situation is complicated somewhat by augmenting points.

Box and Hunter [14] derive the general rules for experimental blocking, which they illustrate. Possible blocking for Box Behnken designs is given in

[10]. For orthogonal blocking of a rotatable design, the following must hold.

1. Within each block the design columns must be orthogonal,

$$\sum_{u=1}^{n_b} x_{iu}x_{ju}=0 \qquad i\neq j \tag{3.103}$$

2. The total contribution to the sums of squares of each block must be proportional to the number of points allocated to the block. Thus

$$\frac{\sum_{u=1}^{n_b} x_{iu}^2}{\sum_{u=1}^{N} x_{iu}^2}=\frac{n_b}{N} \tag{3.104}$$

The second rule adjusts for different size blocks (e.g., between factorial and axial points) and the allocation of center points to blocks.

Design Criteria. Two sets of factors affect the quality of estimates obtained from designed experiments for regression models. One set is under the complete control of the experimenter, the design itself. Thus, before data taking is begun, it is possible to specify certain matters concretely: the shape and type of bias, the precision of estimates, what estimates will be correlated, and so on. Because of this, the experimenter will oftentimes be in a position beforehand to determine whether his experimental goals can be reached and at what approximate cost. A second set of factors is generally unknown. These factors include the assumptions of constant variance, adequacy of the chosen model, appropriate scale factors, and so on. The choice of experimental design naturally includes consideration of factors that may be present in a given situation.

Our general goals for a good design, therefore, include [14]:

1. The ability to estimate the approximating model within the region of interest at an acceptable level of accuracy.
2. The ability to detect significant departures from our assumptions.
3. A limited number of experimental points to accomplish (1) and (2).
4. The ability to be blocked.
5. Should be built about a design of lesser order in case a higher order model becomes necessary.

Because of these criteria we have considered the effects of bias, the efficiency of estimates, and the variance and precision of estimates. Since the orientation of the response surface is generally unknown, we have favored designs that are rotatable so that estimates are equally precise, regardless of orientation. Uniform precision estimates are used both to even out the variance of estimates and to limit bias, since, as shown in (3.98), bias is a function of λ_4, which increases as the precision at the center of the design does. Since the models considered so far are essentially interpolating polynomials, no design can prevent the variance of the estimate from increasing rapidly as the design region is quit.

The criteria discussed so far are by no means exhaustive. For instance, although tradeoffs in precision and bias have been mentioned, no explicit method has been given. This can be done by considering the average mean squared error

$$J=\frac{Nk}{\sigma^2}\int_R E[\hat{y}(\mathbf{x})-g(\mathbf{x})]^2\,d\mathbf{x} \tag{3.105}$$

where

$$k=\left[\int_R d\mathbf{x}\right]^{-1}$$

and both multiple integrals are integrated over the entire experimental region R. The integral in (3.105) can be expanded and divided into two portions representing the average variance of y and the average square of the bias,

$$\begin{aligned} J&=\frac{Nk}{\sigma^2}\int_R E[\hat{y}(\mathbf{x})-g(\mathbf{x})]^2\,d\mathbf{x}\\ &=\frac{Nk}{\sigma^2}\left\{\int_R E\{\hat{y}(\mathbf{x})-E[\hat{y}(\mathbf{x})]\}^2+\int_R[E(\hat{y}(\mathbf{x})-g(x)]^2\,d\mathbf{x}\right\}\\ &=V+B \end{aligned} \tag{3.106}$$

The designer can then investigate the effect of minimizing V, B, or J in choosing a design. An introduction to this approach is given in Myers [45] and Box and Draper [12]. Designs in which V is minimized are called minimum variance designs, and those in which B is minimized are minimum bias designs (minimum mean-squared bias).

In the absence of direct knowledge about a response surface, it is of course difficult to determine the proper choice of a design. This problem will be faced again in Chapter 5 and following, when applications are considered. Montgomery and Evans [43] have considered this empirically for several surfaces and levels of error. They considered second-order designs and compared estimates of the optimum and its location. They report generally that,

1. Rotatable designs generally outperformed nonrotatable designs.

2. Both the central composite and hexagonal designs performed well over all surfaces.
3. Since the hexagonal designs contain fewer points than the central composites, they may well be the most efficient of the designs considered.

The test surfaces were all two dimensional and covered a broad range of surfaces, both smooth, rough and highly biased. Designs considered were an orthogonal 3^2, orthogonal, uniform precision and minimum bias central composites, and orthogonal and uniform precision hexagonal designs.

Other criteria for evaluating designs based upon the determinant of the moment matrix, the trace of the precision matrix and properties of the correlation matrix R of the residual are presented by Hahn et al. [32]. These criteria relate to confidence intervals for model estimates, total average variance of all coefficients, and so forth. Of particular interest is their method for evaluating the effects of missing data or analyzing nonstandard designs. The analysis depends only upon the design so the analysis can be made prior to experimentation.

ADVANCED REGRESSION TOPICS

The preceding two sections have introduced the essentials of standard linear regression models and the design of experiments for such models. By standard linear models we have meant polynomial models linear in the estimated parameters. Our method of solution has been the least squares criteria, which has assumed that any errors were distributed NID $(0,\sigma^2)$. In this section we will extend these techniques to any implicitly linear (or linearizable) model and examine some specialized model structures as well as appıoaches to other error distributions. Treatment of nonlinear models is postponed to the next section.

Linearizing Transformations

As pointed out above, so far we have been approximating functions with low-order polynomials linear in the unknown parameters, β_i. We can extend these techniques directly to models which are implicitly linear, that is, models that can be transformed into a form which is linear in its unknown parameters. Consider the following model

$$y=\alpha e^{\beta_0+\beta_1 x_1+\varepsilon} \tag{3.107}$$

where ε is a normally distributed random variable. Then, taking the natural log of both sides,

$$\ln y = \ln \alpha + \beta_0 + \beta_1 x_1 + \varepsilon$$

to which we could fit a linear model with

$$E(\hat{y}) = \ln y \qquad E(\hat{\beta}_0) = \ln \alpha + \beta_0 \qquad E(\hat{\beta}_1) = \beta_1$$

using our standard techniques and the model

$$\hat{y} = \hat{\beta}_0 + \hat{\beta}_1 x_1$$

Similarly, other models can be shown to be implicitly linear such as the multiplicative model

$$y = \alpha x_1^{\beta} x_2^{\gamma} x_3^{\delta} e^{\varepsilon}$$

since

$$\ln y = \ln \alpha + \beta \ln x_1 + \gamma \ln x_2 + \delta \ln x_3 + \varepsilon$$

Note that the model $y = \alpha x_1^{\beta} \varepsilon$ is only implicitly linear with regard to distribution if $\ln \varepsilon$ is normally distributed. Other models include reciprocal

$$y = (\beta_0 + \beta_1 x_1 + \beta_2 x_2 + \varepsilon)^{-1}$$

where

$$\frac{1}{y} = \beta_0 + \beta_1 x_1 + \beta_2 x_2 + \varepsilon$$

Also, if

$$y = [1 + e^{\beta_0 + \beta_1 x_1 + \varepsilon}]^{-1}$$

several transformations are used to obtain

$$\ln\left(\frac{1}{y} - 1\right) = \beta_0 + \beta_1 x_1 + \varepsilon$$

Thus any model that is implicitly linear can be fit using our standard regression techniques using the transformed variables.

Often, of course, we will not know in advance that a transformation will be suitable for a given set of data. We do so, however, whenever a transformation is indicated from a theoretical consideration, when a transformation will simplify the model, or when an examination of the fitted residuals suggests that either the untransformed data is not properly distributed or that some systematic variation is not being accounted for by the fitted model.

As suggested above, there are a wide range of transformations to consider. Fortunately, a large number such as

$$x^{-1/2}, 1/x, \sqrt{x}, \ln x, e^{x}, x^{2}, \text{etc.}$$

can be summarized with

$$(x+c)^p$$

and the problem becomes one of deciding what c and p to use in fitting the linearized model,

$$(y+b)^q = \beta_0 + \beta_1(x+c)^p$$

where x or y or both may be transformed. In the case that b, q or c, p are unknown, the problem is nonlinear and can be approached using the techniques of the next section. Various general approaches are also given by Box and Tidwell [16], Tukey [47], Draper and Hunter [22], and Dolby [21]. The Dolby approach is based upon the observation that the family of transformations given by $(x+c)^p$ is the solution to a differential equation

$$\frac{d}{dx}\frac{y'}{y''} = \frac{1}{p-1}$$

when $y=(x+c)^p$. A finite difference approximation of the differential equation based upon the data allows the strength of the transformation to be estimated from the data as a first guess of a transformation. The paper of Draper and Hunter proposes investigating various transformations with a view toward reducing various mean-square ratios corresponding to interaction, error mean squares, etc. The approach, therefore, is essentially solving a series of linear regression models as a function of the transformation constants. Some statistical criteria for limiting the range of transformations in a particular problem are also discussed. The methods of Box and Tidwell are somewhat more involved. They use derivatives of power transformations (similar to those above) to construct a sequence of transformed variables until a convergence is achieved.

With the availability of nonlinear modeling codes, transformation parameters can become part of the general model-fitting problem and the preselection of transformations to allow a linear analysis with the transformed data is less common. The primary exception to this rule would be when either a particular theoretic model might reasonably be presumed or when the object of a study is modeling rather than approximation. The technique suggested by Draper and Hunter might then be favored since a trade-off between transforming and not transforming, and over various transformation levels, can be graphically depicted.

Explicitly Constrained Least Squares

In the context of Chapter 4 on Optimization, the linear least-squares problem is an unconstrained minimization of the residual sum of squares $S(\hat{\boldsymbol{\beta}})$

$$\min S(\hat{\boldsymbol{\beta}}) = (X\hat{\boldsymbol{\beta}} - y)^T(X\hat{\boldsymbol{\beta}} - y) \tag{3.108}$$

whose solution can be given by the normal equations (3.14). However, in some cases restrictions upon the $\hat{\beta}$ do exist and this constraint information must be incorporated into the standard model for solution.

Consider the following situations:

1. An engineer is interested in characterizing the frictional head loss in a fluid through an experimental flow configuration. At this early stage in the project, the engineer believes a linear model will be adequate to give him a feel for the situation. Each of his parameters is a quantitative description of some test factor (fluid flow rate, wetted diameter, etc.), and the fitted coefficients $\hat{\beta}_i$ are therefore per unit contributions of head loss due to each factor. The minimum unit head loss is of course zero, and he would like to constrain his parameters to be nonnegative. Negative values could occur due to bias terms and, if the experiment is not designed properly, due to interaction effects omitted from the model. His regression model is of the form

$$\min S(\hat{\boldsymbol{\beta}})$$

such that

$$\hat{\beta}_i \geqslant 0 \qquad i = 1, 2, \ldots, k$$

2. An engineer is involved in setting up a simple process control unit that will depend upon estimates of concentrations within several pieces of process equipment. It is proposed that these concentrations be estimated from observations of process stream temperatures, pH, and resistivity, all of which can be measured continuously in the process. The control processor can solve algebraic equations, so a least-squares model is proposed to estimate each of the concentrations, from transformations of the monitored variables. The engineer observes actual concentrations of each of p components as a function of the monitor variables to relate the two together. Since he is monitoring concentrations, he may be in a position to specify what certain combinations of the predicted values should equal. Since he wishes to include this within his model, he can proceed in the following manner. Letting

$$\begin{bmatrix} X_1 & & & 0 \\ & X_2 & & \\ & & X_3 & \\ 0 & & & X_p \end{bmatrix} \begin{bmatrix} \boldsymbol{\beta}_1 \\ \boldsymbol{\beta}_2 \\ \\ \boldsymbol{\beta}_p \end{bmatrix} \cong \begin{bmatrix} \mathbf{C}_1 \\ \mathbf{C}_2 \\ \\ \mathbf{C}_p \end{bmatrix} \tag{3.109a}$$

$$X\boldsymbol{\beta} \cong \mathbf{y} \tag{3.109b}$$

he can solve the p least-squares problems as a single problem. Any known restrictions on the predicted concentrations can be included as constraints. For instance, if it is known that

$$E(\hat{y}_1)+3E(\hat{y}_2)=100$$

for certain subsets of the observations, then

$$\begin{bmatrix} W_1 & 0 \\ 0 & 3W_2 \end{bmatrix}\begin{bmatrix} \beta_1 \\ \beta_2 \end{bmatrix}=100.1$$

would be an appropriate constraint. W_1 and W_2 would consist of those portions of x_1 and x_2 which the constraint applied. The general problem would be

$$\min S(\hat{\beta})=(X\hat{\beta}-\mathbf{y})^T(X\hat{\beta}-\mathbf{y})$$

subject to

$$W\hat{\beta}=\mathbf{Z}$$

3. In a similar vein, consider an economist using least squares to estimate contributions to the GNP from dollar activities in certain industries from observations taken from historical records, and so forth. In certain cases he too may feel able to restrict combinations of predictions due to specialized knowledge or aggregate statistics available for entire industries. These restrictions could be equalities, as in 2, or inequalities of the form

$$\min S(\hat{\beta})=(X\hat{\beta}-\mathbf{y})^T(X\hat{\beta}-\mathbf{y})$$

subject to

$$W\hat{\beta}\leqslant\mathbf{Z}$$

We will show in the next chapter that these problems form a special class of optimization problems called Quadratic Programming problems. Their special feature is that as long as the constraints are linear, the overall problem can be solved via linear matrix calculations. In essence, the constraints restrict the degrees of freedom within the regression problem without altering its essential features otherwise. Lawson and Hanson [41] may be consulted for a description of this problem and computer codes with which to solve it.

Although the essential features of linear regression are unchanged by constraints, it is not right to suppose the restrictions can be ignored or satisfied after the unconstrained problem is solved. For instance, in the first problem it would be incorrect to set any negative coefficients obtained from the unconstrained case to zero, for it is likely that a better overall fit could be obtained by the constrained problem. It should also be noted that the minimum from the constrained problem may be greater than that obtained from an unconstrained formulation. If the constraints are accurate, the unconstrained minimum can be considered merely an artificial result.

Implicitly Constrained Least Squares

In the last section attention was focused upon explicit constraints upon the least-squares estimation problem. We will show in the next chapter that at the solution, equality and active inequality constraints are satisfied exactly, which reduces the overall degrees of freedom left for the residual mean squares and very likely increases the size of S. We will be willing to put up with this as long as the fit of model is still adequate and other features of the model—implied by the constraints—are now satisfied. A potentially more serious problem occurs when dependencies within the data occur without our knowledge. The results we obtain are likely to be quite arbitrary, with high variances for the permissable range of the actual value of our estimates. As with explicit constraints, the identification of these implicit constraints may imply an increase in our squared error, and the result will be more likely to reflect reality and will generally possess shorter confidence intervals.

Dependencies and general ill-conditioning can arise in various ways. Earlier we showed that poor scaling could give rise to arbitrary estimates. One of the reasons is that poorly scaled data makes it difficult to distinguish independent vectors. Linear dependencies can occur from data that is estimated using other data columns and from normalizing the results of several columns to fit some expectation about the data. In addition, when dependencies among expected values do exist among variables, they will be obscured by error somewhat; enough, in general, to allow a solution to be calculated, but enough also to obscure the fact that the solution is likely to be both artificial and unstable. Illustrations of these problems in practice can be obtained in Box et al. [15]. Near dependency can also become a factor when regression is performed upon unplanned data. It is common that one or more variables will be strongly correlated (positively or negatively) or changing slowly or not at all over the course of experiments.

The problem with linear dependencies is that the normal equations involve inversion of a singular or nearly singular matrix. Other regression methods, such as maximum likelihood estimation to be introduced later, involve the minimization of determinants of nearly singular matrices. We would know there was trouble if the determinant was exactly zero, but short of that, how do we determine a "really" small result from an artificial one? Fortunately there is help. We will survey three general approaches, all of which are fairly recent developments.

Box et al. [15] has already been cited. They are dealing with data that is obtained by measurement of chemical systems for fitting kinetic models. Since some of the data is hard or impossible to measure directly, there is a temptation, particularly when the experimenter is not familiar with linear estimation, to calculate one set of data from other data. Also, because of general mass laws, there is a temptation to normalize certain results so that the data will conform exactly. Finally, certain ranges of data selection or the results from certain processes will give rise to dependencies due to equilibrium conditions obscured by experimental errors.

The analysis involves a comparison of the eigenvalues of X^TX where the columns of X have been corrected for the mean. The eigenvalues are the λ_i, which satisfy the characteristic equation of X^TX,

$$|X^TX-\lambda_i I|=0 \tag{3.110}$$

In the case where dependencies within the data exist, m_1 of the λ_i will differ from zero due to roundoff errors, another m_2 for the dependencies among expected values, which will differ from zero because of measurement errors, and the remaining $k-m_1-m_2$ λ_i, which will be significantly larger than either, representing independent results. Note that in an orthogonal design scaled according to our normal convention $\lambda_i=N$ for all i. The first set of λ_i are expected to be determined by the precision of calculations, generally estimated as ± 0.5 in the last significant digit calculated,and uniformly distributed. Then the standard deviation for calculations of d significant digits is

$$\sigma_{re}^2 \approx \frac{0.5\times 10^{-5}}{12}$$

for a total of N data items, and $i\in I_{m1}$,

$$E(\lambda_i)=(N-1)\sigma_{re}^2$$

The eigenvalues for dependencies $i\in I_{m_2}$ in the data are likewise given by

$$E(\lambda_i)=(N-1)\sigma^2=(N-1)S^2$$

Once these eigenvalues are identified, their eigenvectors $\boldsymbol{v}_i$ are calculated according to

$$X^TX\boldsymbol{v}_i=\lambda_i\boldsymbol{v}_i \tag{3.111}$$

Since eigenvectors are not unique [$\alpha\boldsymbol{v}$ is obviously a solution to (3.111)], we generally stipulate the normalized form of $\boldsymbol{v}_i$,

$$\boldsymbol{v}_i^T\boldsymbol{v}_i=1$$

The eigenvectors $\boldsymbol{v}_i$, $i\in I_{m_1}$, I_{m_2} can then be used to identify the inherent dependencies within the data columns. A linearly independent set of the data vectors is then selected and used to solve the regression problem of reduced rank. There is no hard and fast procedure in picking this set, although the combination should be orthogonal to the dependencies outlined by the eigenvectors. In the paper it is noted that remaining eigenvectors (after the dependent eigenvalues are eliminated) define directions orthogonal to the discarded "singularity plane," and linear combinations of these vectors can be used in solving the least-squares problem.

A method similar to the above that is more explicit in generating the independent subset of X to use is the singular value decomposition referred to earlier. This method utilizes the decomposition of a poorly conditioned X via

$$X = USV^T$$

where S is a diagonal matrix containing the singular values of X, U is an $m \times m$ orthogonal matrix and V is an $n \times n$ orthogonal matrix. The singular values given in S are the positive square roots of the eigenvalues of X^TX. We eliminate those singular values that are of the range of uncertainty in our precision or data due to measurement error. The least squares problem is then solved using the USV^T with reduced rank.

The orthogonal matrices U and V may be obtained from the standard decomposition of X

$$X = HRK^T \tag{3.112}$$

where H is an orthogonal $m \times m$ matrix and K is orthogonal $n \times n$, and R is of the form

$$R = \begin{bmatrix} R_{11} & 0 \\ 0 & 0 \end{bmatrix}$$

where R_{11} is a nonsingular triangular matrix. R_{11} can itself be expressed by a composition

$$R_{11} = \tilde{U}\tilde{S}\tilde{V}^T$$

where $\tilde{U}$, $\tilde{S}$, and $\tilde{V}$ are all $k \times k$ and $\tilde{S}$ contains the nonsingular values of S arranged in nonincreasing order. Then, by defining

$$R = USV^T$$

where

$$\hat{U} = \begin{bmatrix} \tilde{U} & 0 \\ 0 & I_{m-k} \end{bmatrix}$$

$$\hat{V} = \begin{bmatrix} \tilde{V} & 0 \\ 0 & I_{n-k} \end{bmatrix}$$

and

$$S = \begin{bmatrix} \tilde{S} & 0 \\ 0 & 0 \end{bmatrix}$$

then if

$$U = H\hat{U}$$

and

$$V = K\hat{V}$$

(3.112) can be given by

$$X = USV^T$$

as indicated. The new regression problem becomes

$$USV^T\boldsymbol{\beta} \cong \mathbf{y}$$

when $\cong$ signifies the least-squares solution. This can be represented equivalently as

$$S\mathbf{p} \cong \mathbf{g}$$

where

$$\boldsymbol{\beta} = V\mathbf{p}$$

and

$$\mathbf{g} = U^T\mathbf{y}$$

The solution obtained via this decomposition can be defined by

$$\boldsymbol{\beta}^{(k)} = \sum_{j=1}^{k} p_j \boldsymbol{v}_j$$

where the $\boldsymbol{v}_j$ are columns from V, including those associated with the singular values discarded in the analysis. A convenient definition for our solutions will be

$$\boldsymbol{\beta}^{(0)} = \mathbf{0}$$
$$\boldsymbol{\beta}^{(k)} = \boldsymbol{\beta}^{(k-1)} + p_k \boldsymbol{v}_k$$

A property of these solutions is that it is possible to calculate the residual norm p_k associated with the candidate solution $\beta^{(k)}$, since

$$\rho_k = \|X\boldsymbol{\beta}^{(k)} - \mathbf{y}\|$$

is also given by

$$\rho_k^2 = \|\mathbf{g}^{(2)}\|^2 + \sum_{i=k+1}^{n} (g_i^{(1)})^2$$

where $\mathbf{g}^T=[\mathbf{g}^{(1)^T},\mathbf{g}^{(2)^T}]$, with the $\mathbf{g}^{(1)}$ as the n values associated with the n nonsingular values of S, and $\mathbf{g}^{(2)}$ of length $m-n$ associated with the missing singular values. These residuals can also be given by

$$\rho_n^2=\|\mathbf{g}^{(2)}\|^2$$

$$\rho_k^2=\rho_{k+1}^2+\left(g_{k+1}^{(1)}\right)^2 \qquad k=n-1,\ldots,1,0$$

We can express the standard deviation of the error as a function of the residual and the number of terms, k, used in the model

$$S_k=\left(\frac{\rho_k^2}{m-k}\right)^{1/2}$$

Clearly, the least residual occurs when all n possible terms are included in the model and it increases as terms are removed from the model. However, the parameter estimates ρ_i each have an uncertainty proportional to their inverse singular value, so that as each term is added to the model, the uncertainty associated with that vector $\boldsymbol{v}_i$ increases. Selection of the optimum k thus is a tradeoff between the uncertainty in $\boldsymbol{\beta}^{(k)}$ and the residual ρ_k^2 associated with that solution. The null coefficients mentioned earlier consist, therefore, of the linear combination of the $\boldsymbol{v}_i$ associated with great uncertainty. They may be considered arbitrary, and their use will not significantly increase the residual.

The two methods discussed so far arise out of primarily computational considerations, and consist, in general, of procedures for eliminating dependencies and near dependencies within the moment matrix X^TX and its inverse $(X^TX)^{-1}$. A third method proceeds by augmenting the moment matrix by adding a constant C to all the diagonal terms, so that the normal equations become

$$(X^TX+CI)\hat{\boldsymbol{\beta}}^*=X^T\mathbf{y} \tag{3.113}$$

Several interpretations may be attached to the constant C, as we will discuss. Our primary motivation is to provide an estimate $\hat{\boldsymbol{\beta}}^*$ that will have smaller variance than $\hat{\mathbf{g}}$ when X^TX is poorly conditioned. The estimate $\hat{\boldsymbol{\beta}}^*$ will also differ from $\hat{\boldsymbol{\beta}}$ in that it will be a biased estimate of $\boldsymbol{\beta}$ since for $C>0$

$$\begin{aligned} E(\hat{\beta}^*)&=(X^TX+CI)^{-1}X^TX\boldsymbol{\beta} \\ &\neq\boldsymbol{\beta} \end{aligned} \tag{3.114}$$

Note that $\hat{\boldsymbol{\beta}}^*$ reduces to $\hat{\boldsymbol{\beta}}$ when $C=0$.

The estimates $\hat{\boldsymbol{\beta}}^*$ given by (3.114) are called the *Ridge Regression* estimates of $\boldsymbol{\beta}$ [see Hoerl and Kennard, 35 and 36]. They arise from a statistical consideration of the expected difference between $\hat{\boldsymbol{\beta}}$ and $\boldsymbol{\beta}$, given by

$$L_1^2 = (\hat{\boldsymbol{\beta}} - \boldsymbol{\beta})^T (\hat{\boldsymbol{\beta}} - \boldsymbol{\beta}) \tag{3.115}$$

We can show that

$$E(L_1^2) = \sigma^2 \operatorname{Tr}(X^T X)^{-1} \tag{3.116}$$

where Tr is "trace," and that when the errors are normally distributed

$$\operatorname{Var}(L_1^2) = 2\sigma^4 \operatorname{Tr}(X^T X)^{-2} \tag{3.117}$$

Now, the trace of a matrix is constant for any orthogonal transformation of the matrix so $\operatorname{Tr}(X^T X)$

$$\operatorname{Tr}(X^T X)^{-1} = \operatorname{Tr} P X^T X P = \operatorname{Tr} \Lambda = \sum \lambda_i \tag{3.118}$$

where P is the canonical transformation and Λ is the diagonal matrix containing the eigenvalues λ_i of $X^T X$. By substituting (3.118) in (3.116), we see that

$$E(L_1^2) = \sigma^2 \sum \lambda_i^{-1} \tag{3.119}$$

and

$$\operatorname{Var} L_1^2 = \sigma^4 \sum \lambda_i^{-2} \tag{3.120}$$

If we order the eigenvalues λ_i according to

$$\lambda_{\max} = \lambda_1 \geqslant \ldots \geqslant \lambda_r = \lambda_{\min} > 0$$

it is clear that the lower bound for (3.119) is given by $\sigma^2/\lambda_{\min}$ and equation (3.120) by $\sigma^4/\lambda_{\min}^2$.

Another property of the Ridge Regression estimates $\hat{\boldsymbol{\beta}}^*$ is that for $C > 0$

$$\hat{\boldsymbol{\beta}}^{*}T\hat{\boldsymbol{\beta}}^{*} < \hat{\boldsymbol{\beta}}^T \hat{\boldsymbol{\beta}}$$

and in the limit as $C \to \infty$, $\hat{\boldsymbol{\beta}}^{*T} \hat{\boldsymbol{\beta}}^{*} \to 0$, so that the length of the estimates is being reduced. This is not unexpected, of course, since we have remarked earlier that ill conditioning tends to produce estimates of large magnitude, so that as we correct the conditioning, we expect that the size of the estimates will decrease. In fact, one approach to interpreting the constant C is as a solution to the optimization problem

$$\min \phi = (y - X\boldsymbol{\beta})^T (y - X\boldsymbol{\beta})$$

subject to

$$\boldsymbol{\beta}^T \boldsymbol{\beta} - R^2 = 0$$

The value of R^2 is a limit on the size of $\hat{\boldsymbol{\beta}}$. We will show in the next chapter that a Lagrangian function

$$l(\hat{\boldsymbol{\beta}},C)=(y-X\hat{\boldsymbol{\beta}})^T(y-X\hat{\boldsymbol{\beta}})-\frac{1}{C}(\hat{\boldsymbol{\beta}}^T\hat{\boldsymbol{\beta}}-R^2)$$

can be minimized as an unconstrained problem to provide a solution to the constrained problem. Taking the derivatives of $l(\hat{\boldsymbol{\beta}},C)$ with respect to $\hat{\boldsymbol{\beta}}$ and C yields

$$(X^TX+CI)\hat{\boldsymbol{\beta}}=X^T\mathbf{y} \tag{3.121}$$

and

$$\hat{\boldsymbol{\beta}}^T\hat{\boldsymbol{\beta}}-R^2=0 \tag{3.122}$$

The value of C is obtained by substituting (3.122) into (3.121). Since R^2 is unknown, what is normally done is to investigate R^2 versus C or to plot ϕ and the β_i versus C. This latter plot is called the *Ridge Trace* and is discussed extensively by Hoerl and Kennard [36].

A final interpretation for the parameter C can also be supplied. The biased estimate $\hat{\boldsymbol{\beta}}^*$ is given by

$$\begin{aligned}\hat{\boldsymbol{\beta}}^* &= Z\hat{\boldsymbol{\beta}}\\ &=\left[I-(X^TX+CI)^{-1}\right]\hat{\boldsymbol{\beta}}\end{aligned} \tag{3.123}$$

where the matrix Z is a function of C and can be thought of as a variance reducer.

The three methods discussed in this section are obviously related. Both the eigenvalue analysis of Box et al. and the Singular Value Decomposition (SVD) are dealing with the orthogonal subspace corresponding to the larger singular values (eigenvalues) of X^TX. And Lawson and Hanson provide an example that suggests that the stabilized Ridge Regression (or Marquardt stabilization) values are similar to those obtained by SVD.

Computationally SVD is likely to be the preferable method. SVD and eigenvalue analysis differ primarily in their representation of X^TX (SVD using the simpler orthogonal decomposition), whereas the Ridge Trace will require the solution of many linear problems to obtain the Ridge Trace. Where this computation is not onerous, however, the graphical Ridge Trace may be preferred.

Weighted Least Squares

The general linear least squares problem can be generalized to the form

$$\min(\mathbf{y}-X\boldsymbol{\beta})^TW(\mathbf{y}-X\boldsymbol{\beta}) \tag{3.124}$$

where W is a weighting matrix which assigns relative importance to the squared residuals $\hat{\varepsilon}_i^2$ and sometimes to products of residuals $\hat{\varepsilon}_i\hat{\varepsilon}_j$. Equation (3.124) is the weighted least-squares problem.

Taking the usual derivatives, we now have

$$\hat{\beta}=(X^TWX)^{-1}X^TW\mathbf{y} \tag{3.125}$$

as the solution of normal equations. The new estimates are unbiased when $\text{Var}(\varepsilon)\sim \text{NID}(\mathbf{0}, W^{-1}\sigma^2)$ and $E(\mathbf{y})=X\boldsymbol{\beta}$. The variance of the estimates is now

$$\text{Var}\,\hat{\beta}=(X^TWX)^{-1}\sigma^2 \tag{3.126}$$

The total sum of squares is given by $\mathbf{y}^TW\mathbf{y}$ and the sum of squares due to the regression, SS_R,

$$SS_R=\mathbf{y}^TWX(X^TWX)^{-1}X^TW\mathbf{y} \tag{3.127}$$

The weighted least-squares problem arises principally when the precision of certain estimates is known to differ in some way so that instead of

$$\text{Var}(\varepsilon)=I\sigma^2 \tag{3.128}$$

as in the unweighted case,

$$\text{Var}(\varepsilon)=V\sigma^2=W^{-1}\sigma^2 \tag{3.129}$$

The matrix V is a general variance matrix for the errors ε_i. The minimum variance estimates for the $\hat{\boldsymbol{\beta}}$ are obtained when weights are given by $W=V^{-1}$.

The matrix elements σ_{ij}^2 may be known from several sources:

1. Suppose several replications are taken at each experimental point, from which variances can be estimated. Then the appropriate σ_{ii}^2 are assigned the estimate S_i^2.
2. Experimental results can be obtained from several pieces of equipment which are known from historical data, maintenance records, or whatever, to have various levels of precision in results. These disparities are included in the estimation through the variance matrix by relative levels of variability.
3. In simulation runs it is typical for the variance of a response to be available from the output, so each experiment can be assigned its own variance. This is clearly preferable to running the simulation until a set variance level is obtained.

Of course it is also possible that the variance will be a known function of the experimental point x_{iu}, $i=1,2,\ldots,k$, or the magnitude of y_u. This will sometimes be indicated from a study of the residuals obtained from the unweighted least squares estimates.

Weighted least squares supposes knowledge of the variance matrix $V\sigma^2$. In the cases where V is unknown or the approximation $I\sigma^2$ is inappropriate, the method of maximum likelihood can be employed to estimate both V and $\hat{\boldsymbol{\beta}}$ given V.

Maximum Likelihood Methods

Weighted least squares is a generalization of our standard linear regression technique, which is the special case of equal variance in all observations. We can, however, carry this generalization further with the introduction of maximum likelihood estimates, both to illustrate how the error distribution can be included in the model, and what types of criteria will apply to distributions other than the normal. The maximum likelihood estimators will prove especially useful when we begin treating nonlinear models.

Our observations $\mathbf{y}$ are random variables whose expected value is assumed to be given by the linear model $X\boldsymbol{\beta}$. The exact values y_i differ from this value by a random value ε_i, which is unknown, although it is *likely* that most of the time ε_i will be small, or y_i will be approximately equal to $\mathbf{x}_i^T\boldsymbol{\beta}$. If the errors are normally distributed with known variances σ^2, we can put this more concretely by giving the marginal probability of y given $\mathbf{x}_i$ as

$$f(y_i|\mathbf{x}_i)=\frac{1}{\sigma(2\pi)^{1/2}}\exp\left(-\frac{\varepsilon_i^2}{2\sigma_i^2}\right) \tag{3.130}$$

The most likely values of y_i, therefore, are those for which the residual ε_i are least. For a set of points, we consider the likelihood function $L(\mathbf{y}|X)$ given by

$$L(\mathbf{y}|X)=f(y_n|y_{n-1},y_{n-2}\ldots y_i,X)f(y_{n-1}|y_{n-2}\ldots y_1,X)\ldots f(y_1|X)$$

which is

$$L(\mathbf{y}|X)=\prod_{i=1}^{n} f(y_i|\mathbf{x}_i) \tag{3.131}$$

when the observations are all independent. Substituting (3.131) into (3.130), we obtain

$$L(\mathbf{y}|X)=\frac{1}{(2\pi)^n/2}\prod_{i=1}^{n}\left(\frac{1}{\sigma_i}\right)\exp\left(-\frac{1}{2}\sum_{L=1}^{n}\frac{\varepsilon_i^2}{\sigma_i^2}\right) \tag{3.132}$$

In maximizing $L(\mathbf{y}|X)$, we generally use the log likelihood function $\log L(\mathbf{y}|X)$, which is equivalent, since the logarithm is a monotonic increasing function of its argument. This is generally easier to treat. Equation

(3.132) becomes

$$\log L = -\log\left[\prod_{i=1}^{n}(\sigma_i)\right] - \frac{n}{2}\log 2\pi - \frac{1}{2}\sum\frac{\varepsilon_i^2}{\sigma_i^2} \tag{3.133}$$

The first two terms on the right are constant and the maximization of $\log L$ is clearly obtained via the minimization of

$$\min \sum_{i=1}^{n}\frac{\varepsilon_i^2}{\sigma_i^2}$$

This is the weighted least-squares criteria. Equation (3.133) reduces to

$$\log L = -n\log\sigma - \frac{n}{2}\log 2\pi - \frac{1}{2\sigma^2}\sum\varepsilon_i^2 \tag{3.134}$$

when the $\sigma_i^2 = \sigma^2$ are all equal. This is the form of our standard least-squares problem, since its maximum will obviously be obtained with the minimization of the sum of the squared residuals

$$\min \sum_{i=1}^{n}\varepsilon_i^2$$

We will generalize these methods further in the next section with application to nonlinear models and multiple responses. In all cases the overall technique remains the same: residuals are formed as functions of the unknown parameters and will be used to estimate the likelihood functions L or $\log L$. The optimal $\boldsymbol{\beta}^*$ will be obtained that satisfy

$$L(\mathbf{y};\hat{\boldsymbol{\beta}}) = \max_{\boldsymbol{\beta}} L(\mathbf{y};\boldsymbol{\beta})$$

or, equivalently

$$\log L(\mathbf{y};\hat{\boldsymbol{\beta}}) = \max_{\boldsymbol{\beta}} \log L(\mathbf{y};\boldsymbol{\beta}) \tag{3.135}$$

Using the likelihood function we may also obtain the criteria for distributions other than the normal. Consider errors which come from the double exponential distribution,

$$f(\varepsilon) = \frac{1}{2\sigma}e^{-|\varepsilon|/\sigma} \tag{3.136}$$

The general form of the log likelihood function for n observations is given by

$$\log L = -\log\left(2\prod_{i=1}^{n}\sigma_i\right) - \sum_{i=1}^{n}\frac{|\varepsilon_i|}{\sigma_i} \tag{3.137}$$

The maximum likelihood estimator for such an error distribution would be given by

$$\min_{\beta} \sum_{i=1}^{n} \frac{|\varepsilon_i|}{\sigma_i} \tag{3.138}$$

This estimate is also known as the l_1 or *robust estimator* since (3.138) is a minimization of the 1 norm given by

$$\sum \left| \frac{\varepsilon_i}{\sigma_i} \right|$$

Just as the least-squares estimator corresponds to the l_2 or 2-norm

$$\left\{ \sum \left(\frac{\varepsilon_i}{\sigma_i} \right)^2 \right\}^{1/2}$$

The l_1 estimator is "robust" in the sense that it is less sensitive to large residuals than the least-squares estimate. Large residuals are ignored by the l_1 estimate. This reflects the fact that large residuals are more likely with the exponential distribution and undue weight is not attached to them.

For errors obtained from uniform distributions bounded by $\pm b_i$, the likelihood functions f_i are equally likely when $|\varepsilon_i / b_i| \leqslant 1$ and impossible outside of that range. (The impossibility of outliers with this distribution is a strong characteristic of this distribution.) The appropriate maximum likelihood function is given by

$$L(y|X) = \max \left| \frac{\varepsilon_i}{b_i} \right|$$

and the maximum likelihood estimates $\hat{\boldsymbol{\beta}}$ are given by the minimax problem

$$\min_{\beta} \max \left| \frac{\varepsilon_i}{b_i} \right|$$

This criteria is sometimes known as the l_∞ estimate since, for $x^* = \max(x_i)$ and $x_i = |\varepsilon_i / b_i|$

$$\lim_{P \to \infty} \left(\sum x^p \right)^{1/p} = \lim_{P \to \infty} \frac{x^*}{x^*} \left(\sum x^p \right)^{1/p}$$

$$= x^* \lim_{P \to \infty} \left(\sum \frac{x^p}{x^{*p}} \right)^{1/p}$$

for all $x < x^*$, $\lim_{p \to \infty} (x^p / x^*) = 0$ so that this last line becomes

$$\lim_{p \to \infty} \left(\sum x^p \right)^{1/p} = x^*$$

The l_∞ estimate is thus given by the maximum component in the vector.

Both l_1 and l_∞ estimations differ from that of standard least squares in that the $\hat{\boldsymbol{\beta}}$ can be obtained from an analog of the normal equations. Robust or l_1 estimates can be obtained from a linear programming formulation. The l_∞ estimate is a nonlinear problem which could be solved using a random or direct search technique, using the solution to the l_1 problem as an initial guess. This estimate poses special problems, since the gradient of the objective function given by (3.138) is very likely to be discontinuous.

NONLINEAR REGRESSION AND MODELING

The linear regression models we have considered so far have definite strengths and some strict restrictions as well. As approximating functions, linear models can be usefully employed in a wide variety of situations. The statistical and computational properties of linear models and their experimental designs are well known. In most situations their solution is straightforward, and, if it exists, it can always be found. Nevertheless, their application is limited to models that are linear in the parameters or can be suitably transformed so that they will be linear in the parameters, *and* the experimental error for the linear model should be additive, with constant variance and of known distribution. If any of these assumptions are violated, our knowledge of the results is seriously impaired.

In addition, linear regression is often adopted for approximating purposes, as when the mechanism is unknown and it is hoped that at least locally a linear model of some form will provide a suitable, essentially empirical approximation. Such models are only suitable for *interpolation*: function estimation within the experimental region. Empirical models are unsuited to either extrapolation (prediction) or to study of the underlying mechanism of a function.

Oftentimes, therefore, we are forced to abandon linear estimation for its nonlinear counterpart. For nonlinear models we can relax our stipulations on the form of the model, the homogeneity of the error variance, and the independence among observations. We will still expect that error be additive and of known distribution. The price to be exacted for this flexibility includes computational difficulty, a smaller body of knowledge concerning the statistical properties of nonlinear models, and much less prior knowledge of what constitutes a good nonlinear experimental design.

The simplest nonlinear situation is given by a single response function of constant variance and errors distributed normally and uncorrelated. Let our observations now be given by

$$y_u(\mathbf{x};\boldsymbol{\theta}) = \eta_u(\mathbf{x};\boldsymbol{\theta}) + \varepsilon_u \tag{3.139}$$

where, as usual,

$$E(\varepsilon_u)=0 \qquad E(\varepsilon_u^2)=\sigma_u^2 \tag{3.140}$$

and

$$E(\varepsilon_u \varepsilon_v)=0$$

Now η_u can be modeled by any function in the parameters θ and independent variables X. In kinetics modeling, for instance, it is not uncommon for the general model to be given as a rate equation of the form

$$\frac{d\eta}{dt}=f(\mathbf{x},t;\boldsymbol{\theta}) \tag{3.141}$$

so that the function η is given by an integral equation

$$\eta(t)=\eta(0)+\int_0^t f(\mathbf{x},\tau;\boldsymbol{\theta})\,d\tau \tag{3.142}$$

If the form of f is not simple, it will be necessary to numerically integrate (3.142) for each set of $(\boldsymbol{\theta},t)$.

It is not even necessary that η be an explicit function, such as in (3.141). For instance, (3.141) could be given as

$$\frac{d\eta}{dt}=f(\mathbf{x},\eta,t;\boldsymbol{\theta}) \tag{3.143}$$

so that

$$\eta(t)=\eta(0)+\int_0^t f[\mathbf{x},\eta(\tau),t;\boldsymbol{\theta}]\,d\tau \tag{3.144}$$

or

$$t_u=\int^{\eta_u}\frac{d\alpha}{f(\mathbf{x}_u,\alpha,t_u;\boldsymbol{\theta})} \tag{3.145}$$

which can be given generally by

$$x_u=g(\mathbf{x}_u,\eta_u,\boldsymbol{\theta}) \tag{3.146}$$

In the simple case we employ nonlinear least squares directly, minimizing

$$S(\boldsymbol{\theta})=\sum_{u=1}^{n}\left[y_u-\hat{y}(\mathbf{x}_u,\boldsymbol{\theta})\right]^2 \tag{3.147}$$

and in the implicit case it is necessary to also minimize

$$T(\boldsymbol{\theta})=\sum_{u=1}^{n}\left[x_u-g(\mathbf{x}_u,\eta_u,\boldsymbol{\theta})\right]^2 \tag{3.148}$$

Since x_u is an independent variable, minimization of (3.148) alone would be incorrect, as it would imply not (3.139) but

$$x_u=g(\mathbf{x}_u,\eta_u,\boldsymbol{\theta})+\tilde{\varepsilon}_u \tag{3.149}$$

Use of equation (3.148) alone has been shown to significantly bias the results and to inflate the variance of the estimates.

With linear models it was possible to solve directly for the optimum of the least-squares problem with the normal equations. The minimum of (3.147) is by no means so simple, given the nonlinearity of $y(\mathbf{x}, \boldsymbol{\theta})$. We postpone until the next chapter a discussion of this problem, except to note that the solution will require an iterative solution technique and a converged solution may well depend upon a good initial estimate for the $\boldsymbol{\theta}$.

We derived criteria for linear models in special situations using the maximum likelihood approach in the previous section. Maximum likelihood is especially useful for nonlinear modeling. A particular application is given in system identification and kinetic modeling where observations from several response variables y_{iu}, $i=1,2,\ldots,m$, $u=1,2,\ldots,n$, is available and a common set of parameters links these functions together. Consider the rate reaction given by

$$A \rightarrow B \rightarrow C$$

which could be modeled by

$$\begin{aligned} \dot{\eta}_1 &= -\beta_1 \eta_1 \\ \dot{\eta}_2 &= \beta_1 \eta_1 - \beta_2 \eta_2 \\ \dot{\eta}_3 &= \beta_2 \eta_2 \end{aligned} \tag{3.150}$$

where η_1, η_2, and η_3 represent the mole fractions of A, B, and C during the reaction. The solution to (3.150) for initial conditions $\eta_1 = 1$, $\eta_2 = \eta_3 = 0$ at $t=0$ is

$$\begin{aligned} \eta_1 &= e^{-\beta_1 t} \\ \eta_2 &= \frac{\beta_1}{(\beta_2 - \beta_1)(e^{-\beta_1 t} - e^{-\beta_2 t})} \\ \eta_3 &= 1 + \frac{-\beta_2 e^{-\beta_1 t} + \beta_1 e^{-\beta_2 t}}{\beta_2 - \beta_1} \end{aligned} \tag{3.151}$$

We proceed by measuring y_1, y_2, and y_3 at several points in the reaction independently, so that even though

$$E(y_1 + y_2 + y_3) = 1$$

$$y_{1u} + y_{2u} + y_{3u} \neq 1$$

in general. We expect also that the errors in these observations could be correlated and given by

$$\operatorname{cov}(\mathbf{y}) = V = \{\sigma_{ij}\} \tag{3.152}$$

The appropriate method of estimation is maximum likelihood. Defining the quantities

$$r_{ij}=\sum_{u=1}^{n}\left[y_{iu}-\hat{y}_i(\mathbf{x}_u,\theta)\right]\left[y_{ju}-\hat{y}_j(\mathbf{x}_u,\theta)\right] \tag{3.153}$$

then the likelihood function is given by

$$p(\mathbf{y}|\boldsymbol{\theta},w_{ij})=(2\pi)^{-(1/2)nk}|V^{-1}|^{(1/2)n}\exp\left\{-\frac{1}{2}\sum_{i=1}^{k}\sum_{j=1}^{k}w_{ij}r_{ij}\right\} \tag{3.154}$$

where, as before, $V^{-1}=\{w_{ij}\}$, and the w_{ij} are the weights derived from the covariance matrix V. We can derive the appropriate criteria for estimation depending upon our knowledge of V.

If the σ_{ij} are known, and $\sigma_{ij}=0$ for $i\neq j$, then the $\boldsymbol{\theta}$ should be chosen such that

$$\min\sum_{i=1}^{k}w_{ii}r_{ii}(\boldsymbol{\theta}) \tag{3.155}$$

is solved. This is nonlinear weighted least squares.

If the σ_{ij} are all known and some $\sigma_{ij}\neq 0$, $i\neq j$, then $\boldsymbol{\theta}$ should be chosen such that

$$\min\sum_{i=1}^{k}\sum_{j=1}^{k}w_{ij}r_{ij}(\theta) \tag{3.156}$$

is solved.

In the general case the σ_{ij} are not known at all and must be estimated as well. Box and Draper [11] have shown, using Bayes Theorem, that the appropriate criteria then becomes

$$\min|R|=\sum_{i=1}^{k}r_{ij}R_{ij}=\sum_{i=1}^{k}\sum_{j=1}^{k}\frac{r_{ij}R_{ij}}{k} \tag{3.157}$$

where $R=\{r_{ij}\}$ and R_{ij} is the cofactor of r_{ij}. The minimizing estimated weights w_{ij} are given by

$$w_{ij}=\frac{R_{ij}}{k} \tag{3.158}$$

which can be seen when (3.157) and (3.156) are compared.

To illustrate, consider the example suggested earlier. If V is unknown, the appropriate criteria in estimating $\boldsymbol{\beta}^T=(\beta_1,\beta_2)$ is minimization of

$$|R|=\begin{vmatrix}\sum_{u=1}^{n}(y_{1u}-\hat{y}_{1u})^2 & \sum_{u=1}^{n}(y_{1u}-\hat{y}_{1u})(y_{2u}-\hat{y}_{2u}) & \sum_{u=1}^{n}(y_{1u}-\hat{y}_{1u})(y_{3u}-\hat{y}_{3u})\\ \sum_{u=1}^{n}(y_{1u}-\hat{y}_{1u})(y_{2u}-\hat{y}_{2u}) & \sum_{u=1}^{n}(y_{2u}-\hat{y}_{2u})^2 & \sum_{u=1}^{n}(y_{2u}-\hat{y}_{2u})(y_{3u}-\hat{y}_{3u})\\ \sum_{u=1}^{n}(y_{1u}-\hat{y}_{1u})(y_{3u}-\hat{y}_{3u}) & \sum_{u=1}^{n}(y_{2u}-\hat{y}_{2u})(y_{3u}-\hat{y}_{3u}) & \sum_{u=1}^{n}(y_{3u}-\hat{y}_{3u})^2\end{vmatrix} \tag{3.159}$$

where $\hat{y}_{iu}=y_i(\mathbf{x}_u,\boldsymbol{\beta})$.

Box and Draper demonstrate that this approach improves the estimates of $\boldsymbol{\beta}$ compared to the results obtained when $\boldsymbol{\beta}$ are fit using the y_i separately by producing estimates of less variance. The maximum likelihood method is a way of utilizing all the information that the k sets of observations provide to provide a combined estimate.

SCREENING EXPERIMENTS

At the beginning of many experimental projects, there is considerable uncertainty about which of many candidate factors are significant enough to be included in the study. Two suspicions occur to us. On one hand, just about any factor probably influences some interesting response. On the other, if everything were included in the model, the number of experiments required would be prohibitive. It is often necessary to construct a preliminary design over the possible factors that can screen out the most important factors for further detailed study.

Factor screening is also used when the object of our study is well known but complicated, or too expensive to investigate directly. Consider the case with a large simulation model. The modeler can show, "in the small," how each factor affects the total response, but will be hard put to speculate about which factors, "on the whole," are significant. If our object is to control the system, or to discover optimum settings, we will be willing to sacrifice the precise approximation of the process which the simulation provides, for a reduced model that contains the main features of system behavior.

Our first screening technique is direct analysis and thorough planning. Possible factors should be identified prior to any experimentation, and those that can neither be assumed significant nor trivial must be considered further. Can groups of variables be identified, whose effect should be similar, or whose use will be correlated? Can groups be eliminated, simplified, or reparameterized prior to experimental study? In the large, we can also ask what models might suit the system under study or provide insight into the identity of the main variables or the mode of their effect.

The significance of proper planning and analysis as a screening technique cannot be overemphasized. This point is illustrated in several ways. In the practical realm, Box points out [cited in Box and Hunter, "The Experimental Study of Physical Systems," *Technometrics*, vol. 7, p. 1] that the design of experiments depends upon knowledge of the system being studied, for "if nothing is known about the experimental situation then strictly speaking no experiment can be planned." The experimenter must at the least estimate the general level of factor effects versus experimental variance if he is to code the variables so that significant effects can be distinguished from the noise.

Analysis is discussed in Ignall [39] in his note about experimental designs and simulation models by Hunter and Naylor [37]. Ignall describes how proper analysis can significantly reduce the number of candidate factors beforehand without impairing the quality of the results.

Once the analysis is complete, we are left with a list of factors that should contain the most significant factors, and a rationale for choosing design spacings. We will be examining main effects, and our overall model is therefore linear, although we are primarily more interested in the ranking in the size of the effects than in their actual values. The screening experiment will also be useful in refining our estimate of experimental error.

Several general classes of designs are commonly used for screening studies: Plackett-Burman [46], fractional factorial [13a, 13b], and saturated designs [Booth and Cox, 8], and group screening methods [Watson, 49]. Several of these methods will be discussed. A general discussion of the simulation screening problem is also given in Eldredge [27], and tables are included with designs and analysis information.

Plackett-Burman designs are balanced designs of size $4m$, $m=3,4,\ldots$, which are suitable for linear models with up to $4m-1$ factors. The designs are orthogonal, but interaction effects alias most terms and cannot be separately estimated by these designs. The estimate of the variance is likewise biased by interaction (when degrees of freedom are available) so that the significance of the main effects can be seen. The Plackett Burman designs are generated from a column generator for each design, $m>3$. For instance, the generator for the 12 point design is given by the coded variables (+ for 1, $-$ for -1).

$$++-+++---+-$$

The design consists of the generator in the first column, with each successive column starting one position further right, the passed portion of the generator being appended to the end. A column of all minuses is added as the last row of the design. For instance, for 4 factors the design is

$$D=\begin{bmatrix} 1 & 1 & -1 & 1 \\ 1 & -1 & 1 & 1 \\ -1 & 1 & 1 & 1 \\ 1 & 1 & 1 & -1 \\ 1 & 1 & -1 & -1 \\ 1 & -1 & -1 & -1 \\ -1 & -1 & -1 & 1 \\ -1 & -1 & 1 & -1 \\ -1 & 1 & -1 & 1 \\ 1 & -1 & 1 & 1 \\ -1 & 1 & 1 & -1 \end{bmatrix} \tag{3.160}$$

Plackett-Burman designs are relatively easy to use and analyze and are relatively efficient estimators. Correlations between two-factor interactions and main effects can be eliminated by running a reflected design (Plackett-Burman design of $-D$) and analyzing the two designs together.

Another screening technique is the 2^{k-p} fractional factorial introduced earlier. These orthogonal designs consist of successive fractions of the full factorials. Now, in the full factorials all first-order interactions are orthogonal to the main effects and to each other. As fractions are taken of this full design various groups become aliased together, as defined by the contrast aliased to the mean. The smallest fraction is a completely saturated design with all factors aliased to some interaction. We can control the structure of these aliases, however, as shown earlier, and separate first-order factors from two-factor interactions, and so on. This leads to a convenient way of classifying fractional designs: We define the resolution of a design as the degree in which results must be interpreted. As the resolution is increased, the less various factors and interactions are confounded together. At the lower resolutions factors are greatly confounded, with the least resolution provided by the saturated fractions. A resolution N design is a fraction in which main effects are confounded with interactions of $N-1$ factors and two-factor interactions are confounded with $N-2$ factor interactions. Thus saturated designs are of Resolution III and the half fraction will be of Resolution N. Plackett-Burman designs for which $4m=2^k$ are saturated designs of Resolution III.

To illustrate the resolution of a design, consider the 2^{4-1}_{IV} design given by

$$D=\begin{bmatrix} -1 & -1 & -1 & -1 \\ 1 & -1 & -1 & 1 \\ -1 & 1 & -1 & 1 \\ 1 & 1 & -1 & -1 \\ -1 & -1 & 1 & 1 \\ 1 & -1 & 1 & -1 \\ -1 & 1 & 1 & -1 \\ 1 & 1 & 1 & 1 \\ 0 & 0 & 0 & 0 \\ 0 & 0 & 0 & 0 \\ 0 & 0 & 0 & 0 \\ 0 & 0 & 0 & 0 \end{bmatrix}$$

The defining contrast here is $ABCD$, and the alias structure is

$$\begin{array}{ll} A \equiv BCD & I = ABCD \\ B \equiv ACD & AB \equiv CD \\ C \equiv ABD & AC \equiv BD \\ D \equiv ABC & BC \equiv AD \end{array}$$

Four center points have been included to estimate error variance. Notice that since this is a design of Resolution IV, the main effects are confounded only with 3-factor interactions and the two-factor interactions are confounded with each other.

The design given above allows us great flexibility with regard to what can follow it. The confounded effects can be estimated using this design and the remaining 8 of the full factorial if they are found significant. If any main factor is not significant, the 8 factorial points define a 2^3 full factorial design in the remaining three factors, and if only two main effects are significant, the points define two replications of the 2^2 full factorial.

As the number of potential factors increases, sequential screening strategies, such as the group screening procedure of Watson, become necessary. In the group screening procedure several factors are aliased together as group factors in an orthogonal design. Groups which are not found to be significant are not further pursued, but additional experiments are run to distinguish the active factors within a significant group.

Various strategies have been proposed for supersaturated designs for the detection of significant factors, although obviously not all factors can be independently determined. The designs of Booth and Cox are constructed so as to be almost orthogonal and compare favorably with the *random balance experiments*. Random balance experiments are generated randomly, sacrificing orthogonality and efficiency for economy and simplicity. Analysis of such designs is not simple, and the ridge regression and singular value decomposition methods might be useful in this regard.

REFERENCES

1. Anderson, T. F., D. S. Abrams, and E. A. Grens, II, "Evaluation of Parameters for Nonlinear Thermodynamic Models," *American Institute of Chemical Engineering Journal*, vol. 24, no. 1, January 1978, pp. 20–29.
2. Anscombe, F. J., "Rejection of Outliers," *Technometrics*, vol. 2, no. 2, May 1960, pp. 123–147.
3. Anscombe, F. J., and J. W. Tukey, "The Examination and Analysis of Residuals," *Technometrics*, vol. 5, no. 2, May 1963, pp. 141–160.
4. Ball, W. E., and L. C. D. Groenweghe, "Determination of Best-Fit Rate Constants in Chemical Kinetics," *I and EC Fundamentals*, vol. 5, no. 2, May 1966, pp. 181–184.
5. Bard, Y., *Nonlinear Parameter Estimation*, Academic Press, New York, 1974.
6. Bard, Y., and L. Lapidus, "Kinetics Analysis by Digital Parameter Estimation," *Catalysis Reviews*, vol. 2, no. 1, 1968, pp. 67–112.
7. Bevington, P. R., *Data Reduction and Error Analysis for the Physical Sciences*, McGraw-Hill, New York, 1969.
8. Booth, K. H. V., and D. R. Cox, "Some Systematic Supersaturated Designs," *Technometrics*, vol. 4, no. 4, pp. 489–495.

9. Box, G. E. P., "Multifactor Designs of First Order," *Biometrika*, vol. 39, 1952, pp. 49–57.

10. Box, G. E. P., and D. W. Behnken, "Some New Three Level Designs for the Study of Quantitative Variables," *Technometrics*, vol. 2, no. 4, Nov. 1960, pp. 455–475.

11. Box, G. E. P., and N. R. Draper, "The Bayesian Estimation of Common Parameters from Several Responses," *Biometrika*, vol. 52, 1965, pp. 355–365.

12. Box, G. E. P., and N. R. Draper, "A Basis of Selection of a Response Surface Design," *Journal of the American Statistical Association*, vol. 54, Sept. 1959, p. 287.

13. Box, G. E. P., and J. S. Hunter, "The 2^{k-p} Fractional Factorial Designs—Part I," *Technometrics*, vol. 3, no. 3, pp. 311–351; Part II, *Technometrics*, vol. 3, no. 4, pp. 449–478.

14. Box, G. E. P., and J. S. Hunter, "Multi-factor Experimental Designs for Exploring Response Surfaces," *Annals of Mathematical Statistics*, vol. 28, 1957, pp. 195–241.

15. Box, G. E. P., W. G. Hunter, J. F. MacGregor, and J. Erjavec, "Some Problems Associated with the Analysis of Multiresponse Data," *Technometrics*, vol. 15, no. 1, Feb. 1973, pp. 33–51.

16. Box, G. E. P., and P. W. Tidwell, "Transformation of the Independent Variables," *Technometrics*, vol. 4, 1962, pp. 531–550.

17. Box, G. E. P., and K. B. Wilson, "On the Experimental Attainment of Optimum Conditions," *Journal of the Royal Statistical Society*, Ser. B, vol. 13, 1951, pp. 1–45.

18. Brooks, S. H., and M. R. Mickey, "Optimum Estimation of Gradient Direction in Steepest Ascent Experiments," *Biometrics*, vol. 17, no. 1, March 1961, pp. 48–56.

19. Davies, O. L., *Design and Analysis of Industrial Experiments*, 2nd Ed., Hafner Publishing Co., New York, 1956.

20. Debaun, R. M., "Block Effects in the Determination of Optimum Conditions," *Biometrics*, vol. 12, 1956, pp. 20–22.

21. Dolby, J. L., "A Quick Method for Choosing a Transformation," *Technometrics*, vol. 5, no. 3, August 1963, pp. 317–325.

22. Draper, N. R., and W. G. Hunter, "Transformations: Some Examples Revisited," *Technometrics*, vol. 11, no. 1, Feb. 1969, pp. 23–40.

23. Draper, N. R., and H. Smith, *Applied Regression Analysis*, John Wiley and Sons, New York, 1966.

24. Eakman, J. M., "Strategy for Estimation of Rate Constants from Isothermal Reaction Data," *Industrial and Engineering Chemical Fundamentals*, vol. 8, no. 1, Feb. 1969, pp. 53–58.

25. Efron, B., and C. Morris, "Stein's Paradox in Statistics," *Scientific American*, May 1977, pp. 119–127.

26. Efron, B., and C. Morris, "Stein's Estimation Rule and Its Competitors—An Empirical Bayes Approach," *Journal of the American Statistical Association*, vol. 68, no. 341, March 1973, pp. 117–130.

27. Eldredge, D. L., "Experimental Optimization of Statistical Simulation," *Proceedings of the 1974 Winter Simulation Conference*, Elmont, New York, 1974.

28. Forsythe, G. E., M. A. Malcolm, and C. B. Moler, *Computer Methods for Mathematical Computations*, Prentice-Hall, Englewood Cliffs, NJ, 1977.

29. Forsythe, G. E., and C. B. Moler, *Computer Solution of Linear Algebraic Systems*, Prentice-Hall, Englewood Cliffs, NJ, 1967.

30. Gorman, J. W., and J. E. Hinman, "Simplex Lattice Designs for Multicomponent Systems," *Technometrics*, vol. 4, no. 4, pp. 463–487.

31. Graybill, F. A., *An Introduction to Linear Statistical Models*, Vol. I, McGraw-Hill, New York, 1961.
32. Hahn, G. J., W. Q. Meeker, Jr., and P. I. Feder, "The Evaluation and Comparison of Experimental Designs for Fitting Regression Relationships," *Journal of Quality Technology*, vol. 8, no. 3, July 1976, pp. 140–157.
33. Henrici, P., *Elements of Numerical Analysis*, John Wiley and Sons, New York, 1964.
34. Hicks, C. R., *Fundamental Concepts in the Design of Experiments*, Holt, Rinehart and Winston, New York, 1973.
35. Hoerl, A. E., and R. W. Kennard, "Ridge Regression: Biased Estimation for Nonorthogonal Problems," *Technometrics*, vol. 12, no. 1, Feb. 1970, pp. 55–67.
36. Hoerl, A. E., and R. W. Kennard, "Ridge Regression: Applications to Nonorthogonal Problems," *Technometrics*, vol. 12, no. 1, Feb. 1970, pp. 69–82.
37. Hunter, J. R., and R. H. Naylor, "Experimental Designs for Computer Simulation Experiments," *Management Science*, vol. 16, pp. 422–434.
38. Hunter, W. G., "Estimation of Unknown Constants from Multiresponse Data," *Industrial and Engineering Chemical Fundamentals*, vol. 6, p. 461.
39. Ignall, E. J., "On Experimental Designs for Computer Simulation Experiments," Notes, *Management Science*, vol. 18, no. 7, April 1972, p. 384.
40. Isaacson, E., and H. B. Keller, *Analysis of Numerical Methods*, John Wiley and Sons, New York, 1966.
41. Lawson, C. L., and R. J. Hanson, *Solving Least Squares Problems*, Prentice-Hall, Englewood Cliffs, NJ, 1974.
42. Montgomery, D. C., *Design and Analysis of Experiments*, John Wiley and Sons, New York, 1976.
43. Montgomery, D. C., and D. M. Evans, "Second-Order Response Surface Designs in Digital Simulation," 41st ORSA Meeting, New Orleans (1972).
44. Montgomery, D. C., J. J. Talavage, and C. J. Mullen, "A Response Surface Approach to Improving Traffic Signal Settings in a Street Network," *Transportation Research*, vol. 6, 1972, pp. 69–80.
45. Myers, R. H., *Response Surface Methodology*, distributed by Edwards Brothers, Inc., Ann Arbor, MI, 1976.
46. Plackett, R. L., and J. P. Burman, "The Design of Multifactorial Experiments," *Biometrika*, vol. 33, 1946, pp. 305–325.
47. Tukey, J. W., "On the Comparative Anatomy of Transformations," *Ann. Math. Stat.*, vol. 28, pp. 602–632.
48. Walpole, R. E., and R. H. Myers, *Probability and Statistics for Engineers and Scientists*, 2nd Ed., Macmillan Publishing Co., New York, 1978.
49. Watson, G. S., "A Study of the Group Screening Method," *Technometrics*, vol. 3, no. 3, pp. 371–388.
50. Whitwell, J. C., and G. K. Morbey, "Reduced Designs of Resolution five," *Technometrics*, vol. 3, no. 4, Nov. 1961, p. 459.
51. Wilkinson, J. H., *Rounding Errors in Algebraic Processes*, Prentice-Hall, Englewood Cliffs, NJ, 1963.
52. Wilkinson, J. H., *The Algebraic Eigenvalue Problem*, Oxford University Press, New York, 1965.

Chapter 4

Fundamentals of Optimization

INTRODUCTION

Our strategic objective is optimization, the attainment of the best solution for the system under study. At each step of the process of moving toward the optimum, promising input values will be determined by the optimization methodology according to some algorithm. The uncertainty of measurement will affect the exact choice of points for experiments, for at the tactical level the statistical considerations discussed in the two previous chapters will predominate.

An illustration of the strategic use of optimization is the response surface or gradient search approach described in this chapter. A linear or quadratic estimate of a response is determined, and the next experiments are deployed along an algorithmically determined direction. For an unconstrained system this is often the steepest ascent direction. In constrained systems, the search direction is a feasible direction estimate, that is, a direction that gains improvement in the primary response y_i while moving along estimated bounds.

Because the systems of interest are experimentally determined, optimization methods will often be used tactically to solve specific subproblems during the drive to the optimum solution. The least-squares techniques of the last chapter are solutions to the optimization problem of minimizing the squared error of the estimate. In the linear case this solution is the normal equations. In the nonlinear case there is no longer a general solution and each problem must be solved separately.

It will sometimes prove useful to optimize an approximation of the system rather than the system itself (as is the case with response surface methods). This can be done when the estimate is quadratic or higher and will be especially useful when constraints are encountered. "Experiments" made with the approximating system are relatively cheap compared to the actual system, of course, and a more detailed look for constrained solutions can be made. In feasible direction methods, an optimization model is proposed to solve for an approximate best experimental direction.

Because of the different uses for optimization techniques, several different techniques will be described in this chapter and their properties discussed. These properties determine their application in the various problems. Strategic techniques, those to be applied to experimental measurements to determine the next set points, must be above all relatively insensitive to error. Techniques that are applied to approximating problems can be error sensitive, but should converge to solutions consistently and with reasonable efficiency. Techniques for solving nonlinear least-squares problems should be robust and efficient.

Just as there are several different applications for optimization problems, this chapter will also discuss several types of optimization problems. The simplest are unconstrained problems, including the line search

$$\max f(\lambda)$$

and the scalar valued vector optimization

$$\max f(\mathbf{x})$$

Since there are no constraints, optimization consists of finding a point where the function obtains a maximum. The optimization problem could just as well have been formulated as a minimization, since the two are equivalent; that is,

$$\max[-f(\mathbf{x})] = \min f(\mathbf{x})$$

The value at which $f(\mathbf{x})$ obtains its minimum is the point at which $-f(\mathbf{x})$ is maximum.The simplest constraints are called *explicit* and consist simply of bounds on the input vector **x**. Solutions to the unconstrained problem lying outside the explicit constraints are rejected. Explicit constraints are easy to handle because the feasible set is simply defined and there is no uncertainty about the inputs and their feasibility.

Implicit constraints are given by functions of two forms, termed inequality and equality constraints, as follows:

$$g_i(\mathbf{x}) \leqslant 0 \qquad i = 1, \ldots, p \tag{4.1}$$

$$h_j(\mathbf{x}) = 0 \qquad j = 1, \ldots, n \tag{4.2}$$

Inequality constraints at the boundary are called active constraints, as in (4.1). *Equality* constraints, as in (4.2), are of course always active. Unless the constraint functions are especially simple, the feasible region can be quite complicated and the constrained optima quite difficult to obtain.

Constrained problems with implicit constraints are called *multiple response* problems when more than one response must be measured.

Another type of multiple response problem is the multiple objective problem, where several functions are to be optimized simultaneously. Such problems may be constrained or unconstrained.

SINGLE-VARIABLE OPTIMIZATION

This simplest of optimization problems is also among the more ubiquitous of optimization techniques. It is simple because with one variable there are no complications due to interactions among variables, models can be constructed with a minimum of terms, and there is no "curse" of dimensionality. The problem is quite common, however, as line searches are an integral part of ascent (or descent) methods, such as the steepest ascent, Newton-Raphson, Fletcher-Reeves, Marquardt, and feasible direction methods to be introduced in later sections.

The general single-variable problem is of the form

$$\max f(x)$$

subject to $x \varepsilon \Omega$ where in general Ω is simply some closed interval of the real line. The set Ω represents the set of feasible x. A typical single variable function $f(x)$ is illustrated in Figure 4.1. A line search has a related problem in several dimensions of the form

$$\max f(\mathbf{x}_0 + \lambda \mathbf{s})$$

subject to $\mathbf{x} \varepsilon \Omega$. Even though the search now takes place in a multidimensional space, the single variable λ, giving the step along the search direction $\mathbf{s}$, defines the point in that space so that the two problems are equivalent.

Single-variable searches are popular because of their simplicity and because of their relative economy. A direction estimating experiment in n

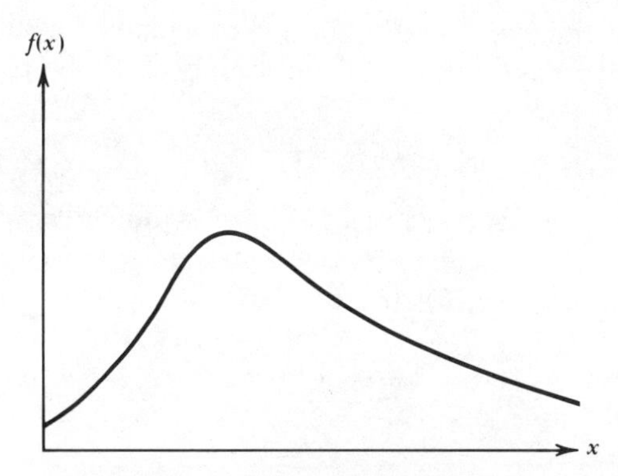

Figure 4.1. A single-variable search in x.

dimensions requires at least $n+1$ measurements of the objective function, whereas the number of measurements in a line search is independent of the actual dimensionality of the problem. However, since the movement in the test region often occurs as a result of the line search, it is also important that its accuracy be good.

Most line search methods assume *unimodality* in the search area. This considerably simplifies the problem, since points on the far side of the worst point can be eliminated from further consideration. Of the techniques to follow, the interval reducing methods will depend upon this implicitly, and the polynomial techniques perhaps not at all.

Interval Reduction Methods

In intervals of known limits, the assumption of *unimodality* is exploited to reduce the interval of uncertainty about the optimum as quickly as possible. This "interval of uncertainty" is the interval known to contain the optimum, but within which the exact location of the optimum is not known with certainty. As this interval decreases, the uncertainty regarding the position of the optimum correspondingly decreases.

For two points placed in a known interval, the interval of uncertainty is minimized for two points placed symmetrically about the midpoint. This is the basis of a *dichotomous search.* In the limit the points are placed "side by side." In practice, they must be placed far enough apart for the two measurements to be *distinguishable*. This distance is called the *resolution distance* and is denoted by ε. The optimal placement of points is given by

$$x_1 = b - \left(\frac{b-a+\varepsilon}{2}\right) \tag{4.3a}$$

$$x_2 = a + \left(\frac{b-a+\varepsilon}{2}\right) \tag{4.3b}$$

b is the upper end point and a the lower endpoint in the initial search interval. The expected interval of uncertainty after the placement of two search points is

$$L_2 = \frac{b-a}{2} + \frac{\varepsilon}{2} \tag{4.4}$$

The search proceeds with the placement of symmetrical points within the remaining interval of uncertainty. For instance, if $y(x_2) > y(x_1)$, the interval of uncertainty is now (x_1, b), and two new measurements are placed at

$$x_3 = x_1 + \frac{b - x_1 + \varepsilon}{2}$$

$$x_4 = b - \frac{b - x_1 + \varepsilon}{2}$$

so that the interval of uncertainty after the placement of four points is given by

$$L_4 = b - x_3 = \frac{b-a}{4} + \left(1 - \tfrac{1}{4}\right)\varepsilon$$

In general, after k pairs of search points have been placed, the remaining interval of uncertainty is

$$L_{2k} = \frac{b-a}{2^k} + \left(1 - \frac{1}{2^k}\right)\varepsilon \tag{4.5}$$

Since ε limits the resolution of measurements, the interval of uncertainty must be greater than ε. In practice the interval is reduced to 2ε and a final measurement is placed in the center of the final interval. Therefore, the number of experimental points will be

$$(2^k + 1)\varepsilon \leqslant (b-a)$$

and k is the largest integer satisfying the relation. For instance, consider the problem

$$\max y(x) = xe^{-x}$$

for x in the interval $0 \leqslant x \leqslant 2$. Here $a = 0$ and $b = 2$. We would like to estimate x^* to within 0.05 of its true position. The number of point sets required to obtain this precision with a dichotomous search is given by

$$(2^k + 1)(0.05) \leqslant 2$$

$k = 5$ is the greatest value of k satisfying this equation. The first two points are given by (4.3a) and (4.3b),

$$x_1 = 2 - \left(\frac{2 - 0 + 0.05}{2}\right) = 1.025 \qquad y(x_1) = 0.367766$$

$$x_2 = 0 + \left(\frac{2 - 0 + 0.05}{2}\right) = 0.975 \qquad y(x_2) = 0.367763$$

Since $y(x_1) > y(x_2)$, the new search interval is $0.975 \leqslant x \leqslant 2$. Successive applications of the algorithm yields

i	x_i	y_i	a	b
1	1.025	0.367766		
2	0.975	0.367763	0.975	2
3	1.5125	0.333292		
4	1.4625	0.338797	0.975	1.5125
5	1.26875	0.356751		
6	1.21875	0.360262	0.975	1.26875
7	1.146875	0.364279		
8	1.096875	0.366261	0.975	1.146875
9	1.085938	0.366396		
10	1.035938	0.36747	0.975	1.085938

The remaining interval of uncertainty exceeds 2ε, and the best estimate, at x_1, can be used as the optimum or a final point can be placed between the last interval, "just to make sure."

$$x_{11} = \tfrac{1}{2}(a+b) = 1.030469 \qquad y(x_1) = 0.367712$$

and x_1 is still the best estimate of the optimum:

$$x^* = x_1 = 1.025 \qquad y^* = 0.367766$$

The actual optimum can be solved analytically, since at the optimum

$$\frac{d}{dx}(x\,e^{-x}) = 0 = e^{-x} - x\,e^{-x}$$

$$= (1-x)e^{-x}$$

so that $x^* = 1$ and $y^* = 0.367879$. Our estimate was within $\varepsilon/2$ of the optimum x and within 0.031% in estimating the optimum y.

This problem can also illustrate the practical significance of the resolution distance ε. Suppose $\varepsilon = 0.01$ and calculations were limited to six significant digits. Then x_1 and x_2 would have been given by

$$x_1 = 1.005 \qquad y(x_1) = 0.36787486$$

$$x_2 = 0.995 \qquad y(x_2) = 0.36787483$$

and, to six significant digits the relative magnitude of $y(x_1)$ and $y(x_2)$ could not have been resolved.

Fibonacci Search

Dichotomous search minimizes the interval of uncertainty at each step of measurement, which requires two experiments. However, what is often of ultimate importance is the *final interval*, not each interval along the way.

An alternative strategy is to place search points far enough apart so that one point will fall within the new interval of uncertainty. In this way only one additional experiment per step is required to reduce the interval of uncertainty.

Consider the symmetric arrangement of points shown in Figure 4.2. The interval of uncertainty is constant on both sides, and larger than in the case of the dichotomous search case. For simplicity, say that $y(x_2) > y(x_1)$. Then a third point is located in the new search interval (x_1, b) symmetric with respect to x_2, and so on, through $x_4, x_5, \ldots$.

It can then be shown that for such an arrangement of search points, the intervals of uncertainty are related by

$$L_i = L_{i+1} + L_{i+2} \qquad i = 1, 2, \ldots, n-2 \tag{4.6}$$

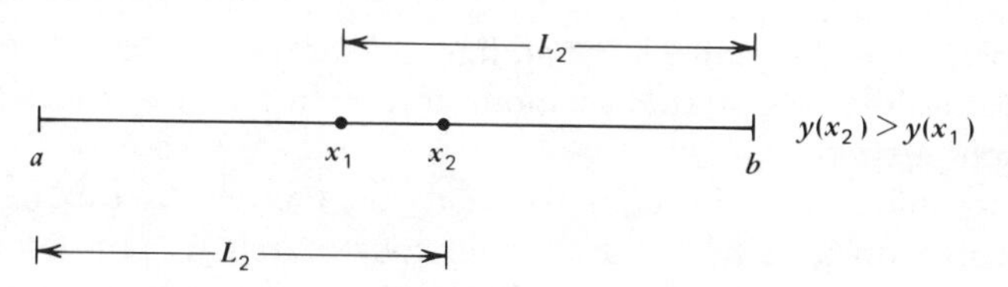

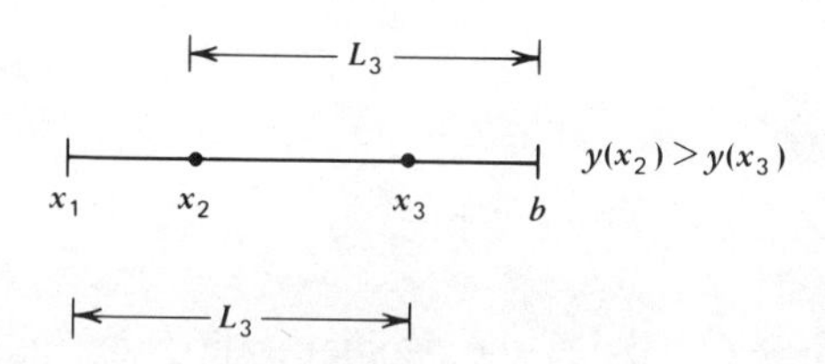

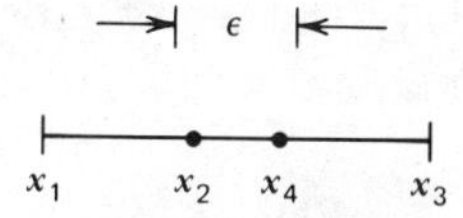

Figure 4.2. Single variable search by interval elimination.

and that the general relationship over the entire search is given by

$$L_n = \frac{1}{F_n} L_1 + \frac{F_{n-2}}{F_n} \varepsilon \tag{4.7}$$

where L_1 is the initial search area $(b-a)$ and the F_i are the Fibonacci numbers satisfying the difference equation

$$F_{k+1} = F_k + F_{k-1} \qquad k = 1,2,3\ldots \tag{4.8}$$

when

$$F_0 = F_1 = 1 \tag{4.9}$$

Once the final search interval L_n is chosen, the successive search intervals can be obtained from

$$L_{n-k} = F_{k+1} L_n - F_{k-1} \varepsilon \qquad k = n-2, n-3, \ldots, 1 \tag{4.10}$$

New points are determined by adding or subtracting the new interval length from the appropriate base point to obtain a new symmetric pair in the search interval.

This technique is quite effective, and for a given L_n can be reached faster than with a comparable dichotomous search. For instance, $n=16$ measurements are necessary in dichotomous search to reduce the search interval to below 0.5%, whereas $n=12$ will suffice with the Fibonacci sequence.

Ignoring ε for a moment, simplifying indices in (4.10), and eliminating L_n, yields the relation

$$L_{i+1}=\frac{F_{n-i}L_i}{F_{n-i+1}} \tag{4.11}$$

As i increases the ratio, F_{n-1}/F_{n-i+1} approaches a constant and

$$L_{i+1}=\frac{1}{\rho}L_i \tag{4.12}$$

where $\rho=1.618034\ldots$ the "golden ratio," or "golden section," a constant of certain aesthetic importance in architecture. For $(n-i)>4$ the approximation is good, and from 12 on it is quite good, so that search intervals approach those of the full Fibonacci treatment. This near-Fibonacci search method is called *golden section* search, and is highly recommended for its simplicity and search efficiency.

Consider the problem given earlier for the dichotomous search procedure,

$$\max xe^{-x}$$

in the interval $0\leqslant x\leqslant 2$. If we use a golden section search, the interval of uncertainty after n trials is ρ^n. The number of search points necessary to obtain an uncertainty of 0.05 is therefore given by the smallest value of k such that $\rho^{-k}\leqslant 0.05$, which is 7.

$$L_0=2$$

$$x_1=0+\frac{L_0}{\rho}=1.236068 \qquad y(x_1)=0.359108$$

$$x_2=2-\frac{1}{\rho}L_1=0.763932 \qquad y(x_2)=0.355863$$

The new region of exploration is $0.763932\leqslant x\leqslant 2$, since the region given by $x<0.763932$ has been eliminated. A third point is given by

$$L_3=\frac{1}{\rho}L_2=1.236068$$

$$x_3=a+\frac{1}{\rho}L_2=1.527864$$

i	x_i	y_i	L_{i-1}	a	b
1	1.236068	0.359108	2	0	2
2	0.763932	0.355863	1.236068	0.763932	2.00
3	1.527864	0.331544	0.763932	0.763932	1.527864
4	1.055728	0.367329	0.472136	0.763932	1.236068
5	0.944272	0.367287	0.291796	0.944272	1.236068
6	1.124602	0.365250	0.180330	0.944272	1.124602
7	1.013158	0.367847	0.111456	0.944272	1.055728

Thus, our optimum is $X^* = 1.013$, $y^* = 0.3679$. Golden section search is very simple and is easily applied to line searches for an optimal step-size λ in multidimensional search. We shall also illustrate golden section search in Chapter 6.

MULTIVARIABLE OPTIMIZATION

Direct Search Techniques

Search techniques that make no assumptions about the continuity of derivatives are often called *direct search* algorithms. The algorithms of the previous section are of this type: the dichotomous, Fibonnaci, and golden section searches assumed only continuity and unimodality in the objective function, whereas only continuity and smoothness were required in univariate curve fitting. The algorithms of this section will extend these ideas to the optimization of multivariate functions. The basic algorithms will be presented in terms of the unconstrained optimization problem. Later in this section, penalty and barrier function methods will be introduced for converting constrained problems into equivalent unconstrained problems suitable for optimization by direct search.

Random Search As the name implies, random search is the random investigation of the solution space for the optimum. It is a brute force method of almost minimal assumptions: no foreknowledge of the likely position of the optimum is assumed, so that its position is equally likely throughout the experimental region. In unconstrained problems it is not even necessary that the objective function be continuous.

The general optimization problem for random search algorithms is given by

$$\max y_0(\mathbf{x}) \tag{4.13}$$

subject to

$$\mathbf{a} \leqslant \mathbf{x} \leqslant \mathbf{b}$$

where the limits on the $\mathbf{x}$ define the experimental region. A simple random search generates search points $\mathbf{x}^{(i)}$ according to

$$\mathbf{x}^{(i)} = \mathbf{a} + \mathbf{r}_i^T(\mathbf{b} - \mathbf{a}) \tag{4.14}$$

where the $\mathbf{r}_i$ are uniform random numbers in the interval $(0, 1)$. Each observation $y_0[\mathbf{x}^{(i)}]$ is compared to the current maximum. If it is greater than the current maximum, the latest point becomes the newest estimate of the optimum location. The search can be terminated at a set number of evaluations or after some set number of evaluations passes without improving the current estimate of the optimum.

The likelihood that the simple random search will locate the optimum increases with the number of trials. If the position of the optimum is equally likely to be anywhere in the experimental region, then the probability of placing a point in the region of the uncertainty of the optimum is equal to the relative volume of uncertainty,

$$P(\mathbf{x} \in s^*) = \frac{\prod_{j=1}^{n} \Delta x_j}{\prod_{j=1}^{n} (b_j - a_j)} \tag{4.15}$$

If we convert the coordinate uncertainties, Δx_i, in (4.15) to relative uncertainties,

$$\Delta x_j = \rho(b_j - a_j)$$

then (4.15) reduces to

$$P(\mathbf{x} \in s^*) = \rho^n \tag{4.16}$$

Now, the probability that a given point falls outside the optimal region is given by

$$P(\mathbf{x}_i \in s^*) = 1 - P(\mathbf{x}_i \notin s^*) \tag{4.17}$$

and the joint probability that all m trials fall outside the true optimum is

$$P(\mathbf{x}_i \notin s^*, i = 1, \ldots, m) = [1 - P(\mathbf{x}_i \in s^*)]^m, \tag{4.18}$$

Therefore, the probability that at least a single trial falls within the optimum region s^* is given by

$$P(\mathbf{x} \cap s^* \neq \varnothing) = 1 - [1 - P(\mathbf{x}_i \in s^*)]^m \tag{4.19}$$

which we may rewrite as

$$P(i \geqslant 1) = 1 - [1 - P(i = 0)]^m \tag{4.20}$$

solving for n, the number of trials, yields

$$n = \frac{\log[1 - P(i \geqslant 1)]}{\log[1 - P(i = 0)]} \tag{4.21}$$

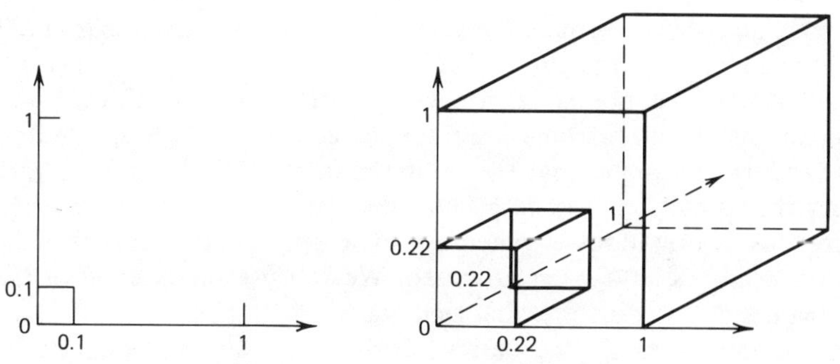

Figure 4.3. The "curse of dimensionality."

If $P(i=0)=0.01$ and $P(i \geqslant 1)=0.95$, a total of 299 trials will be necessary. To boost the probability of at least one successful trial to 0.99 will require a further 162 search points.

It is important to remember that the probabilities used in these equations solve for the *volume* region of uncertainty. As the dimensionality increases this volume can be a small fraction of the total volume while the factor uncertainty remains fairly high. For instance, let $P(\mathbf{x} \in s^*)=0.01$. For two dimensions, using (4.16), $\rho=0.10$. For three dimensions, ρ becomes 22%. This situation, shown graphically in Figure 4.3, is sometimes called the "curse of dimensionality," since the search volume is a product of each factor interval.

If we desired that the factor level uncertainties be reduced to a certain level, than (4.21) would be replaced with

$$m=-\frac{1}{\rho^n}\log(1-P(i \geqslant 1)) \tag{4.22}$$

For factor uncertainties of 10% and a probability 0.95 of having at least one point within the optimum region for a problem of 2, 3, and 5 dimensions would require a total of 13,010, 130,0000 and 1.3×10^{10} observations respectively. This is a clear indication of the difficulty of accurately locating the optimum by simply searching the solution space. To guarantee that points are spread over the entire volume of the solution space, the space is sometimes divided into subsections and several points deployed randomly within the subsection. These *stratified random searches* are efficient variations of simple random searches when the total number of experiments is limited. As the number of experiments increases, the two techniques become equivalent. For the stratified random search (4.13) is modified to

$$\mathbf{x}^{(i)}=\mathbf{x}_{\min}^{(i)}+\mathbf{r}_i^T(\mathbf{x}_{\max}^{(i)}-\mathbf{x}_{\min}^{(i)}) \tag{4.23}$$

The $\mathbf{x}_{max}^{(i)}$ and $\mathbf{x}_{min}^{(i)}$ determine the subregion being explored for trial i. They are varied to cover the entire solution space.

So far we have not acted upon any information obtained from our observations. If we assume that the objective function is continuous, however, we can argue that the optimum is more likely to be near large values than small ones. Acting upon this we should adapt our search to utilize this information in generating new points. Several such adaptive random searches have been proposed. We will use the method of Gaddy and Gaines [20] to illustrate this technique.

Let us denote as $\mathbf{x}^*$ the current position of the optimum. Under an adaptive technique new estimates are generated according to

$$x_i^{(j)} = x_i^* + \frac{R_i}{k}(2r_j - 1)^k \tag{4.24}$$

The r_j are uniform random numbers, the R_i are the ranges, $R_i = x_{i\,max} - x_{i\,min}$ on each component direction. Values of x_i which fall outside the experimental region are replaced by feasible values. The factor k is a search factor, which increases as the search progresses. As k increases, the distribution of new points becomes strongly clustered about the current optimum.

Adaptive random procedures are more efficient than simple random searches because points tend to be concentrated in the most likely region for success. Where there is a possibility for several solutions, the stratified random procedure can be used to survey the entire search region, and the most likely candidates can be used to initiate adaptive random searches.

We will demonstrate more systematic procedures for searching the solution space, but random search techniques have certain properties that make them useful in certain situations. Their strongest claim is that they are extremely likely to converge to the region of an optimum regardless of the shape of the surface or discontinuities in derivatives, although they will converge slowly to actual optimum. This can be useful for problems that are relatively easy to evaluate but are so complicated or poorly behaved that the assumptions of the more sophisticated routines are violated. A comparison of several techniques and an adaptive random search called the *controlled random search* is given by Goulcher and Long [22]. In situations where mathematical models are available for evaluation, random search methods can be competitive with other techniques.

Sequential Simplex Search. We can make random searches more efficient by exploiting the assumed continuity of the response surface. This was the approach taken in adaptive procedures, since, given continuity, points near the current maximum were more likely to contain the optimum. In this manner the curse of dimension is alleviated somewhat, since implicitly this assumption reduces the search volume to the immediate vicinity of the best-observed point.

Several techniques formalize this approach still further. A systematic approach is given by the sequential simplex of Spendley et al. [41], which we now consider. Recalling from the previous chapter, a simplex is a regular figure of $n+1$ vertices in n space. A property of such designs is that the reflection of any point with respect to the remaining n points forms a simplex with those points. The two simplexes share n points, the place across which the one point was reflected. Now, consider: If the point being reflected has the least value, then the function values on the plane will be higher. In the local region the function is likely to be unimodal so this side of the plane can be rejected as a search region, and the other side of the plane, the reflection point for example, is the logical point to place a next trial. This is equivalent to the dichotomous and Fibonacci search logics.

In the sequential simplex search, movement is made via reflection and contraction. After the functions are evaluated at each point, the worst point is reflected across the remaining points. If this reflected point results in improvement, the old point is discarded and the new point is added to the simplex, and the new worst point is selected for further reflections.

Reflection points are easily obtained. Let the simplex be given by the $n+1$ vectors $\mathbf{x}_i$. We locate the worst point, $\mathbf{x}_q$, and calculate the centroid $\mathbf{x}_c$ of the remaining points, where

$$x_{jc} = \frac{1}{n+1} \sum_{\substack{i=1 \\ i \neq q}}^{n+1} x_{ij} \tag{4.25}$$

$\mathbf{x}_c$ lies in the plane of the remaining n points. The reflection point $\mathbf{x}_r$ is given by

$$\mathbf{x}_r = \mathbf{x}_q + 2(\mathbf{x}_r - \mathbf{x}_q) \tag{4.26}$$

The progress of a typical simplex search is illustrated in Figure 4.4.

Several pathological cases can occur in simplex searches. For instance, the reflected point $\mathbf{x}_r$ in (4.26) may itself be a new worst value. Its reflection would of course not be an improving point, so instead the *next* worst value is used for reflection in equations (4.25) and (4.26).

As the optimum is approached, the technique will either reach a point where no reflections are possible, or the design will pivot on one of the points and cycle around the optimum. In the first case the design can be contracted toward the centroid of the complete simplex and the search continued on a smaller scale. The situation in the second case can be detected by keeping track of the *age*, in iterations, of each simplex point. A point that is too old is a pivot point and the design is cycling. The optimization can be stopped there or the design contracted and resumed on a smaller scale.

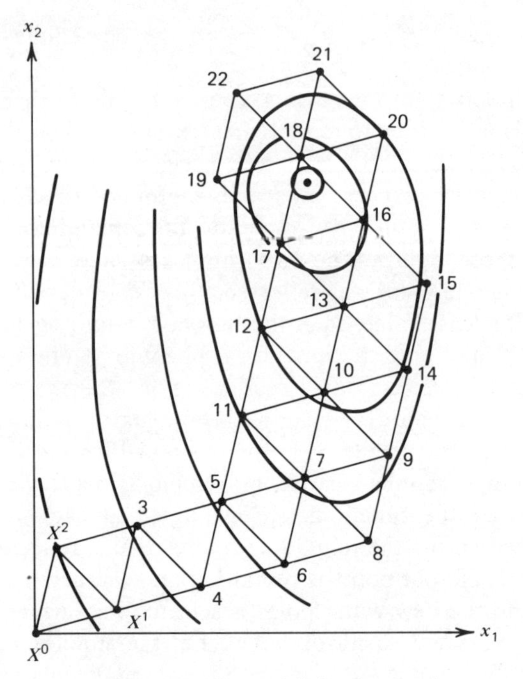

Figure 4.4. Sequential simplex search.

Relative to standard optimization problems, sequential simplex lacks the ability to accelerate its progress when a promising search direction occurs. An adaptation of the sequential simplex algorithm by Nelder and Mead [36] includes provision for acceleration. A computer code of this algorithm is given in Kuester and Mize [27].

In the Nelder-Mead algorithm, if the reflected point $\mathbf{x}_r$ given in (4.26) is the best point, an expansion point given by

$$\mathbf{x}_E = \mathbf{x}_c + \gamma(\mathbf{x}_r - \mathbf{x}_c), \tag{4.27}$$

where $\gamma > 1$, is used to accelerate past $\mathbf{x}_r$ in search of rapid improvement in the objective function. Contraction points given by $\mathbf{x}_b$ in

$$\mathbf{x}_b = \mathbf{x}_c - \beta(\mathbf{x}_c - \mathbf{x}_q) \tag{4.28}$$

and

$$\mathbf{x}_b = \mathbf{x}_c - \beta(\mathbf{x}_c - \mathbf{x}_r) \tag{4.29}$$

with $0 < \beta < 1$ when the reflected point $\mathbf{x}_r$ has a worse function value than the current points or when the reflected point has a better value than $\mathbf{x}_q$ but it is not the best value.

Because of the acceleration and contraction procedures, this version of the simplex algorithm does not maintain its original shape throughout the progress of the search. Instead, it adapts to the terrain being explored to expand along long ridges and contract amid curving ones and near optima. The search is terminated, in fact, when the design has contracted about a single best function value.

The simplex has limited adaptability and can be thwarted by steep curving ridges, so that it is sometimes a good idea to perform a post optimal adaptive random search to see whether a true optimum has been located. In addition, since the simplex algorithm assumes local unimodality, multiple starting points are necessary to guard against multiple local optima.

Complex Search. Most of the calculations in the sequential simplex algorithm deal with the centroid poiint x_c to define the reflected, expansion and contraction points. Aside from the geometric interpretation given earlier, however, there is no reason to limit the overall design to $n+1$ points. A generalization of the Nelder-Mead simplex along these lines is provided by the complex search of Box [10] using any number of randomly generated starting points. The algorithm is quite similar to that of Nelder-Mead except that the summations now run over more points. A version of Box complex search adapted for inequality constraints is given by Kuester-Mize [27].

Pattern Search. The direct search procedures given so far primarily have been interval elimination procedures, although with simplex expansion something like direction estimation is being attempted. It is the approximation of an ascent direction that is the basis of the pattern search of Hooke and Jeeves [23]. In this algorithm perturbations about a base point $\mathbf{x}_1$ called explorations are given by

$$\mathbf{x}_1^* = \mathbf{x}_1 \pm \Delta\mathbf{x}_i^* \tag{4.30}$$

to obtain a point $\mathbf{x}_1^*$ satisfying

$$y(\mathbf{x}_1^*) > y(\mathbf{x}_1) \tag{4.31}$$

In most versions of the algorithm each coordinate direction Δx_i is changed one at a time and the solution to (4.31) is the result of n sequential coordinate maximizations [see, for example, 27]. The result of the local exploration is used to attempt a new base point $\mathbf{x}_2$ using

$$\mathbf{x}_2 = \mathbf{x}_1 + 2(\mathbf{x}_1^* - \mathbf{x}_1) \tag{4.32}$$

Local exploration is conducted about $\mathbf{x}_2$ to yield $\mathbf{x}_2^*$. We use the two points $\mathbf{x}^*\mathrm{b}_1$ and $\mathbf{x}_2^*$ to make a pattern move

$$\mathbf{x}_3 = \mathbf{x}_1^* + 2(\mathbf{x}_2^* - \mathbf{x}_1^*) \tag{4.33}$$

If this pattern move is successful, $\mathbf{x}_3$ is tentatively set to $\mathbf{x}_3^*$ and more pattern moves are attempted until improvement cannot be made, according to

$$\mathbf{x}_1+_1=\mathbf{x}_{i-1}+2(\mathbf{x}_i^*-\mathbf{x}_{i-1}) \tag{4.34}$$

These pattern moves accelerate the search along a promising direction in two ways. In the first, pattern moves are continued until a pattern move fails to produce an improvement. Local exploration is not resumed until this point. But there is acceleration in the sense that each successive pattern move is bolder than the last. For instance, if $\boldsymbol{\Delta}$ is defined by

$$\boldsymbol{\Delta}=\mathbf{x}_1^*-\mathbf{x}_1 \tag{4.35}$$

then using (4.32)

$$\mathbf{x}_2-\mathbf{x}_1=2\boldsymbol{\Delta} \tag{4.36}$$

and this with (4.33)

$$\mathbf{x}_3-\mathbf{x}_1=5\boldsymbol{\Delta} \tag{4.37}$$

and therefore

$$\mathbf{x}_3-\mathbf{x}_2=3\boldsymbol{\Delta} \tag{4.38}$$

In general, then, for step i of successive pattern moves

$$\mathbf{x}_i-\mathbf{x}_{i-1}=i\boldsymbol{\Delta} \tag{4.39}$$

For pattern search, clearly nothing succeeds so well as success!

The pattern procedure makes local explorations and pattern moves until local exploration fails to yield a feasible new base point. At that point the Δx_i may be reduced in size to reduce the scale of the search. If this fails to yield any improvement, the procedure is terminated.

As with simplex, pattern can get stuck on sharp ridges, so a post optimal analysis by a random procedure is unwarranted. Several improvements in the basic algorithm have been suggested. One is to vary the Δx_i independently of each other. Another is to rotate the search directions to align with canonical directions of the problem under consideration. Both techniques are aimed at reducing the directional limitations of the coordinate searches. An evolutionary step in this direction is the employment of finite difference estimates of the derivatives at a base point using the central difference formula,

$$\frac{dy}{dx_i}\doteq\frac{y(\mathbf{x}+\Delta x_i/2)-y(\mathbf{x}-\Delta x_i/2)}{\Delta x_i} \tag{4.40}$$

Such techniques belong to the borderline between direct search and derivative methods. Equation (4.40) is applicable to direct search since derivatives are being estimated from function observations on the $y(\mathbf{x})$, but since the equation implies the existence of a derivative, techniques that use such estimates are classed as derivative methods.

Modifications for Constrained Direct Search. Direct search techniques can be readily modified to solve constrained optimization problems of the type

$$\max y(\mathbf{x}) \tag{4.41}$$

such that

$$\mathbf{h}(\mathbf{x})=\mathbf{0} \tag{4.42}$$

$$\mathbf{g}(\mathbf{x})\leqslant\mathbf{0} \tag{4.43}$$

One method is to transform the constrained problem given by equations (4.41)–(4.43) into an equivalent unconstrained problem of the form

$$\max y(\mathbf{x})-p_1[\mathbf{h}(\mathbf{x}),\mathbf{w}_1]-p_2[\mathbf{g}(\mathbf{x}),\mathbf{w}_2] \tag{4.44}$$

in which p_1 and p_2 are appropriate *penalty functions* and the $\mathbf{w}_i$ are weighting parameters associated with the penalty functions. We will examine several classes of penalty functions for use in optimization problems of the type given in (4.44)

Some methods can be applied almost unaltered to the optimization in which equation 4.42 does not occur. For instance, in the complex algorithm of Kuester and Mize [27], points that violate any constraints given by (4.43) are contracted until they are feasible. To start the algorithm, all initial values, however, must themselves be feasible. As Noh [37] points out, the feasible region may not always be that obvious. He outlines a two-phased approach for obtaining an initial set of feasible points. In the first phase the optimization problem is

$$\min \tilde{g}_i(\mathbf{x}) \tag{4.45}$$

where

$$\tilde{g}_i(\mathbf{x})=\begin{cases}0 & g_i(\mathbf{x})\leqslant 0\\ g_i(\mathbf{x}) & g_i(\mathbf{x})>0\end{cases} \tag{4.46}$$

The first phase terminates when the objective function is zero, which occurs when all constraints are satisfied. The second phase uses these points as its initial values to solve the optimization problem given by (4.41) and (4.43).

A similar approach for dealing with the pattern search solution for inequality constrained problems has been suggested by Mugele [33]. When exploration points are infeasible but show improvement in the objective function, the observation is stored and used to estimate local feasible values by interpolating the constraint position. The constrained problem terminates when the optimum is on the constraint and no better feasible point can be found.

Both of these approaches are called *interior* algorithms, since the solution at all stages is in the interior of the feasible region. Interior methods are especially suited for inequality constrained problems. There is a class of penalty functions applicable to inequality constraints which are interior penalties as well. These functions are most often called *barrier functions*, since their form is

$$p[g_i(\mathbf{x}), w_i] = \begin{cases} M + g_i(\mathbf{x}) & g_i(\mathbf{x}) \geqslant 0 \\ \dfrac{w_i}{|g_i(\mathbf{x})|} & g_i(\mathbf{x}) < 0 \end{cases} \tag{4.47}$$

M is a very large constant which is applied whenever the constraint is violated. The barrier consists of the increasing value of the penalty as it is approached and the absolute value of the constraint function decreases. This is illustrated for several values of W in Figure 4.5.

The solution procedure is iterative. A series of problems of the form

$$\max_{\mathbf{w} \to \mathbf{0}} y(\mathbf{x}) - p[\mathbf{g}(\mathbf{x}), \mathbf{W}] \tag{4.48}$$

are solved with decreasing values of the penalty weights, allowing the barrier to become increasingly sharp.

Barrier functions are not applicable to equality constraints. Instead, exterior penalties are used to penalize deviations from the constraint.

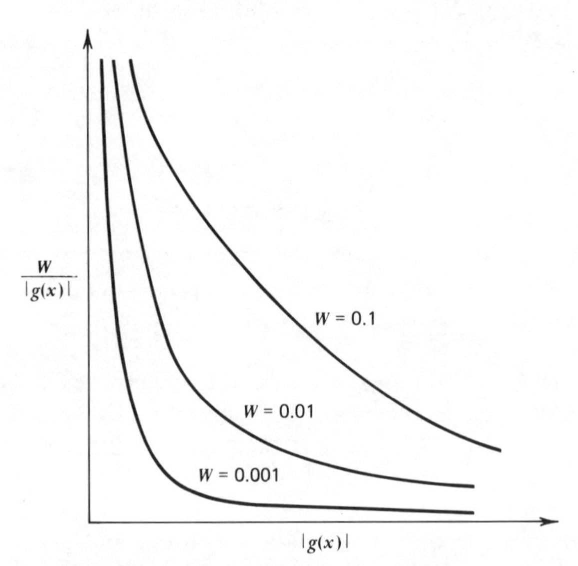

Figure 4.5. $W/|g(x)|$ versus $|g(x)|$ for various W.

These generally take the form of either

$$p[h_i(\mathbf{x}), w_i] = w_i h_i^2(\mathbf{x}) \tag{4.49}$$

or

$$p[h_i(\mathbf{x}), w_i] = w_i(e^{h^2(\mathbf{x})} - 1) \tag{4.50}$$

In this case the weights increase during iterations toward the optimum. Exterior penalties are also sometimes applied to inequality constraints in the form

$$p[g_i(\mathbf{x}), w_i] = \begin{cases} 0 & g_i(\mathbf{x}) \leqslant 0 \\ w_i g_i^2(\mathbf{x}) & g_i(\mathbf{x}) > 0 \end{cases} \tag{4.51}$$

Gradient Based Unconstrained Optimization

We turn now from direct search methods to the classical derivative methods of optimization. We shall use the calculus to provide insight into the nature of optimum solutions and for algorithms in obtaining those solutions to the unconstrained optimization problem. The constrained optimization problem will be considered in the next section. We shall be concerned in both sections with the generalizations of the scalar calculus applied to multivariate functions.

In the scaler calculus a function is said to have a local extreme point x^*, if, for any small h

$$f(x^*) \geqslant f(x^* + h) \tag{4.52}$$

Such an extreme is a local extreme point unless

$$f(x^*) \geqslant f(x) \tag{4.53}$$

for all x, in which case x^* is a global extreme point. A *necessary* condition for a point to be an extreme point is that the derivative at that point be zero:

$$\left.\frac{df}{dx}\right|_{x^*} = f'(x^*) = 0 \tag{4.54}$$

This, however, is not a sufficient condition for a point to be an extreme point. For the point to be an extreme point the second derivative must also be nonzero; $f''(x^*) < 0$ for the extreme point to be a local maximum, and $f''(x^*) > 0$ for the extreme point to be a local minimum. If $f''(x^*) = 0$, the point may be an extreme point or an inflection point, and the higher derivatives can be consulted further to classify it.

The multivariate equivalent of the derivative is the *gradient vector*

$$\nabla f(\mathbf{x}) = \left[\frac{\partial f(\mathbf{x})}{\partial x_1}, \frac{\partial f(\mathbf{x})}{\partial x_2}, \ldots, \frac{\partial f(\mathbf{x})}{\partial x_n} \right] \tag{4.55}$$

(pronounced "del f" or "grad f") and consists of the row vector of partial derivatives of the function $f(\mathbf{x})$. For functions twice differentiable, the analog to the second derivative is the *Hessian matrix*

$$H(\mathbf{x}) = \nabla^2 f(\mathbf{x}) = \left| \begin{matrix} \frac{\partial^2 f}{\partial x_1^2} & \frac{\partial^2 f}{\partial x_2 \partial x_1} & \cdots & \frac{\partial^2 f}{\partial x_n \partial x_1} \\ & \frac{\partial^2 f}{\partial x_2^2} & \cdots & \frac{\partial^2 f}{\partial x_n \partial x_2} \\ & & & \cdots \\ \text{sym} & & & \frac{\partial^2 f}{\partial x_n^2} \end{matrix} \right| \tag{4.56}$$

where $\nabla^2 f(\mathbf{x}) = \nabla[\nabla f(\mathbf{x})^T]$. The Hessian is the gradient of the transposed gradient of f.

With the aid of the gradient and the Hessian, it is easy to construct the Taylor series approximation for a point near an extreme point $\mathbf{x}^*$.

$$f(\mathbf{x}) = f(\mathbf{x}^*) + \nabla f(\mathbf{x}^*)\Delta\mathbf{x} + \frac{1}{2}\Delta\mathbf{x}^T \nabla^2 f(\mathbf{x}^*)\Delta\mathbf{x} + 0(\|\Delta\mathbf{x}\|^3) \tag{4.57}$$

where $\mathbf{x} = \mathbf{x}^* + \Delta\mathbf{x}$. The last term $\mathbf{0}(\|\Delta x\|^3)$ represents all terms of order 3 and higher in the $\Delta\mathbf{x}$. These have the property that for small enough $\Delta\mathbf{x}$ the first two terms in $\Delta\mathbf{x}^T\Delta\mathbf{x}$ dominate:

$$\lim_{\|\Delta x\| \to 0} \frac{0(\|\Delta\mathbf{x}\|^3)}{\|\Delta\mathbf{x}\|^2} \tag{4.58}$$

For small enough $\Delta\mathbf{x}$ the (4.57) can be rearranged to yield

$$f(\mathbf{x}) - f(\mathbf{x}^*) = \nabla f(\mathbf{x}^*)\Delta\mathbf{x} + \ldots$$

Consider what happens if $\nabla f(\mathbf{x}^*) \neq 0$. For small enough $\Delta\mathbf{x}$

$$f(\mathbf{x}^* + \Delta\mathbf{x}) - f(\mathbf{x}^*) = \nabla f(\mathbf{x}^*)\Delta\mathbf{x} > 0 \tag{4.59}$$

and

$$f(\mathbf{x}^* - \Delta\mathbf{x}) - f(\mathbf{x}^*) = -\nabla f(\mathbf{x}^*)\Delta\mathbf{x} < 0 \tag{4.60}$$

Equation (4.59) contradicts the assertion that $\mathbf{x}^*$ is an extreme point. The two equations taken together indicate that an extreme point $\nabla f(\mathbf{x}^*) = \mathbf{0}$.

This is the necessary condition for extreme points of multivariate functions.

The gradient can be used to generate n simultaneous equations in n unknowns and solved for possible extreme points. For instance, consider the unconstrained problem,

$$y = g(x_1, x_2) = 30 + 2x_1 + 4x_2 - 5x_1^2 + 6x_1x_2 - 7x_x^2$$

Taking the gradient of g and setting it equal to the zero vector gives the two equations

$$2 - 10x_1 + 6x_2 = 0$$

$$4 + 6x_1 - 14x_2 = 0$$

whose solution is given by $(x_1^*, x_2^*) = (\frac{1}{2}, \frac{1}{2})$ where $y = 31\frac{1}{2}$. A check of neighboring values shows this point to be a local maximum.

In deriving the necessary conditions for extreme points, use was made of the fact that for small $\mathbf{\Delta x}$ near the extreme points the $\mathbf{\Delta x}$ terms dominated those of order two and higher. At the extreme point the gradient vanishes and the second-order terms are dominant.

Consider again the first terms of the Taylor's Series near an optimum

$$f(\mathbf{x}^* + \mathbf{\Delta x}) - f(\mathbf{x}^*) = \tfrac{1}{2}\mathbf{\Delta x}^T \nabla^2 f(\mathbf{x}^*)\mathbf{\Delta x} + \mathbf{0}(\|\mathbf{\Delta x}\|^3) \tag{4.61}$$

As $\mathbf{\Delta x}$ becomes small the second order terms dominate the right hand side. If the right hand side is always negative, then

$$f(\mathbf{x}^* + \mathbf{\Delta x}) - f(\mathbf{x}^*) \leqslant 0$$

or

$$f(\mathbf{x}^*) \geqslant f(\mathbf{x}^* + \mathbf{\Delta x})$$

Therefore, $f(\mathbf{x}^*)$ is a local *maximum*. Similarly, if the right hand side is always positive, $f(\mathbf{x}^*)$ is a local *minimum*.

The Hessian $\nabla^2 f(\mathbf{x}^*)$ is a matrix (4.57). For the right-hand side of (4.61) to be negative for all $\mathbf{\Delta x}$, it must have the property of being negative definite,

$$\mathbf{\Delta x}^T \nabla^2 f(\mathbf{x}^*)\mathbf{\Delta x} < 0 \tag{4.62}$$

for all $\mathbf{\Delta x}$ at the point $\mathbf{x}^*$. It is a sufficient condition for $\mathbf{x}^*$ to be a local maximum when $\nabla f(\mathbf{x}^*)$ vanishes and $\nabla^2 f(\mathbf{x}^*)$ is *negative definite*. For $\mathbf{x}^*$ to be a local minimum point, the gradient must vanish and the Hessian must be *positive definite*.

The problem in the previous example can be represented in matrix notation as

$$\max g(\mathbf{x}) = 30 + (2,4)\begin{bmatrix} x_1 \\ x_2 \end{bmatrix} + (x_1, x_2)\begin{bmatrix} -5 & 3 \\ 3 & -7 \end{bmatrix}\begin{bmatrix} x_1 \\ x_2 \end{bmatrix}$$

The gradient is

$$\begin{bmatrix} 2 \\ 4 \end{bmatrix} + 2\begin{bmatrix} -5 & 3 \\ 3 & -7 \end{bmatrix}\begin{bmatrix} x_1 \\ x_2 \end{bmatrix}$$

The singular point can be solved, as in the last example, from this equation:

$$\begin{bmatrix} -5 & 3 \\ 3 & -7 \end{bmatrix}\begin{bmatrix} x_1 \\ x_2 \end{bmatrix} = \begin{bmatrix} -1 \\ -2 \end{bmatrix}$$

to yield $\mathbf{x}^* = (\frac{1}{2}, \frac{1}{2})^T$. The Hessian is

$$2\begin{bmatrix} -5 & 3 \\ 3 & -7 \end{bmatrix} = \begin{bmatrix} -10 & 6 \\ 6 & -14 \end{bmatrix}$$

Using this matrix in (4.62) and substituting $\mathbf{x}$ for $\Delta\mathbf{x}$ yields

$$-10x_1^2 + 12x_1x_2 - 14x_2^2 < 0$$

or

$$5x_1^2 + 7x_2^2 > 6x_1x_2$$

This is true for any nonzero x_1 and x_2 so the matrix is clearly negative definite and our solution a local maximum. In this example g is a quadratic function so the Hessian is constant and negative definite everywhere. Thus the extreme point $\mathbf{x}^* = (\frac{1}{2}, \frac{1}{2})^T$ is a unique, global maximum point.

When the matrix of the Hessian is semidefinite, that is, when there are some directions $\Delta\mathbf{x}_0$ for which

$$\Delta\mathbf{x}_0^T H(\mathbf{x}^*)\Delta\mathbf{x}_0 = 0 \qquad (4.63)$$

and all others in which

$$\Delta\mathbf{x}^T H(\mathbf{x}^*)\Delta\mathbf{x} < 0 \qquad (4.64)$$

then less can be said about the stationary point $\mathbf{x}^*$. If the Hessian matrix is constant, the direction $\Delta\mathbf{x}_0$ points to another $\mathbf{x}_0$ in the neighborhood of $\mathbf{x}^*$ such that $f(\mathbf{x}_0) = f(\mathbf{x}^*)$. When the Hessian is not constant, more analysis of the situation is necessary.

In the case where $H(\mathbf{x}^*)$ is *indefinite*, some directions suggest that it is a local maximum and others that it is a minimum. Such a point is called a *saddle*, and this is indeed the case: the saddle is a point that is a minimum in one direction and a maximum in another, as illustrated in Figure 4.6. In cases where the Hessian vanishes altogether, the higher order terms must be consulted for the nature of the stationary point.

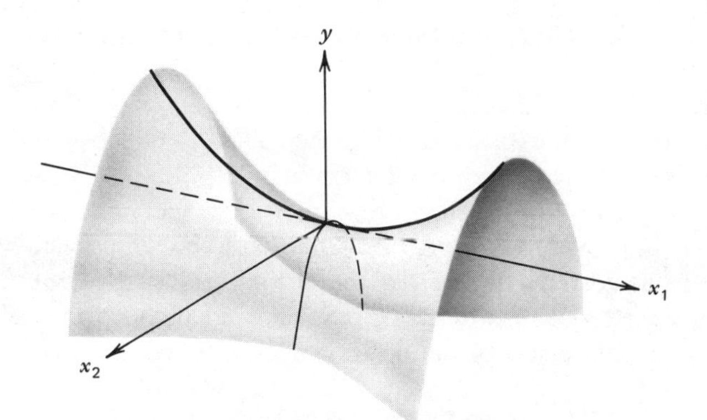

Figure 4.6. A saddle system.

"Ridge" Analysis An alternative method to (4.64) can be used to determine whether a matrix is definite, semidefinite or indefinite. Because of the interaction terms $x_i x_j$, it is difficult to tell whether a matrix is definite or indefinite. However, because the Hessian is a real symmetrical matrix, it is possible to obtain a canonical representation of the matrix in which the interactions vanish. Only the quadratic terms will be nonzero in this representation and the matrix can be calculated according to the sign of the individual terms.

Since $H(\mathbf{x})$ is a real symmetric matrix, there exists an orthogonal transformation P that will diagonalize $H(\mathbf{x})$:

$$P^T H(\mathbf{x}) P = \Lambda \tag{4.65}$$

such that

$$P^T P = P^{-1} P = I \tag{4.66}$$

The matrix Λ is the canonical form of $H(\mathbf{x})$ and contains as its only nonzero elements the eigenvalues λ_i of $H(\mathbf{x})$ along its diagonal. Equation (4.66) results from the orthogonality of P. The canonical coordinates w_i are related to the factor coordinates Δx_i according to

$$\Delta\mathbf{x} = P\mathbf{w} \tag{4.67}$$

Using this equation

$$\begin{aligned} \Delta\mathbf{x}^T H(\mathbf{x}) \Delta\mathbf{x} &= \mathbf{w}^T P^T H(\mathbf{x}) P\mathbf{w} \\ &= \mathbf{w}^T \Lambda \mathbf{w} \\ &= \sum_{i=1}^{n} w_i^2 \lambda_i \end{aligned} \tag{4.68}$$

The form of (4.68) which is most useful is that given in the last line. The terms w_i^2 are all positive so the sign will be determined by the λ_i, the eigenvalues of $H(\mathbf{x})$. If these are all negative, then clearly $H(\mathbf{x})$ is negative definite. If some of the λ_i are missing then $H(\mathbf{x})$ is negative semidefinite. When the signs of the λ_i are mixed $H(\mathbf{x})$ is indefinite.

Besides being useful in classifying matrices, the eigenvalues of the Hessian matrix provide additional information about the sensitivity of the stationary value to deviations from the optimum. Using (4.68) and (4.63) and ignoring the terms of order 3 and higher, we can study the behavior of the function in the region of the optimum

$$\Delta f(\mathbf{x}^*) = f(\mathbf{x}^*) - f(\mathbf{x}^* + \Delta\mathbf{x}) = -\sum_{i=1}^{n} \lambda_i w_i^2 \tag{4.69}$$

Then $f(\mathbf{x}^*)$ is a maximum for any combination of w_i when $\lambda_i < 0$. The relative magnitude of the eigenvalues indicate the relative rate of change in f for each of the canonical directions w_i. Those with the largest magnitude change quickest and it is these directions that are the most sensitive to deviation from the actual stationary point. The smaller eigenvalues indicate directions of lesser sensitivity. Equation (4.67) can be used to determine the directions corresponding to these eigenvalues.

In the previous example the eigenvalues $\lambda_1 = -6 + \sqrt{10}$ and $\lambda_2 = -6 - \sqrt{10}$ were given. The eigenvectors form the columns of the transformation matrix P and can be used with (4.67) to give the corresponding directions

$$\begin{bmatrix} \Delta x_1 \\ \Delta x_2 \end{bmatrix} = \begin{bmatrix} -3 & -3 \\ 1+\sqrt{10} & 1-\sqrt{10} \end{bmatrix} \begin{bmatrix} w_1 \\ w_2 \end{bmatrix}$$

The first eigenvector $(-3, 1+\sqrt{10})$ gives the direction of least sensitivity, that of the small eigenvalue. The second eigenvector is given for $w_2 = 1$ and gives the most sensitive direction, $(-3, 1-\sqrt{10})$.

This canonical analysis seen in Figure 4.7, provides a picture of the behavior of the function g about the stationary point and suggests, when the eigenvalues are unequal, that the level curves are elliptical. The major axis belongs to the least sensitive direction (given by an eigenvector) and the minor axis is that of the greater eigenvalue. In the deterministic case this is of only cursory value. In the experimental realm, where the stationary point is only an estimate, this Ridge Analysis provides further knowledge about which directions must be estimated most precisely and which are less important. In addition, recalling the problem of scaling in the last chapter, it is obvious that then natural scaling is related to the eigenvalues of the Hessian.

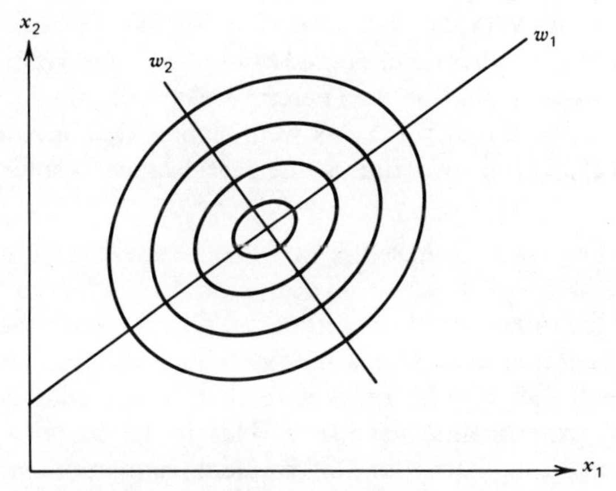

Figure 4.7. A canonical representation of a system.

Gradient Search: Hill Climbing

The gradient provides useful information about whether a point is an extreme point or not, since it is necessary for the gradient to vanish at that point. However, the gradient is extremely useful in many algorithms for searching for such points. Consider first that the components of the gradient are rates of change in f with respect to each of the component directions. To a first-order approximation, returning to the Taylor's Series,

$$f(\mathbf{x}+\mathbf{\Delta x})-f(\mathbf{x})=\nabla f(\mathbf{x})\mathbf{\Delta x} \tag{4.70}$$

The gradient provides an estimate of the expected change in $f(\mathbf{x})$ for a given $\mathbf{\Delta x}$. What $\mathbf{\Delta x}$ should be used to maximize this change? The product in (4.70) can be rewritten as

$$\nabla f(\mathbf{x})\ \mathbf{\Delta x}=\|\nabla f(\mathbf{x})\|^{1/2}\|\mathbf{\Delta x}\|^{1/2}\cos(\theta) \tag{4.71}$$

and θ is the angle between the two vectors. For arbitrary (but small) $\mathbf{\Delta x}$ the equation is clearly maximized for a θ of 0°, that is, when $\mathbf{\Delta x}$ and $\nabla f(\mathbf{x})$ are parallel. The gradient vector is thus the direction of greatest change.

Obviously, a direct way of maximization is to estimate or calculate the gradient and to proceed for a small step s in that direction. The gradient can be reevaluated at this new point and the process continued until the stationary point is attained. Since the function $f(\mathbf{x})$ can be considered a "hill" in several dimensions, this technique is often called "hill climbing," or, after the properties of the gradient, "steepest ascent."

In practice, steepest ascent is performed somewhat differently than the above suggests. Obviously, because the gradient is the direction of steepest ascent, the product $\nabla f(\mathbf{x})\Delta\mathbf{x}$ is the greatest for the $\Delta\mathbf{x}$ parallel to $\nabla f(\mathbf{x})$. However, since the gradient is assumed continuous, the gradient at $\mathbf{x}+\Delta\mathbf{x}$ will be approximately that at $\mathbf{x}$. Therefore, $\Delta\mathbf{x}$ will still be close to the steepest ascent direction at $\mathbf{x}+\Delta\mathbf{x}$ as well. Typically, therefore, the direction given $\nabla f(\mathbf{x})$ is used to define a line search at $\mathbf{x}$ to solve the optimal step problem

$$\max f[\mathbf{x}+\rho\nabla f(\mathbf{x})] \tag{4.72}$$

The primary advantage of this method is that a good line search will require fewer function evaluations of $f(\mathbf{x})$ than each estimation of $\nabla f(\mathbf{x})$, and progress will therefore be quicker relative to the number of function evaluations (or experiments) necessary. This is the basis of the steepest ascent algorithm, an alternation of gradient estimation steps and line searches. Actual "movement" occurs during the line search part of the procedure. The algorithm terminates when the gradient becomes small enough or actually vanishes.

The steepest ascent algorithm has the general property of rapid movement when far from the optimum, but suffers from convergence problems near the optimum as the gradient decreases. In addition, overall pattern of movement, especially near the optimum, is characterized by excessive zig-zagging. This is indicative of the breakdown of the linear approximation in estimating the changes in $f(\mathbf{x})$. Near the optimum, of course, the second-order terms begin to dominate. Because of this steepest ascent is primarily useful for starting points far from the optimum and where only an approximate estimate of the stationary point is necessary.

Newton Algorithm: Second-Order Methods

The primary defects of the steepest ascent method have to do with the local property of the gradient. The gradient is a local property of a function, and as the search progresses, the gradient typically changes direction and decreases in magnitude, approaching zero, of course, near the optimum. It is near the optimum that steepest ascent performs the worst and it is the rapid change in direction more so than decrease in the gradient magnitude which limits the efficiency of the search progress. This is because the search begins to zigzag, with each step progressively smaller. The zigzagging can be considered a measure of the ellipticalness of a system, since the gradients of a circular system all point to the optimum. A simple way to improve a steepest ascent search is to have it scaled in its natural scaling where the level curves are circular. The scale factors can be estimated from the eigenvalues.

Several other first-order techniques are advanced for solving this problem. Relaxation methods attempt to take advantage of the fact that the line search passes the point where the gradient points towards the optimum better than the gradient at the end of the line search. Unfortunately, it is not possible to know *a priori* whether relaxation or even over-relaxation in a given situation is superior. The relaxation technique selects a new base point according to the formula:

$$\mathbf{x}_r = \mathbf{x}_0 + \beta\rho^* \nabla f(\mathbf{x}_0) \tag{4.73}$$

where $0<\beta<1$ for relaxation and $\beta>1$ for over-relaxation; ρ^* is a solution to (4.72).

A more promising technique is that of parallel tangents (PARTAN), which keeps track of past experience to construct the tangents that presumably lead more directly to the optimum point. This is actually a quasi-second-order technique and will be discussed later in this section.

Steepest ascent was based upon the first-order Taylor's approximation. To obtain information about the change in the gradient, it is necessary to use a second-order approximation. The second-order term, the Hessian, describes changes in the gradient that should lead to a better search direction by deflecting the best local direction to a less localized best direction.

Consider the approximation formulas at a point near the optima $\mathbf{x}^*$ and at it,

$$f(\mathbf{x}+\Delta\mathbf{x}^*) = f(\mathbf{x}) + \nabla f(\mathbf{x})\Delta\mathbf{x}^* + \tfrac{1}{2}\Delta\mathbf{x}^{*T}\nabla^2 f(\mathbf{x})\Delta\mathbf{x}^*$$

and

$$(\mathbf{x}) = f(\mathbf{x}+\Delta\mathbf{x}^*) - \nabla f(\mathbf{x}+\Delta\mathbf{x}^*)\Delta\mathbf{x}^* + \tfrac{1}{2}\Delta\mathbf{x}^{*T}\nabla^2 f(\mathbf{x}^*)\Delta\mathbf{x}^*$$

where $\mathbf{x}^* = \mathbf{x} + \Delta\mathbf{x}^*$ is the optimum point. The object is to estimate a second-order $\Delta\mathbf{x}^*$ for an arbitrary point $\mathbf{x}$. At the optimum the gradient is zero. If it is assumed that the Hessian is nearly constant from $\mathbf{x}$ to $\mathbf{x}^*$, then the two equations can be added to produce

$$\left[\nabla f(\mathbf{x}) + \Delta\mathbf{x}^*\nabla^2 f(\mathbf{x})\right]\Delta\mathbf{x}^* = 0$$

or

$$\Delta\mathbf{x}^* = -\left[\nabla^2 f(\mathbf{x})\right]^{-1}\nabla f^T(\mathbf{x}) \tag{4.74}$$

This equation can be recast as an iterative formula,

$$\mathbf{x}_{n+1} = \mathbf{x}_n - \left[\nabla^2 f(\mathbf{x}_n)\right]^{-1}\nabla f^T(\mathbf{x}_n) \tag{4.75}$$

It is assumed that the Hessian is not singular at $\mathbf{x}_n$. This formula is the basic Newton algorithm, and is quite similar to the Newton-Raphson algorithm for the scalar case,

$$x_{n+1}=x_n-\frac{f'(x_n)}{f''(x_n)}$$

It can be seen that the Hessian acts to deflect the steepest ascent deflection in order to speed up its convergence. The Newton algorithm is in fact exact for the quadratic case and of higher order convergence than steepest ascent for higher order cases.

Although the Newton algorithm is superior to that of steepest ascent, it has some drawbacks. The first is its expense. Each iteration requires the evaluation of the Hessian, which must also be inverted. This requires function evaluations on the order of $n^2/2$ plus matrix inversion. In addition there is no guarantee that the Hessian is nonsingular at points far from the optimum (where it should be definite).

Several alternative strategies are proposed to reduce the average cost of the technique. For instance, far from the optimum the steepest ascent direction can be used without modification. This can be done by approximating the Hessian with the negative of the identity matrix, which yields the steepest ascent formula directly from the Newton equation

$$\begin{aligned}\mathbf{x}_{n+1}&=\mathbf{x}_n-(-I)^{-1}\nabla f^T(\mathbf{x}_n)\\&=\mathbf{x}_n+\nabla f^T(\mathbf{x}_n)\end{aligned}$$

As the optimum is approached, the Hessian can be calculated and used. Another alternative is to save the inverse Hessian and to reuse it for several steps. The convergence of such techniques would be between that of steepest ascent and exact Newton procedures.

Conjugate Gradient, Fletcher Reeves, and Partan Algorithms

One class of approximate Newton algorithms are related to conjugate direction considerations, and are based upon a quadratic model adequately approximating the objective function, at least locally. For quadratic problems the n steps are required for an exact solution, and the technique is equivalent to taking n steps to construct the Hessian in the quadratic case from gradient information.

Given the approximating problem of

$$\max\frac{1}{2}\mathbf{x}'Q\mathbf{x}+\mathbf{b}'\mathbf{x} \tag{4.76}$$

it is possible to express the solution $\mathbf{x}^*$ in terms of the Q-orthogonal, or conjugate directions given by

$$\mathbf{d}_i' Q \mathbf{d}_j = 0 \qquad i \neq j \tag{4.77}$$

The entire set of these conjugate directions can be shown to be linearly independent when Q is symmetric and definite [31]. Therefore the solution vector $\mathbf{x}^*$ can be given as a linear combination of conjugate directions,

$$\mathbf{x}^* = \alpha_1 \mathbf{d}_1 + \alpha_2 \mathbf{d}_2 + \ldots + \alpha_n \mathbf{d}_n \tag{4.78}$$

Since the problem is quadratic the solution is already known,

$$Q\mathbf{x}^* = -\mathbf{b} \tag{4.79}$$

The two equations can be combined to calculate the α_i's. Equation (4.78) is premultiplied by Q and the inner product with $\mathbf{d}_i$ yields,

$$\alpha_i = \frac{\mathbf{d}_i' Q \mathbf{x}^*}{\mathbf{d}_i' Q \mathbf{d}_i} = \frac{-\mathbf{d}_i' \mathbf{b}}{\mathbf{d}_i' Q \mathbf{d}_i} \tag{4.80}$$

since all the other terms drop out. The solution is thus given by

$$\mathbf{x}^* = \sum_{i=1}^{n} \frac{\mathbf{d}_i' \mathbf{b}}{\mathbf{d}_i' Q \mathbf{d}_i} \mathbf{d}_i \tag{4.81}$$

where prime also denotes the transpose.Luenberger [31] shows that given a set of $\mathbf{d}_i$, $i = 1, \ldots, n$, it is possible to converge to a solution to the linear equation (4.79) in n steps. This is the solution of the quadratic maximum problem. In addition, it is shown that the conjugate directions can be generated from the gradients at each step and do not need to be known in advance. Letting $\mathbf{d}_1 = \nabla f_1 = Q\mathbf{x}_1 + \mathbf{b}$, where $f_1 = f(\mathbf{x}_1)$, the algorithm proceeds according to,

$$\mathbf{x}_{k+1} = \mathbf{x}_k + \alpha_k \mathbf{d}_k \tag{4.82}$$

$$\alpha_k = -\frac{\nabla f_k \mathbf{d}_k}{\mathbf{d}_k' Q \mathbf{d}_k} \tag{4.83}$$

$$\mathbf{d}_{k+1} = \nabla f_{k+1} + \beta_k \mathbf{d}_k \tag{4.84}$$

$$\beta_k = -\frac{\nabla f_{k+1} Q \mathbf{d}_k}{\mathbf{d}_k' Q \mathbf{d}_k} \tag{4.85}$$

Figure 4.8 illustrates the progress of conjugate gradient search, shown with the dashed line, as compared to gradient search.

The solution in the conjugate gradient algorithm is exact for the quadratic case. The technique is related at every step to the gradient direction, however, so the technique is applicable to nonquadratic problems as well, as the gradient promises ascent for every step. In the nonquadratic case, Q is replaced by $H(\mathbf{x}_k)$ and updated at each step. The algorithm then becomes,

$$\mathbf{d}_1 = \nabla f(\mathbf{x}_1) \tag{4.86}$$

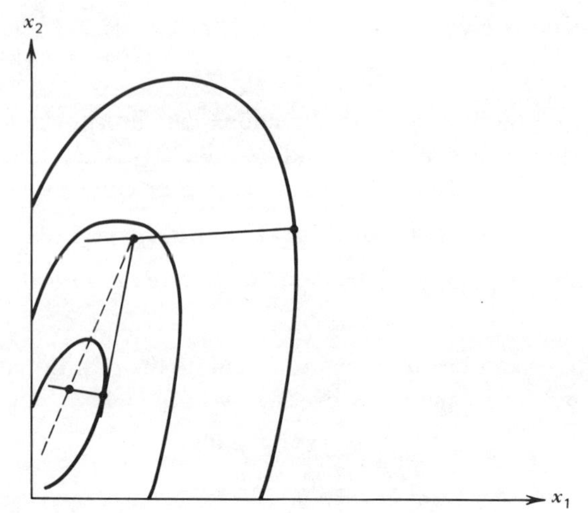

Figure 4.8. Gradient search versus conjugate gradient search.

And for $k=2,\ldots,n$

$$\mathbf{x}_{k+1}=\mathbf{x}_k+\alpha_k\mathbf{d}_k \tag{4.87}$$

$$\alpha_k=-\frac{\nabla f_k\mathbf{d}_k}{\mathbf{d}_k' H(\mathbf{x}_k)\mathbf{d}_k} \tag{4.88}$$

$$\nabla f_{k+1}=\nabla f(\mathbf{x}_{k+1}) \tag{4.89}$$

$$\mathbf{d}_{k+1}=\nabla f_{k+1}^T+\beta_k\mathbf{d}_k \tag{4.90}$$

and

$$\beta_k=\frac{\nabla f_{k+1}H(\mathbf{x}_k)\mathbf{d}_k}{\mathbf{d}_k^T H(\mathbf{x}_k)\mathbf{d}_k} \tag{4.91}$$

The algorithm is restarted every n steps and is identical to the conjugate gradient algorithm whenever $f(\mathbf{x})$ is quadratic.

The disadvantage of the method is that the Hessian is required at every step, although its inverse is not. No line search is required.

The Fletcher Reeves algorithm is associated with the conjugate gradient algorithm and essentially trades a line search for the Hessian evaluation. The equation for α_k gives the optimal step along $\mathbf{d}_k$, which could be estimated directly without the Hessian matrix. The Hessian information from the line search is also sufficient to determine the β_k without explicit calculation of the Hessian. The Fletcher Reeves algorithm becomes, letting,

$$f_k=f(\mathbf{x}_k) \tag{4.92}$$

$$\mathbf{d}_1=\nabla f_1^T \tag{4.93}$$

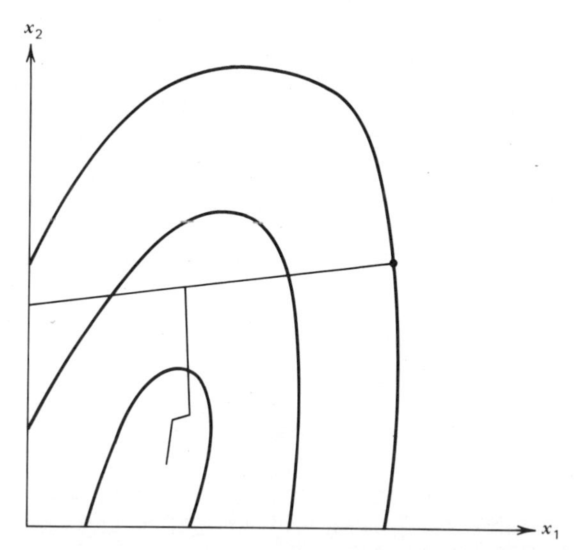

Figure 4.9. Fletcher-Reeves conjugate gradient search.

and let α_k^*, $k=1,\ldots,n$ solve

$$\max f(\mathbf{x}_i+\alpha_i^*\mathbf{d}_i) \tag{4.94}$$

Then

$$\mathbf{x}_{k+1}=\mathbf{x}_k+\alpha_i^*\mathbf{d}_i \tag{4.95}$$

$$\mathbf{d}_{k+1}=\nabla f_{k+1}^t+\beta_k\mathbf{d}_k \tag{4.96}$$

$$\beta_k=\frac{\nabla f_{k+1}\nabla f_{k+1}^T}{\nabla f_k\nabla f_k^T} \tag{4.97}$$

The algorithm is restarted every n steps. Figure 4.9 depicts a Fletcher-Reeves search.

A final method of utilizing gradient information is the method of parallel tangents, shown in Figure 4.10. As mentioned earlier, the method derives from the observation that steepest ascent tends to approach an optimum by a zigzag pattern which appears to be contained within two lines not equal to the gradient directions. By utilizing the results of previous steps it is possible to estimate these lines to speed the converge of search. Luenberger shows that in the quadratic case this is implicitly the conjugate gradient, for, although the conjugate directions are not obtained, the endpoints at each iteration are the same as if they had been obtained. In addition, although the method is therefore related as well to that of Fletcher-Reeves, this method is less sensitive to error in the line searches.

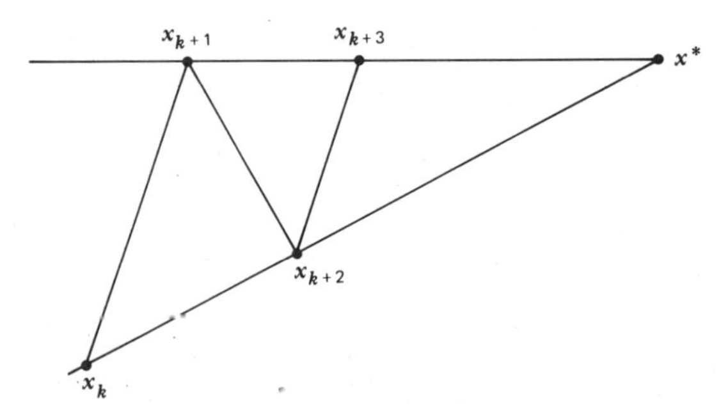

Figure 4.10. Method of parallel tangents (PARTAN).

The Constrained Optimization Problem

Unlike the previous section the concern of the present section is optimization under constraint. The general problem may be given as

$$\max f(\mathbf{x}) \tag{4.98}$$

subject to

$$\mathbf{h}(\mathbf{x}) = \mathbf{0} \tag{4.99}$$

$$\mathbf{g}(\mathbf{x}) \leqslant \mathbf{0} \tag{4.100}$$

The two sets of constraint functions, $\mathbf{h}(\mathbf{x})$ and $\mathbf{g}(\mathbf{x})$ define a constraint set, the elements of which are called feasible points or feasible $\mathbf{x}$. The problem could be summarized as,

$$\max f(\mathbf{x})$$

such that

$$\mathbf{x} \in \Omega$$

where Ω is the set of feasible $\mathbf{x}$.

In the general case the Ω can take any form, including convex, simply connected, and even unconnected sets of points. Some techniques take advantage of the properties of special kinds of sets. For instance, convex sets are a particularly useful type of constraint set. A convex constraint set is any one in which the linear combination of two feasible points are contained within the constraint set. For instance, let $\mathbf{x}_1, \mathbf{x}_2 \in \Omega$ be any two feasible points in Ω. For Ω to be convex, any linear combination of $\mathbf{x}_1$ and $\mathbf{x}_2$ must be contained within Ω

$$\alpha \mathbf{x}_1 + (1-\alpha)\mathbf{x}_2 \in \Omega \qquad 0 \leqslant \alpha \leqslant 1$$

Linear programming is one technique which takes advantage of convex constraint sets.

A general approach to the unconstrained problem could start with the concept of feasible directions. In the unconstrained case a feasible improving direction was any $\mathbf{d}$ such that

$$\nabla f(\mathbf{x})\mathbf{d} \geqslant 0 \tag{4.101}$$

In the last section several directions such as the steepest ascent direction, the Newton "deflected gradient" and the conjugate gradient directions would be feasible ascent directions in this sense. In the constrained problem a feasible direction $\mathbf{d}$ is one in which (4.101) and

$$\mathbf{x} + \alpha\mathbf{d} \in \Omega \qquad \alpha > 0 \tag{4.102}$$

are satisfied.

Before considering the necessary and sufficient conditions for the constrained problem exactly, consider the situation at a general constraint point. At such a point all of the $h_i(\mathbf{x})$ and perhaps some of the $g_j(\mathbf{x})$ are satisfied exactly (they are "active"). In the unconstrained case an ascent direction is any for which (4.101) is true, at least locally. For such a direction to be feasible, the following must be true:

$$h(\mathbf{x} + \alpha\mathbf{d}) = 0 \qquad \alpha \geqslant 0 \tag{4.103}$$

$$g(\mathbf{x} + \alpha\mathbf{d}) \leqslant 0 \qquad \alpha \geqslant 0 \tag{4.104}$$

In the case of (4.103), this means that there is an ascent direction which also satisfies all of the $h_i(\mathbf{x})$ and all of the $g_j(\mathbf{x})$ in (4.104). Ignoring for a moment the $g_j(\mathbf{x})$, consider the situation in (4.103) at a constrained optimum. An approximate feasible direction would be one for which

$$h_i(\mathbf{x}^* + \alpha\mathbf{d}) = h_i(\mathbf{x}^*) + \alpha \nabla h_i(\mathbf{x}^*)\mathbf{d} + \ldots$$

and

$$\alpha \nabla h_i(\mathbf{x}^*)\mathbf{d} = 0$$

since

$$h_i(\mathbf{x}^*) = 0$$

Consider the situation at a single constrained point, as shown in Figure 4.11. The gradient of the constraint defines a tangent that approximates the constrained point. Since there is a non-zero projection of ∇f upon this tangent direction, it defines a feasible direction. In the example the constraint is followed until no improvement is possible in f. At such a constrained point the feasible directions are now orthogonal to both ∇h and ∇f. In the case of ∇h this is because a feasible direction is necessarily a tangent to the constraint gradient. Relative to ∇f it signifies that the

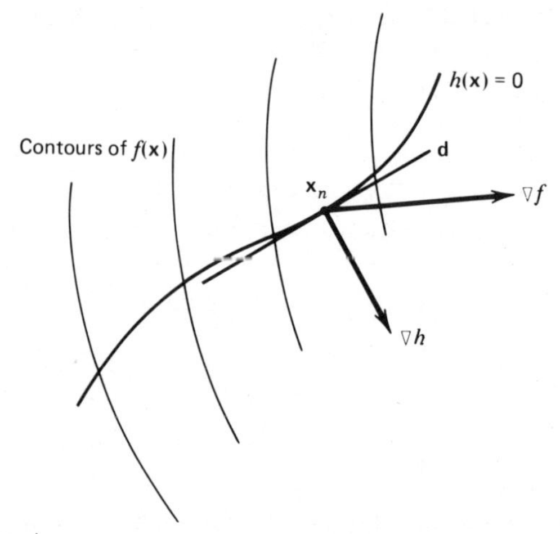

Figure 4.11. **d** is approximating tangent to $h(\mathbf{x})=0$ at $\mathbf{x}_n$.

feasible direction results in no improvement in f. The end result, however, is that ∇f and ∇h are linearly dependent, so that there are two α's not equal to zero such that, indicating

$$\alpha_1 \nabla f + \alpha_2 \nabla h = \mathbf{0} \tag{4.105}$$

and that at the constrained optimum ∇f and ∇h are parallel. This condition is illustrated in Figure 4.12.

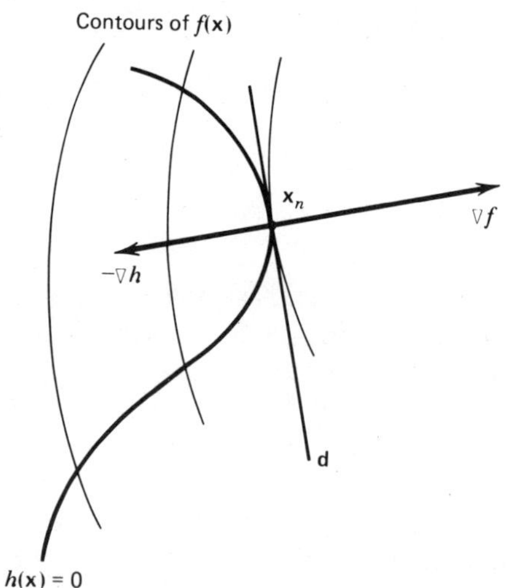

Figure 4.12. Constrained optimum at $\mathbf{x}_n$ with $\nabla f = \nabla h$ and $\nabla f \cdot \mathbf{d} = 0$.

Consider now the more general case with several constraints active. The feasible directions will satisfy several equations and define a

$$\nabla h_1 \mathbf{d} = 0$$

$$\nabla h_2 \mathbf{d} = 0$$

$$\cdots$$

$$\nabla h_m \mathbf{d} = 0$$

tangent space of dimension $n-m$. For $\mathbf{x}^*$ at this point to be a constrained maximum, then the space defined by these $\mathbf{d}$ must be orthogonal to ∇f, or else, as in the first example, there would exist a better feasible point. Now consider that the problem space is of dimension n. The constraint subspace is of dimension $n-m$. It is defined by the m independent gradients of the active constraints, which is itself a space of dimension m. Now ∇f is orthogonal to the constraint space and it has no components there, so if it exists it must be a linear combination of the ∇h (there is nowhere else for it to be!) and in this case,

$$\nabla f = \sum_{i=1}^{m} \alpha_i \nabla h_i \tag{4.106}$$

Therefore, just as in the constrained problem the combination

$$\nabla f \boldsymbol{\Delta} \mathbf{x} = \nabla f \mathbf{d} = 0$$

at the optimum, only the gradient itself does not disappear. This same result can be reached in a different way. Consider the lagrangian function

$$\max l(\mathbf{x}, \boldsymbol{\lambda}) = f(\mathbf{x}) + \boldsymbol{\lambda}^T \mathbf{h}(\mathbf{x}) \qquad \boldsymbol{\lambda} \in R^m \tag{4.107}$$

As an unconstrained problem, the necessary conditions on $l(\mathbf{x}, \boldsymbol{\lambda})$ to be a constrained optima are

$$\frac{\partial l}{\partial \mathbf{x}} = \nabla f + \lambda^T \nabla \mathbf{h}(\mathbf{x}) = \mathbf{0} \tag{4.108}$$

$$\frac{\partial l}{\partial \boldsymbol{\lambda}} = \mathbf{h}(\mathbf{x}) = \mathbf{0} \tag{4.109}$$

Equation (4.108) is a restatement of the earlier argument that the gradient of f will be a linear combination of the constraint gradients. Equation (4.109) is simply the requirement that the constraints are satisfied.

The simultaneous equations thus given are generally nonlinear and difficult to solve directly. Nevertheless, they are useful for determining whether a given solution is a local constrained maximum. The sufficient conditions on $L(\mathbf{x}, \boldsymbol{\lambda})$ require that the Hessian $L(\mathbf{x}, \boldsymbol{\lambda})$ of $l(\mathbf{x}, \boldsymbol{\lambda})$ must be negative semidefinite with respect to $\mathbf{x}$ of feasible points. As it turns out, the lagrangian is a *saddle point*, a maximum with respect to a feasible $\mathbf{x}$, and a minimum with respect to the $\boldsymbol{\lambda}$. Therefore, (4.108) implies (4.106)

$$\nabla f(\mathbf{x}^*) = \sum_{i=1}^{m} \lambda_i \nabla h(\mathbf{x}^*)$$

The active $g_j(\mathbf{x})$ can also be included in this statement since an active $g_j(\mathbf{x})$ is satisfied exactly and can be treated as one of the $h_i(\mathbf{x})$. The complete equation, then, is

$$\nabla f(\mathbf{x}^*) = \sum_{i=1}^{m} \lambda_i \nabla h_i(\mathbf{x}^*) + \sum_{j \in J} \mu_j \nabla g_j(\mathbf{x}) \tag{4.110}$$

where J specifies the set of active inequality constraints.

Consider now the lagrangian function

$$l(\mathbf{x}, \boldsymbol{\lambda}, \boldsymbol{\mu}) = f(\mathbf{x}) + \sum_{i=1}^{m} \lambda_i h_i(\mathbf{x}) + \sum_{j=1}^{l} \mu_j g_j(\mathbf{x}), \qquad \mu_j \geqslant 0 \tag{4.111}$$

The lagrangian can be considered an unconstrained function. Its necessary condition for an optimum would be

$$l_x = \nabla f(\mathbf{x}) + \boldsymbol{\lambda}^T \nabla \mathbf{h}(\mathbf{x}) + \boldsymbol{\mu}^T \nabla \mathbf{g}(\mathbf{x}) = \mathbf{0} \tag{4.112}$$

$$l_\lambda = \mathbf{h}(\mathbf{x}) = \mathbf{0} \tag{4.113}$$

$$l_\mu = \mathbf{g}(\mathbf{x}) = \mathbf{0} \qquad \mu_j > 0 \tag{4.114}$$

$$g_j(\mathbf{x}) < 0 \qquad \mu_j = 0 \tag{4.115}$$

Equations (4.114) and (4.115) can be replaced by the requirement that

$$\boldsymbol{\mu}^T \mathbf{g}(\mathbf{x}) = 0 \qquad \boldsymbol{\mu} \geqslant \mathbf{0} \tag{4.116}$$

Equations (4.112), (4.113), and (4.116) form the *necessary conditions* for $\mathbf{x}^*$ to be a locally constrained optimum when equality and inequality constraints are present. The *sufficient conditions* for the constrained optimum are similar to those of the unconstrained problem, and

$$L(\mathbf{x}, \boldsymbol{\lambda}, \boldsymbol{\mu}) = \nabla^2 l = \nabla^2 f(\mathbf{x}) + \boldsymbol{\lambda}^T \nabla^2 \mathbf{h}(\mathbf{x}) + \boldsymbol{\mu}^T \nabla^2 \mathbf{g}(\mathbf{x}) \tag{4.117}$$

must be negative semidefinite for any y in the tangent subspace of the active ∇h and ∇g.

Equations of this solution are generally nonlinear simultaneous and not amenable to direct solution, as was the similar case in the unconstrained situation.

Applications of the Lagrangian Analysis

This section and the next will concern applications of the lagrangian formulation for analyzing solutions and developing algorithms. More general considerations will lead to other algorithms in the final portion of the section.

A special class of constrained optimization problems with linear objective functions and linear constraints is called linear programming. It enjoys a number of computational advantages over the more general class of nonlinear problems, and is applicable to a large range of practical situations. It has therefore won wide acceptance throughout industry. Problems of several thousand variables are quite routinely solved in this fashion.

The general problem is of this form:

$$\max \mathbf{c}^T\mathbf{x} \tag{4.118}$$

such that

$$A\mathbf{x} \leqslant \mathbf{b} \tag{4.119}$$

$$\mathbf{x} \geqslant \mathbf{0} \tag{4.120}$$

It is convenient to solve this problem by adding a set of slack variables $\mathbf{y}$ to the constraint equation in order to change the inequalities to equalities,

$$A\mathbf{x}+\mathbf{y}=\mathbf{b} \tag{4.121}$$

$$\mathbf{x} \geqslant 0 \qquad \mathbf{y} \geqslant 0 \tag{4.122}$$

Equation (4.121) can be written as

$$(A,I)(\mathbf{x},\mathbf{y})^T=\mathbf{b} \tag{4.123}$$

There will be a single nonnegative slack variable for every equation.

A special property of this problem is the convexity of the constraint set. In general, there will be m constraints and as many as $n+m$ unknowns. Then it is easy to see that any combination of two feasible solutions, $(\mathbf{x}_1^T,\mathbf{y}_1^T)$ and $(\mathbf{x}_2^T,\mathbf{y}_2^T)$ will itself be a feasible solution, since if

$$A\mathbf{x}_1+\mathbf{y}_1=\mathbf{b}$$

$$A\mathbf{x}_2+\mathbf{y}_2=\mathbf{b}$$

then

$$\mathbf{x}^T=\alpha(\mathbf{x}_1^T,\mathbf{y}_1^T)+(1-\alpha)(\mathbf{x}_2^T,\mathbf{y}_2^T)$$

$$(A)\mathbf{x}=A(\alpha\mathbf{x}_1)+\alpha\mathbf{y}_1+A(1-\alpha)\mathbf{x}_2+(1-\alpha)\mathbf{y}_2=\alpha\mathbf{b}+(1-\alpha)\mathbf{b}=\mathbf{b}$$

The general procedure for solving this problem is called the Simplex Algorithm, which takes advantage of this convexity to seek the solution. Note that since $\mathbf{c}^T\mathbf{x}$ is a linear function, there is no solution in the interior of the constraint set. The constraints will always be satisfied exactly. The algorithm starts at a feasible point, $\mathbf{x}=\mathbf{0}$, $\mathbf{y}=\mathbf{b}$ and a value for the objective function of 0. In the general problem a max $\nabla f\mathbf{d}=\mathbf{c}^T\mathbf{d}$ would be sought, but in this problem this can be done such that the best variable, given the present set of constraints, can be brought into the basis. As a new variable enters the basis, one departs, so that the system is exactly determined at any point. As the algorithm proceeds the x_i come into the solution, and the y_j depart until an optimum set of x_i and y_j have been located.

In the previous example the lagrangian variables $\boldsymbol{\lambda}$ were also solved, and as noted earlier, only those λ_j associated with active constraints were nonzero. These nonzero values give the marginal value of increments in the constraints, that is, the expected incremental gain in the objective function for small changes in the constraint value. In one sense these values can be considered the "worth" of the constraint, and any time the actual cost is less than that "worth," the buy could be considered judicious since the anticipated gain in the objective will overshadow the cost. These sensitivity values, however, are local; larger changes may show a decreasing worth as other constraints become more "important" or new constraints become active, decreasing the "worth" of change in the old constraint to zero.

The linear programming problem is a simple but often used approach to considering tradeoffs between sets of constraints and some objective. It will be used later to determine a best feasible direction in the presence of several constraints. More detailed treatments of linear programming and linear programming algorithms can be found in [21,27,31].

The linear programming method is generally limited to problems that are linear in both the objective function and the constraints. Linear programming has been applied to certain aspects of nonlinear programming. One special case is the quadratic programming problem, which is quadratic in the objective function and linear in the constraints,

$$\max y(x)=\mathbf{c}^T\mathbf{x}+\tfrac{1}{2}\mathbf{x}^T Q\mathbf{x} \tag{4.124}$$

subject to

$$A\mathbf{x}\leqslant\mathbf{b} \tag{4.125}$$

Here it is possible to apply linear programming because the equations of the sufficient conditions for an optimum, i.e.,

$$\mathbf{c}+Q\mathbf{x}+\boldsymbol{\lambda}^T A=0 \tag{4.126}$$

and

$$A\mathbf{x}=\mathbf{b} \tag{4.127}$$

are all linear in the x_i and λ_i. With some modification to the simplex method [21], it is possible to solve the quadratic programming problem via linear programming.

EXAMPLE. In experimentation it is not uncommon to estimate the response variable with a quadratic model subject to restrictions on the range of the design and known constraints on the input variables. For instance, consider the following quadratic response function,

$$\min(-0.00834x_1 - 0.04667x_3 + 0.0004058x_1x_3)$$

subject to

$$100 \leqslant x_1 \leqslant 130$$
$$105 \leqslant x_2 \leqslant 175$$
$$9 \leqslant x_3 \leqslant 21$$

and

$$0.34510x_1 + 0.07788x_2 + 0.71527x_3 = 59.265$$

Such a problem would be a well-defined quadratic programming problem. The solution,

$$x_1 = 129.38 \qquad x_2 = 105 \qquad x_3 = 9$$

can be obtained via linear programming methods.

Linear programming can also be used in several other aspects of nonlinear programming. As described later in the section, linear programming is also used to construct a best feasible direction for exploration near constraints. This is especially feasible when the various functions are known only to a linear approximation and the various constraints and objectives can be expressed as linear equations.

In the general experimental situation several response functions are measured at once so that if one can be estimated as a quadratic function, the others may also. However, when the constraints become quadratic it is no longer possible to solve for the constrained optima via quadratic programming. The lagrangian formulation can be used to generate the new nonlinear equations that must be satisfied at the optimum. Although such equations are nonlinear, they will in general be no worse than quadratic when quadratic models are estimated, and should be amenable to solution by the least-squares methods of the next section.

Consider the system given by Umland and Smith [42]. In an experimental system, estimates of process yield and product purity were estimated to be

$$\hat{Y} = 55.84 + 7.31x_1 + 26.65x_2 - 3.03x_1^2 - 6.96x_2^2 + 2.69x_1x_2$$

and

$$\hat{P} = 85.72 + 21.85x_1 + 8.59x_2 - 9.20x_1^2 - 5.18x_2^2 - 6.26x_1x_2$$

The lagrangian equations for 90% purity are given as,

$$\varphi_1(\mathbf{x}) = 7.31 - 6.06x_1 + 2.69x_2 + \lambda(21.85 - 18.40x_1 - 6.26x_2) = 0$$

$$\varphi_2(\mathbf{x}) = 26.65 - 13.92x_2 + 2.69x_1 + \lambda(8.59 - 10.36x_2 - 6.26x_1) = 0$$

$$\varphi_3(\mathbf{x}) = 85.72 + 21.85x_1 + 8.59x_2 - 9.20x_1^2 - 5.18x_2^2 - 6.26x_1x_2 - 90 = 0$$

Noting that $\varphi=0$ will be satisfied when $\varphi^2=0$, the simultaneous equations can be solved as an unconstrained optimization of the least squares type

$$\min \varphi_i^2(\mathbf{x}) = \max - \varphi_i^2(\mathbf{x})$$

Umland and Smith used a steepest descent routine to solve the problem at several purity levels,

Purity:	94.87	92.47	89.995
Yield:	83.66	86.73	88.68
x_1	0.965	1.005	1.075
x_2	1.088	1.316	1.479
λ	1.612	0.971	0.665

In this particular problem, however, the lagrange multiplier can be used to considerably reduce the difficulty of the problem. As noted by Chow [13], the problem of Umland and Smith is linear in the λ, so that

$$\nabla f + \boldsymbol{\lambda}^T \nabla \mathbf{h} = \mathbf{0}$$

becomes

$$\begin{bmatrix} 18.4\lambda+6.06 & 6.26\lambda-2.69 \\ 6.26\lambda-2.69 & 10.36\lambda+13.92 \end{bmatrix} \begin{bmatrix} x_1 \\ x_2 \end{bmatrix} = \begin{bmatrix} 21.85\lambda+7.31 \\ 8.59\lambda+26.65 \end{bmatrix}$$

which yields the solution

$$x_1 = \frac{173.44+236.16\lambda+172.59\lambda^2}{77.12-159.67\lambda-229.81\lambda^2}$$

$$x_2 = \frac{181.16+451.32\lambda+21.28\lambda^2}{77.12-159.67\lambda-229.81\lambda^2}$$

This reduces the problem from one of three dimensions to one of a single dimension. Chow then varies λ to produce several solutions.

λ	x_1	x_2	Purity	Yield
0.00	2.25	2.35	46.85	95.26
−0.10	1.67	1.58	80.68	91.43
−0.20	1.17	0.92	95.46	81.77
−0.15	1.48	1.19	89.80	86.62

Through the lagrangian or *dual variables* the problem is effectively a constrained optimization problem in a single dimension. The method can be generalized. Whenever the number of constraints is (much) lower than the number of variables, the solution of the dual may offer considerable computational savings.

ALGORITHMS FOR SOLVING LEAST-SQUARES PROBLEMS

Least-squares problems arise in a variety of circumstances. The normal equations of linear regression are solutions to a minimization problem involving the squared deviations of responses to estimates. Least-squares used in regression allow a solution to a system of overdetermined equations, that is, where there are more equations than unknowns, and is applicable whether the underlying model is linear or nonlinear. In addition, the least-squares problem also arises in exactly determined systems of equations, such as the simultaneous nonlinear equations at constrained optima. Least squares is then a method of formulating the constrained problem as an unconstrained optimization.

Consider first the general regression problem. Some function $\hat{y}(\mathbf{x},\boldsymbol{\beta})$ of the variables $\mathbf{x}$ and some parameters $\boldsymbol{\beta}$ is defined. This estimated response can be compared to the measured response at that point with the difference called the residuals ε_i

$$\varepsilon_i = y(\mathbf{x}_i) - \hat{y}(\mathbf{x}_i, \boldsymbol{\beta})$$

The general regression problem becomes

$$\min \sum_{i=1}^{N} \varepsilon_i^2 = \max \sum_{i=1}^{N} (-\varepsilon_i^2)$$

In the general case the number of β_j is less than the number of equations, so that it is not possible for all of the ε_i to equal zero. Therefore, the sum of squared residuals is greater than zero,

$$\boldsymbol{\varepsilon}^T\boldsymbol{\varepsilon} = \sum_{i=1}^{N} \varepsilon_i^2 > 0$$

In case of simultaneous equations of the form

$$g(\mathbf{x}_i) = b_i$$
$$\cdots$$
$$g(\mathbf{x}_n) = b_n$$

The errors are given by

$$\varepsilon_i = g(\mathbf{x}_i) - b_i$$

and represent deviations from the solution. If the system is exactly determined and a solution exists, then the requirement

$$\boldsymbol{\varepsilon} = \mathbf{0}$$

is also satisfied when

$$\boldsymbol{\varepsilon}^T\boldsymbol{\varepsilon} = \sum_{i=1}^{n} \varepsilon_i^2 = 0$$

Therefore, nonlinear equations can also be approached as least squares problems.

The errors in the regression problem are functions of the inputs $\mathbf{x}_i$ and the parameters β_j. In the context of this problem the inputs are fixed and β_j are the variables. Similarly, the error functions in the nonlinear equation case are functions of the variables $\mathbf{x}$. The general problem can be given then as

$$\min \mathbf{f}'(\mathbf{x})\mathbf{f}(\mathbf{x}) = \min \sum_{j=1}^{n} f_j^2(\mathbf{x}) = s$$

where $s^* > 0$ for the regression problem and $s^* = 0$ for the nonlinear equations.

Consider the situation at some point $\mathbf{x}$ in the search for the optimum. Presumably $s(\mathbf{x}) > 0$ at this point. Acting upon the approach in the scalar optimization problem, a reasonable first step would be consideration of a linear approximation of $f(\mathbf{x})$ and of $s(\mathbf{x})$. The linear approximation of $\mathbf{f}(\mathbf{x})$ about the point $\mathbf{x}$ is given by

$$\mathbf{f}(\mathbf{x}) = \mathbf{f}(\hat{\mathbf{x}}) + J_f(\hat{\mathbf{x}})\mathbf{\Delta x} \tag{4.128}$$

This is a vector approximation with the Jacobian J_f given by

$$J_f(\hat{\mathbf{x}}) = \begin{bmatrix} \dfrac{\partial f_1}{\partial x_1} & \dfrac{\partial f_1}{\partial x_2} & \cdots & \dfrac{\partial f_1}{\partial x_n} \\ \dfrac{\partial f_2}{\partial x_1} & \dfrac{\partial f_2}{\partial x_2} & \cdots & \dfrac{\partial f_2}{\partial x_n} \\ \dfrac{\partial f_m}{\partial x_1} & \dfrac{\partial f_m}{\partial x_2} & \cdots & \dfrac{\partial f_m}{\partial x_n} \end{bmatrix}_{\mathbf{x}=\hat{\mathbf{x}}} \tag{4.129}$$

The Jacobian takes the place of the gradient in the scalar case. Using this equation it is possible to solve for estimate changes in $s(\mathbf{x})$.

$$\Delta s(\mathbf{\Delta x}) = s(\hat{\mathbf{x}} + \Delta\mathbf{x}) - s(\hat{\mathbf{x}})$$

$$\begin{aligned} \Delta s(\Delta\mathbf{x}) &= \left[\mathbf{f}(\hat{\mathbf{x}}) + J_f(\hat{\mathbf{x}})\Delta\mathbf{x}\right]^T\left[\mathbf{f}(\hat{\mathbf{x}}) + J_f(\hat{\mathbf{x}})\Delta\mathbf{x}\right] - \mathbf{f}^T(\hat{\mathbf{x}})\mathbf{f}(\hat{\mathbf{x}}) \\ &= 2\mathbf{\Delta x}^T J_f^T(\hat{\mathbf{x}})\mathbf{f}(\hat{\mathbf{x}}) + \Delta\mathbf{x}^T J_f^T(\hat{\mathbf{x}}) J_f(\hat{\mathbf{x}})\Delta\mathbf{x} \\ &= 2\sum_{i=1}^{m}\left[f_i(\hat{\mathbf{x}})\sum_{k=1}^{n}\frac{\partial f_i}{\partial x_k}\Delta x_k\right] + \sum_{i=1}^{m}\left[\sum_{k=1}^{n}\frac{\partial f_i}{\partial x_k}\Delta x_k\right]^2 \end{aligned} \tag{4.130}$$

This equation is the basis of several algorithms for minimizing $s(\mathbf{x})$. Consider first the linear portion of equation (4.130),

$$\Delta s(\mathbf{\Delta x}) = 2\mathbf{\Delta x}^T J_f^T(\hat{\mathbf{x}})\mathbf{f}(\hat{\mathbf{x}}) \tag{4.131}$$

The overall object is to pick $\mathbf{\Delta x}$ such that Δs is minimized so that the right-hand side of (4.131) is less than zero, or

$$s(\hat{\mathbf{x}}+\mathbf{\Delta x})<s(\hat{\mathbf{x}}) \tag{4.132}$$

Since (4.131) is linear there is no finite solution. In addition, the approximation is limited to the neighborhood of $\hat{\mathbf{x}}$ so that $\mathbf{\Delta x}$ must be bounded. A simple method of answering both objections (without going to a quadratic approximation in s) is to restrict $\mathbf{\Delta x}$ to a neighborhood of $\mathbf{x}$ with the constraint

$$\mathbf{\Delta x}^T\mathbf{\Delta x}<\alpha \tag{4.133}$$

The constrained problem

$$\min\ 2\mathbf{\Delta x}^T J_f^T(\hat{\mathbf{x}})\mathbf{f}(\hat{\mathbf{x}}) \tag{4.134}$$

subject to

$$\mathbf{\Delta x}^T\mathbf{\Delta x}=\alpha \tag{4.135}$$

is then solved. Forming the lagrangian for this problem,

$$l(x,\lambda)=2\mathbf{\Delta x}^T J_f^T(\hat{\mathbf{x}})\mathbf{f}(\hat{\mathbf{x}})+\lambda(\mathbf{\Delta x}^T\mathbf{\Delta x}-\alpha) \tag{4.136}$$

The derivative of this function must be zero at the optimum,

$$\frac{\partial l}{\partial \mathbf{\Delta x}}=2J_f^T(\hat{\mathbf{x}})f(\hat{\mathbf{x}})+2\lambda\mathbf{\Delta x}=\mathbf{0}$$

or

$$\mathbf{\Delta x}=-\frac{1}{\lambda}J_f^T(\hat{\mathbf{x}})\mathbf{f}(\hat{x}) \tag{4.137}$$

The actual value of λ will depend upon α and can be obtained from substitution into (4.135). This equation is the vector analog of the scalar steepest descent direction for $\lambda>0$.

Consider now the full quadratic approximation of Δs about $\hat{\mathbf{x}}$. This expression is minimized with respect to $\mathbf{\Delta x}$ when

$$\frac{\partial \Delta s}{\partial \mathbf{\Delta x}}=\mathbf{0}=2J_f^T(\hat{\mathbf{x}})\mathbf{f}(\hat{\mathbf{x}})+2J_f^T J_f\mathbf{\Delta x}$$

whose solution is

$$\mathbf{\Delta x}=-\left[J_f^T(\hat{\mathbf{x}})J_f(\hat{\mathbf{x}})\right]^{-1}J_f^T(\hat{\mathbf{x}})\mathbf{f}(\hat{\mathbf{x}}) \tag{4.138}$$

This is the vector Newton-Raphson equation. Notice the similarity to the normal equations of linear regression, with $\mathbf{f}(\hat{\mathbf{x}})$ taking the place of $\boldsymbol{\varepsilon}(\mathbf{x})$ and $J_f(\mathbf{x})$ that of the experimental matrix $\mathbf{X}$; in fact, this equation is the "optimal step" $\mathbf{\Delta x}^*$ from any estimate $\hat{\boldsymbol{\beta}}$ of $\boldsymbol{\beta}$ to $\boldsymbol{\beta}^*$, the least squares optimal value of $\boldsymbol{\beta}$. Defining

$$J_f(\hat{\boldsymbol{\beta}}) = J_f(\boldsymbol{\beta}) = X$$

$$\boldsymbol{\Delta x}^* = \boldsymbol{\beta}^* - \hat{\boldsymbol{\beta}}$$

$$\mathbf{f}(\hat{X}) = X\hat{\boldsymbol{\beta}} - \mathbf{y}$$

so that (4.138) becomes

$$\boldsymbol{\beta}^* - \hat{\boldsymbol{\beta}} = -(X^TX)^{-1}X^T(\mathbf{x}\hat{\boldsymbol{\beta}} - \mathbf{y})$$

$$\boldsymbol{\beta}^* - \hat{\boldsymbol{\beta}} = (X^TX)^{-1}X^T\mathbf{y} - \hat{\boldsymbol{\beta}}$$

or

$$\boldsymbol{\beta}^* = (X^TX)^{-1}X^T\mathbf{y}$$

The exact solution is obtained since the model is linear, and s quadratic, in the $\boldsymbol{\beta}$.

Also note that for J_f of full rank, (4.138) reduces to

$$\boldsymbol{\Delta x} = -J_f^{-1}(\hat{\mathbf{x}})\mathbf{f}(\hat{\mathbf{x}}) \tag{4.139}$$

if $J_f^{-1}(\hat{\mathbf{x}})$ exists. This is the more familiar form of Newton-Raphson.

This quadratic equation can also be constrained to a region about $\hat{\mathbf{x}}$ using the quadratic approximation. Then the lagrangian becomes

$$l(\mathbf{x},\lambda) = 2\boldsymbol{\Delta x}^TJ_f^T(\hat{\mathbf{x}})\mathbf{f}(\hat{\mathbf{x}}) + \boldsymbol{\Delta x}^TJ_f^T(\hat{\mathbf{x}})J_f(\hat{\mathbf{x}})\boldsymbol{\Delta x} + \lambda(\boldsymbol{\Delta x}^T\boldsymbol{\Delta x} - \alpha) \tag{4.140}$$

with the necessary condition that

$$\frac{\partial l}{\partial \boldsymbol{\Delta x}} = \mathbf{0} = 2J_f^T(\hat{\mathbf{x}})\mathbf{f}(\hat{\mathbf{x}}) + 2J_f^T(\hat{\mathbf{x}})J(\hat{\mathbf{x}})\boldsymbol{\Delta x} + 2\lambda\boldsymbol{\Delta x}$$

or

$$\boldsymbol{\Delta x} = -\left(J_f^T(\hat{\mathbf{x}})J_f(\hat{\mathbf{x}}) + \lambda I\right)^{-1}J_f(\hat{\mathbf{x}})\mathbf{f}(\hat{\mathbf{x}}) \tag{4.141}$$

This is the basis of the Marquardt algorithm. Note that for $\lambda = 0$ this equation reduces to Newton-Raphson, and large λ approaches the method of steepest descent. Hence the algorithm is also sometimes called the Marquardt compromise.

Notice that for linear models the method of Newton-Raphson is exact and the optimum point $\boldsymbol{\beta}^*$ can be obtained in a single step from any starting point. This unfortunately is not the case when the model is not linear in the $\boldsymbol{\beta}$. The speed of the algorithms and their ability to converge depends upon the type of the model and initial starting point $\hat{\mathbf{x}}$.

Constraint Handling

We return to the idea of feasible directions to introduce several methods for dealing with inequality constraints. We recall from (4.101) that an ascent direction **d** given by

$$\nabla f(\mathbf{x})\mathbf{d} > 0$$

is a feasible direction whenever, as given by (4.102),

$$\mathbf{x} + \alpha\mathbf{d} \in \Omega \qquad \alpha > 0$$

To a linear approximation these feasible directions can be obtained from the gradients of the active constraints as lying in the tangent plane defined by

$$J_g\mathbf{d} = \mathbf{0} \tag{4.142}$$

where J_g is the Jacobian of all active inequality constraints. The central idea is that feasible ascent directions will lie on the surface of the active constraints or pointing into the interior of the feasible region.

One way to express this idea is to look for directions of the type

$$\mathbf{d} = W_0 \nabla f(\mathbf{x}) - \mathbf{W}^T J_g(\mathbf{x}) \tag{4.143}$$

where the W_i are appropriate weights, and $W_i = 0$ when $g_i(\mathbf{x}) < 0$. The lead weight W_o controls the tendency to include the ascent direction and the W_i control the tendency to head inside the feasible region. One set of weights is advanced by Klingman and Himmelblau [25]. Here the weights take the form

$$W_o = \|\nabla f(\mathbf{x})\|^{-1} \tag{4.144}$$

$$W_i = \begin{cases} 0 & g_i(\mathbf{x}) < 0 \\ \|\nabla g_i(\mathbf{x})\|^{-1} & g_i(\mathbf{x}) = 0 \end{cases} \tag{4.145}$$

Using these in (4.143) yields the Multiple Gradient Summation Technique (MGS)

$$\mathbf{d} = \frac{\nabla f^T(\mathbf{x})}{\|\nabla f(\mathbf{x})\|} - \sum_{i \in I} \frac{\nabla g_i^T(\mathbf{x})}{\|\nabla g_i(\mathbf{x})\|} \tag{4.146}$$

I is the set of constraints $g_i(\mathbf{x})$ active at $\mathbf{x}$. MGS is essentially a compromise direction based on the normalized gradients. As shown in Figure 4.13 the MGS estimate may not be optimal, but provides a feasible ascent direction. For the MGS estimate in (4.146) to be an ascent direction, (4.142) must be satisfied,

$$\nabla f(\mathbf{x})\left(\frac{\nabla f^T(\mathbf{x})}{\|\nabla f(\mathbf{x})\|} - \sum_{i \in I} \frac{\nabla g_i(\mathbf{x})}{\|\nabla g_i(\mathbf{x})\|}\right) > 0$$

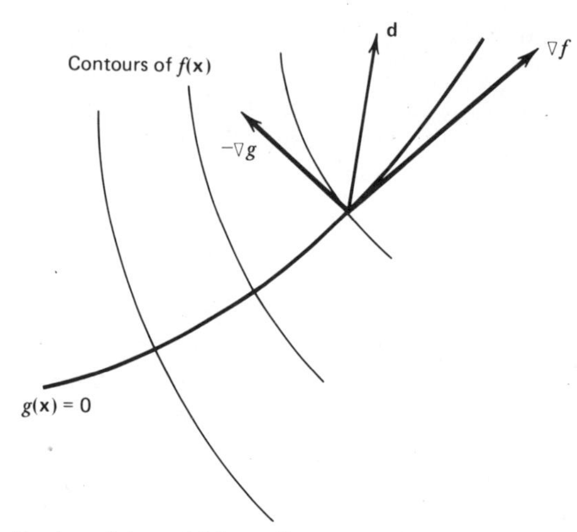

Figure 4.13. Application of the multiple gradient summation technique to a single-constraint system.

so that

$$1 > \sum_{i \in I} \frac{\nabla f(\mathbf{x}) \nabla g_i^T(\mathbf{x})}{\|\nabla f(\mathbf{x})\| \, \|\nabla g_i(\mathbf{x})\|} \tag{4.147}$$

This condition will only be violated if one of the constraints is parallel to $\nabla f(\mathbf{x})$ or the set spans $\nabla f(\mathbf{x})$ so that (4.147) is true. These situations indicate that the set of feasible ascent directions is empty and the constrained point is a local maximum. Equation (4.147) can thus be used as a rough condition for convergence.

Although the direction given by MGS may be a feasible ascent direction, it is unlikely to be a best feasible direction. We can attempt to estimate this direction, however. We can estimate our expected functional gain in $f(\mathbf{x}+\mathbf{d})$ with

$$\varepsilon = f(\mathbf{x}+\mathbf{d}) - f(\mathbf{x}) = \nabla f(\mathbf{x})\mathbf{d} \tag{4.148}$$

We would like to maximize this expected gain subject to remaining feasible. We can formulate this as a linear programming problem in the form

$$\max \varepsilon \tag{4.149}$$

subject to

$$-\nabla f(\mathbf{x})\mathbf{d} + \varepsilon \leqslant 0 \tag{4.150}$$

$$\nabla g_i(\mathbf{x})\mathbf{d} \leqslant 0 \qquad i \in I \tag{4.151}$$

$$\mathbf{1}^T\mathbf{d} \leqslant 1 \tag{4.152}$$

The last constraint limits the length of **d**, which is not important in any case.

The approach for finding the best feasible direction was first suggested by Zoutendijk [45]. The formula given by (4.149)–(4.152) is exact when the $g_i(\mathbf{x})$ are all linear. When some of the constraints are convex we change (4.151) to

$$\nabla g_i(\mathbf{x})\mathbf{d} + \varepsilon\theta_i \leqslant 0 \qquad i \in I \tag{4.151a}$$

The positive constant θ_i is 1 when $g_i(\mathbf{x})$ is convex and 0 when it is linear or concave. This modification repels solution **d** from the tangent plane and into the feasible region. Since we often do not know whether a constraint is convex or not, it is prudent to let all the θ_i be 1.

In discussing the convergence properties of ascent algorithms, Luenberger [31] notes that if an algorithm is not *closed*, convergence can be very slow or fail altogether. This is a particular risk when the algorithm can cycle between two constraint surfaces so that the set I of active constraints changes from iteration to iteration. This cycling phenomena is called *jamming* and tends to slow convergence. One way to prevent jamming is to change (4.151) again so that all constraints are included:

$$\mathbf{g}(x) + J_g(\mathbf{x})\mathbf{d} + \varepsilon\mathbf{u} \leqslant \mathbf{0} \tag{4.151b}$$

For constraints in their interior, $g_i(\mathbf{x})$ is negative and their part of (4.151b)

$$\nabla g_i(\mathbf{x})\mathbf{d} + \varepsilon \leqslant |g_i(\mathbf{x})| \tag{4.153}$$

is looser bound than those $i \in I$ for (4.151a). Notice that (4.151b) reduces to (4.151a) when $\theta_i = 1$ and $i \in I$.

Use of the best feasible direction approach will require the solution of a linear programming problem for each iteration. This may not always be feasible for large nonlinear programming programs, but presents no particular problem in the experimental situation when good linear programming code is available. When all the constraints are linear, much of the linear programming problem given by the constraints in (4.151a) are constant from iteration to iteration. This has been effectively exploited by Wolfe [46] in the reduced gradient technique. Here the old problem is updated each iteration rather than being completely solved every iteration.

We remarked earlier that we can approximate the constraint surface as a plane of dimension $n-m$, where m was the number of active constraints. A feasible ascent direction is given by the projection of the gradient of the objective function upon this plane. This projection is the basis of the gradient projection technique and it is quite easy to calculate.

We know first that the tangent plane is defined in equation (4.142) as

$$J_g(\mathbf{x})\mathbf{d} = \mathbf{0}$$

The vector **d** lies in the tangent subspace of dimension $n-m$. The rest of the space is generated by the rows of $J_g(\mathbf{x})$. Since these two subspaces are orthogonal we can write $\nabla f(\mathbf{x})$ as

$$\nabla f(\mathbf{x}) = \mathbf{d} + J_g^T(\mathbf{x})\boldsymbol{\beta} \tag{4.154}$$

where $\boldsymbol{\beta} \in R^m$ is a real vector of m components. Because of (4.142) we can multiply (4.154) by $J_g(\mathbf{x})$ to obtain

$$\begin{aligned} J_g(\mathbf{x})\nabla f(\mathbf{x}) &= J_g(\mathbf{x})\mathbf{d} + J_g(\mathbf{x})J_g^T(\mathbf{x})\boldsymbol{\beta} \\ &= J_g(\mathbf{x})J_g^T(\mathbf{x})\boldsymbol{\beta} \end{aligned}$$

or

$$\boldsymbol{\beta} = \left[J_g(\mathbf{x})J_g^T(\mathbf{x})\right]^{-1} J_g(\mathbf{x})\nabla f(\mathbf{x})$$

We substitute this back into (4.154) to yield

$$\begin{aligned} \mathbf{d} &= \left\{I - J_g^T(\mathbf{x})\left[J_g(\mathbf{x})J_g^T(\mathbf{x})\right]^{-1} J_g(\mathbf{x})\right\}\nabla f(\mathbf{x}) \\ &= P\nabla f(\mathbf{x}) \end{aligned} \tag{4.155}$$

We call P the *projection matrix* for the active constraints at **x**. It can be shown that a nonzero **d** is an ascent direction and a null **d** is a sufficient condition for a constrained optimum.

The gradient projection and reduced gradient techniques induce a coordinate system on the constraint surface. When the constraints are all linear this induced coordinate system is the same as the original coordinate system. In the nonlinear case they differ somewhat, and the projection may not be feasible, particularly in the case of convex constraint sets. In the course of the algorithm, therefore, if fixed steps are used in place of line searches, it may be necessary to return to the constraint surface from the outside. We then consider the usual linearization about some point **x** to estimate the optimal direction to return to the constraint surface,

$$\mathbf{g}\left[\mathbf{x} + J_g^T(\mathbf{x})\boldsymbol{\alpha}\right] = \mathbf{g}(\mathbf{x}) + J_g(\mathbf{x})J_g^T(\mathbf{x})\boldsymbol{\alpha} \tag{4.156}$$

and solving for the $\boldsymbol{\alpha}$ for which $\mathbf{g}[\mathbf{x} + J_g^T(\mathbf{x})\boldsymbol{\alpha}] = 0$.

$$\boldsymbol{\alpha} = \left[J_g(\mathbf{x})J_g^T(\mathbf{x})\right]^{-1}\mathbf{g}(\mathbf{x}) \tag{4.157}$$

This solution is substituted into $\mathbf{x} + J_g^T(\mathbf{x})\boldsymbol{\alpha}$ to obtain the best search direction away from **x**,

$$\mathbf{d} = J_g^T(\mathbf{x})\left[J_g(\mathbf{x})J_g^T(\mathbf{x})\right]^{-1}\mathbf{g}(\mathbf{x}) \tag{4.158}$$

for returning to the constraint surface.

MULTIPLE OBJECTIVE OPTIMIZATION

We have posed all of our problems thus far as the optimization of a single primary response with or without constraints upon secondary responses. But many times in engineering problems it is impossible to single out a single response as the primary one; instead it is necessary to optimize several responses simultaneously. A case in point occurred in Chapter 3, where the average mean squared error J of an experimental design is shown to consist of two components, a precision term V of the average variance and a bias term B of the average squared bias.

$$J = V + B$$

As demonstrated in Myers [34], it is not always possible to simultaneously optimize the joint criteria, since it depends on terms that are unknown. It is possible to minimize one or the other, but without the missing information there is no way to proceed.

The problem given above is quite general and occurs in most real design problems. Profit and risk are juggled in portfolio selections, mileage emissions and cost could be considered in auto design, and yields of various products and contaminants, and associated costs, are commonly considered in designing chemical systems. But it is far from clear in problems of this sort how much additional gain in mileage balances a gain in emissions or cost, or how much cost will be justified by a loss in emissions.

We shall be considering several methods for dealing with multiple objective problems. Weighting methods assume an appropriate aggregation of the individual responses into a single response. Ridge analysis and other parametric techniques consider study simultaneous changes in several responses so that tradeoffs in responses can be considered. A related approach, goal programming, reformulates the problem in terms of targets or goals, the deviation from the goals is optimized. Finally, the vector optimization problem will be introduced.

The general problem of this section takes the form of (4.159) with U a utility function relating the responses together

$$\max U[\mathbf{y}(\mathbf{x})] \tag{4.159}$$

subject to

$$\mathbf{y}(\mathbf{x}) \leqslant \mathbf{b} \tag{4.160}$$

where any of the $y_i(\mathbf{x})$ may be absent from the objective function or the constraints. For instance, for the auto problem, let y_1 be the design mileage in miles per gallon, y_2 the NOX emissions in milligrams per mile, and y_3 the cost of the design in dollars. Since constraints exist upon some of these terms, the optimization problem might be

$$\max U[y_1(\mathbf{x}), y_2(\mathbf{x}), y_3(\mathbf{x})]$$

subject to

$$y_1(\mathbf{x}) \geqslant 12.5$$
$$y_2(\mathbf{x}) \leqslant 15.$$

We will consider several forms for $U(*)$ and several methods of approaching the two constraints.

Weighted Optimization

Weighted optimization is simply the weighting of several responses together to form a single aggregate response of the form

$$U[\mathbf{y}(\mathbf{x}), \mathbf{w}] = \sum_{i=1}^{m} w_i y_i(\mathbf{x}) \tag{4.161}$$

In the minimization problem positive weights are assigned to responses to be maximized and negative weights to those to be minimized. In our example mileage (y_1) is maximized, and both emissions (y_2) and cost (y_3) are minimized. The equation in the previous example takes the form

$$U[y_1(\mathbf{x}), y_2(\mathbf{x}), y_3(\mathbf{x})] = w_1 y_1(\mathbf{x}) - w_2 y_2(\mathbf{x}) - w_3 y_3(\mathbf{x})$$

The major difficulty with this technique is the familiar "apples and oranges" problem, the comparison of unlike entities. In our example the three responses do not even have the same units. The weights must be chosen to transform the response units to some arbitrary utility scale for direct comparison. For instance, we can assign 1 unit of utility to a 5 mpg gain, a 2 mg/mi decrease in emissions and \$100 decrease in unit cost. Then our objective is

$$U[\mathbf{y}(\mathbf{x})] = 0.20 y_1(\mathbf{x}) - 0.50 y_2(\mathbf{x}) - 0.001 y_3(\mathbf{x})$$

The utility function can then be optimized directly as the primary response variable.

A useful post optimality analysis when using weighted least squares is to test the sensitivity of the solution to changes in the weighting w_i. One then considers the sensitivity in the y_i as the w_i are varied. Ideally the weights should be chosen so that the $y_i(\mathbf{x}^*)$ for each set of w_i do not vary widely. Solutions in which wide variations occur imply a much greater sensitivity to the choice of weights, and alternative methods should be used when weights are highly subjected.

If distributions of the weights can be made, then the expected value of the utility function

$$E\{U[\mathbf{y}(\mathbf{x}),\mathbf{w}]\} \tag{4.162}$$

is maximized in place of (4.161). This formulation allows a quantitative measure of the risk associated with estimated weights to be made.

Threshold Methods

Often enough it proves impossible to weight competing responses to a single utility scale for direct comparison. The simplest alternative is to consider the options of a single response given some minimum allowable performance thresholds on the remaining responses. Once the problem is solved, the solution is finalized and further adjustments to the thresholds made until the solution region vanishes altogether or an acceptable combination of responses is found.

Threshold methods can be treated systematically via the methods of Ridge Analysis, which is itself a special case of the lagrangian dual problem. We have in fact treated a problem of this type in the Ridge Regression and Singular Value Decomposition problems of Chapter 3. There it was discovered that the usual objective function, the sum of squared residuals, is relatively flat about the optimum. To improve the stability of our estimates, we preferred to optimize the confidence interval on our parameter estimates subject to minor deviations in the sum of squared residuals. The ridge analysis equations can in fact be derived as the solution to a constrained optimization problem on the size of the norm of the coefficient estimates $\hat{\boldsymbol{\beta}}$. The standard regression problem is of the form

$$\min\,(\mathbf{y}-X\hat{\boldsymbol{\beta}})^T(\mathbf{y}-X\hat{\boldsymbol{\beta}})$$

to which we can add the constraint,

$$\hat{\boldsymbol{\beta}}^T\hat{\boldsymbol{\beta}}\leqslant R^2$$

A lagrangian formulation is given by

$$\min l(\hat{\boldsymbol{\beta}},\lambda)=(\mathbf{y}-X\hat{\boldsymbol{\beta}})^T(\mathbf{y}-X\hat{\boldsymbol{\beta}})+(\hat{\boldsymbol{\beta}}^T\hat{\boldsymbol{\beta}}-R^2)$$

whose solution is given by

$$\hat{\boldsymbol{\beta}}=(X^TX-\lambda I)^{-1}X^T\mathbf{y} \tag{4.163}$$

subject to the constraint

$$(\boldsymbol{\beta}^T\hat{\boldsymbol{\beta}}-R^2)=0 \tag{4.164}$$

Substitution of (4.163) into (4.164) shows that the Marquardt parameter λ used by ridge regression is equivalent to the vector norm R^2 in (4.164). In fact, using (4.164) we could add R to our ridge trace plots to consider tradeoffs in the residuals and the stability of R.

This parametric approach to multiple objective optimization is that suggested by Umland and Smith [42] and Chow [13] in the last section, where, as in the ridge trace, we consider variations in two response functions with each other. This procedure is generalized by Myer and Carter for quadratic regression models in [35] for systems with two responses. For problems with more than two responses, the analysis is more difficult since the number of parameters increases. For four and more responses more than one plot of responses is necessary and the comparison of solutions is no longer direct.

Goal Programming

With threshold approaches we transform a problem in several responses to a constrained optimization of one response subject to minimum achievements on the remaining response. It falls to the analyst to balance trial constraints to achieve a solution with acceptable tradeoffs in goals. An alternative strategy is employed by goal programming which also depends upon the setting of goals, which correspond to the thresholds just introduced. In this approach deviations from these goals according to an ordinal ranking of preemptive priorities. The original method was developed for linear programming [12], but the development here will be generalized to nonlinear problems. Examples of the method and developments are given in [17, 30].

Let us define the positive and negative deviations d_i^+ and d_i^- of a response y_i from its goal b_i as

$$y_i(\mathbf{x}) + d_i^- - d_i^+ = b_i \tag{4.165}$$

such that

$$d_i^- \geqslant 0 \qquad d_i^+ \geqslant 0$$

Our optimization criteria will depend upon the type of goal we set. For equality attainment our goal is to minimize the sum of deviations, positive and negative,

$$\min d_i^+ + d_i^- \tag{4.166}$$

For minimum attainment of a goal only d_i^- is minimized, since any excess attainment satisfies our minimum. Similarly, for limits on the maximum, we minimize any positive deviations.

Deviations of the type considered in (4.166) correspond to the robust or L_1 regression problem of the form

$$\min |X\boldsymbol{\beta} - \mathbf{y}|$$

which we could write in terms of goal programming as

$$\min \mathbf{1}^T\mathbf{d}^+ + \mathbf{1}^T\mathbf{d}^- = \sum_{i=1}^{n} (d_i^+ + d_i^-) \tag{4.167}$$

such that

$$X\boldsymbol{\beta} + \mathbf{d}^- - \mathbf{d}^+ = \mathbf{y}$$

In the regression problem it is clear that the deviations d_i^+, d_i^- can be considered as a group as in (4.167). However, for incommensurate response functions the deviations will also be incommensurate. Goal programming handles this via preemptive priority classes $P_j, j = 1,\ldots,m$, which may not be commensurate between classes but only within classes. In this formulation all goals in the first class, P_1, are considered first and the optimal solutions obtained. These values are fixed and the next priority class is optimized to compete for the slack left after the higher class has been solved. Because each class is considered separately, it is only within the class that deviations must be commensurate.

We can illustrate the goal programming procedure using the auto design example. Here let us say that fuel economy is our preeminent goal, with emissions limited both by a maximum (15 mg/mi) and targeted to a particular value (12). We would like the design to cost less than $4200. If cost is a greater priority than emissions, we can formulate our problem as

$$\max P_1(d_1^+) + P_2(d_3^-) - P_3(d_2^+ + d_2^-)$$

such that

$$y_1(\mathbf{x}) - d_1^+ = 12.5$$

$$y_2(\mathbf{x}) + d_2^- - d_2^+ = 12$$

$$y_2(\mathbf{x}) \leqslant 15$$

$$y_3(\mathbf{x}) + d_3^- = 4200$$

$$d_1^+ \geqslant 0 \qquad d_2^- \geqslant 0 \qquad d_2^+ \geqslant 0 \qquad d_3^- \geqslant 0$$

The first constraint replaces the old constraint on the minimum mileage since d_1^+ is restricted to nonnegative values. The third constraint could be replaced by a constraint on $d_2^+, d_2^+ \leqslant 3$.

This preemptive priority ranking is an extreme case from the approach of weighted optimization, since incommensurates are now considered completely independently. However, weighting can be included within priority classes for responses thought to be of commensurate utility after weighting.

Geoffrion-Dyer Algorithm

Let us consider the unconstrained multiple response problem given by (4.159),

$$\max U(\mathbf{y}(\mathbf{x}))$$

If the functions $y_i(\mathbf{x})$ and $U(\mathbf{y})$ were known, we could proceed with any of the techniques suggested earlier, such as steepest ascent, using the gradient $\nabla_x U$ of the utility function. We generally know or can estimate the $y_i(\mathbf{x})$ so we can calculate $\nabla_x U$ according to the chain rule

$$\nabla_x U = \sum_{i=1}^{n} \frac{\partial U}{\partial y_i} \nabla_x y_i(\mathbf{x}) \tag{4.168}$$

Our problem reduces to discovering the n derivitives

$$\frac{\partial U}{\partial y_i}, \qquad i=1,\ldots,n$$

One estimation procedure is discussed by Montgomery and Bettencourt [32] as an interactive search technique. Noting that

$$\left(\frac{\partial U}{\partial y_1} \frac{\partial U}{\partial y_2}, \ldots, \frac{\partial U}{\partial y_n} \right)$$

is colinear with

$$\left[1, \left(\frac{\partial U / \partial y_2}{\partial U / \partial y_1} \right), \ldots, \left(\frac{\partial U / \partial y_n}{\partial U / \partial y_1} \right) \right] \tag{4.169}$$

The problem reduces to finding the $n-1$ marginal rates of substitution of f_i with f_1 given by the $n-1$ terms of (4.169). The program confronts the analyst with various perturbations in the responses to determine these approximate rate of substitution. These values can then be used with (4.168) to calculate the gradient of the utility function for a line search. The analyst is then provided with these results to choose a new base point, where the search can either be terminated or a new search direction estimated as before.

This algorithm differs from those given earlier in that no explicit function is assigned to the utility function and there are no effective limits on the number of responses that can be considered. Significant control is left to the analyst in picking starting points in the problem and handling ridges. An approach to the constrained problem is also discussed.

"Nondominated" Vector Optimal Solutions

The techniques considered so far have attempted to reduce, in various ways, multiobjective optimization problems to the standard optimization methodologies of optimizing a single response subject to constraints. This is essentially a tradeoff approach with the implication of some sort of commensurability. As indicated by the discussion of Zeleny [46], there is another approach of simultaneous optimization of all functions such that no solution exists in which all components are better than those at the optimum. Such a solution is called "nondominated," "pareto optimum," or an "efficient set." Optimization of this type does not require commensurability. Algorithmic development in the linear case is discussed in Zeleny with some notes on the nonlinear case.

REFERENCES

1. Abadie, J., and J. Carpentier, "Generalization of the Wolfe Reduced Gradient Method to the Case of Nonlinear Constraints," *Optimization*, R. Fletcher (Ed.), Academic Press, London, 1969, pp. 37–47.
2. Allran, R. R., and S. E. J. Johnson, "An Algorithm for Solving Nonlinear Programming Problems Subject to Nonlinear Inequality Constraints," *The Computer Journal*, vol. 13, no. 2, 1970, pp. 171–177.
3. Bard, Y., *Nonlinear Parameter Estimation*, Academic Press, New York, 1974.
4. Bard, Y., and L. Lapidus, "Kinetics Analysis by Digital Parameter Estimation," *Catalysis Reviews*, vol. 2, no. 1, 1968, pp. 67–112.
5. Bettencourt, V. M., Jr., "An Application of Multiple Response Surface Optimization to the Analysis of Training Effects in Operational Test and Evaluation," unpublished Master's Thesis, School of Industrial and Systems Engineering, Georgia Institute of Technology, 1975.
6. Beveridge, G. S. G., and R. S. Schechter, *Optimization: Theory and Practice*, McGraw-Hill, New York, 1970.
7. Biles, W. E., "A Response Surface Method for Experimental Optimization of Multi-Response Processes," *Industrial and Engineering Chemistry, Process Design and Development*, vol. 14, no. 2, 1975, pp. 152–158.
8. Birta, L. G., "A Parameter Optimization Module for CSSL-based Simulation Software," *Simulation*, vol. 28, no. 4, 1977, pp. 113–121.
9. Box, G. E. P., and K. B. Wilson, "On the Experimental Attainment of Optimum Conditions," *Journal of the Royal Statistical Society*, vol. 13, no. 1, 1951, pp. 1–38.
10. Box, M. J., "A New Method of Constrained Optimization and a Comparison with Other Methods," *Computer Journal*, vol. 8, 1965.
11. Carroll, C. W., "The Created Response Surface Technique for Optimizing Nonlinear Restrained Systems," *Operations Research*, vol. 9, 1961, pp. 169–185.
12. Charnes, A., and W. W. Cooper, *Management Models and Industrial Applications of Linear Programming*, John Wiley and Sons, New York, 1967.

13. Chow, W. W., "A Note on the Calculation of Certain Constrained Maxima," *Technometrics*, Vol. 1, No. 3, 1959.
14. Curtis, P. C., *Multivariate Calculus with Linear Algebra*, John Wiley and Sons, New York, 1972.
15. Dorfman, R., "Steepest Ascent Under Constraints," *Symposium for Simulation Models*, A. C. Hoggart and F. E. Balderson, Eds., Southwestern Publishing Co., Cincinnati, OH, 1963.
16. Draper, N. R., "'Ridge Analysis' of Response Surfaces," *Technometrics*, vol. 5, no. 4, 1963, pp. 469–479.
17. Dyer, J. S., "Interactive Goal Programming," *Management Science*, vol. 19, no. 1, 1973.
18. Fiacco, A. V., and G. P. McCormick, *Nonlinear Programming: Sequential Unconstrained Minimization Techniques*, John Wiley and Sons, New York, 1968.
19. Fletcher, R., and C. M. Reeves, "Function Minimization by Conjugate Gradients," *Computer Journal*, vol. 7, 1964, pp. 149–154.
20. Gaddy, J. L., and L. D. Gaines, "An Examination of the Adaptive Random Search Technique," *AICHE Journal Chemical Engineering Research and Development*, vol. 22, no. 4, 1976.
21. Gottfried, B. S., and J. Weisman, *Introduction to Optimization Theory*, Prentice-Hall, Englewood Cliffs, NJ, 1973.
22. Goulcher, R., and J. J. C. Long, "The Solution of Steady-State Chemical Engineering Optimisation Problems Using a Random-Search Algorithm," *Computers and Chemical Engineering*, vol. 2, 1978, pp. 33–36.
23. Hooke, R., and T. A. Jeeves, "Direct Search Solution of Numerical and Statistical Problems," *Journal of the ACM*, Vol. 18, 1961.
24. Ignizio, J. P., *Goal Programming and Extensions*, Lexington Books, D. C. Heath and Co., Lexington, MA, 1976.
25. Klingman, W. R., and D. M. Himmelblau, "Nonlinear Programming with the Aid of a Multiple Gradient Summation Technique," *Journal of the ACM*, vol. 11, 1964.
26. Kunzi, H. P., W. Krelle, and W. Oettli, *Nonlinear Programming*, trans. by Frank Levin, Blaisdell Publishing Co., Waltham, Mass., 1966.
27. Kuester, J. L., and J. H. Mize, *Optimization Techniques with Fortran*, McGraw-Hill, New York, 1973.
28. Law, V. J., and R. H. Fariss, "Transformational Discrimination for Unconstrained Optimization," *Industrial and Engineering Chemical Fundamentals*, vol. 11, no. 2, 1972, pp. 154–161.
29. Lawson, C. L., and R. J. Hanson, *Solving Least Squares Problems*, Prentice-Hall, Englewood Cliffs, NJ, 1974.
30. Lee, S. M., *Goal Programming for Decision Analysis*, Auerbach, Philadelphia, PA, 1972.
31. Luenberger, D. G., *Introduction to Linear and Nonlinear Programming*, Addison-Wesley, New York, 1973.
32. Montgomery, D. C., and V. M. Bettencourt, "Multiple Response Surface Methods in Computer Simulation," *Simulation*, vol. 29, no. 4, 1977, pp. 113–121.
33. Mugele, R. A., "A Program for Optimal Control of Nonlinear Processes," *IBM Systems Journal*, vol. 2, Sept. 1962.
34. Myers, R. H., *Response Surface Methodology*, distributed by Edwards Brothers, Inc., Ann Arbor, MI, 1976.

35. Myers, R. H., and W. H. Carter, Jr., "Response Surface Techniques for Dual Response Systems," *Technometrics*, vol. 15, no. 2, 1973, pp. 301–317.
36. Nelder, J. A., and R. Mead, "A Simplex Method for Function Minimization," *Computer Journal*, vol. 7, 1964, pp. 308–313.
37. Noh, J. C., "A Two Phase Complex Method for Nonlinear Process Optimization," presented to the 45th National ORSA/TIMS Meeting, Boston, MA, 1974.
38. Pavianni, D. A., and D. M. Himmelblau, "Constrained Nonlinear Optimization by Heuristic Programming," *Operations Research*, vol. 17, no. 5, 1969.
39. Pierre, D. A., and C. H. Dudding, "Constrained Optimization with Gradient Approximations," *IEEE Transactions on Systems, Man and Cybernetics*, vol. 7, no. 2, 1977, pp. 112–116.
40. Pierre, D. A., and M. J. Lowe, *Mathematical Programming Via Augmented Lagrangians*, Addison-Wesley, Reading, MA, 1975.
41. Rosen, J., "The Gradient Projection Method of Nonlinear Programming, I. Linear Constraints," *Journal of the Society of Industrial and Applied Mathematics*, vol. 8, 1960, pp. 181–217; "II. Nonlinear Constraints," *J. Soc. Indust. Appl. Math.*, vol. 9, 1961, pp. 514–532.
42. Seinfeld, J. H., and L. Lapidus, *Mathematical Methods in Chemical Engineering, Vol. 3, Process Modeling, Estimation and Identification*, Prentice-Hall, Englewood Cliffs, NJ, 1974.
43. Spendley, W., G. R. Hext, and R. F. Himsworth, "Sequential Application of Simplex Designs in Optimization and Evolutionary Operations," *Technometrics*, vol. 4, 1962.
44. Umland, A. W., and W. N. Smith, "The Use of Lagrange Multipliers with Response Surfaces," *Technometrics*, vol. 1, no. 3, 1959.
45. Wilde, D. J., and C. S. Beightler, *Foundations of Optimization*, Prentice-Hall, Englewood Cliffs, NJ, 1967.
46. Wolfe, P., "Methods of Nonlinear Programming," Chaper 6, *Nonlinear Programming*, J. Abadie, Ed., North-Holland Publishing Co., Amsterdam, 1970.
47. Zoutendijk, G., *Methods of Feasible Directions*, American Elsevier, New York, 1960.

Chapter 5

Optimization via Experimentation

OVERVIEW OF THE EXPERIMENTAL OPTIMIZATION PROBLEM

We have considered the areas of statistics and statistical inference, experimental design, model building, and optimization as isolated topics so far. Even then it has been apparent that there are essential interrelationships among the topics of statistics, models, and optimization. The idea of a *model* underlies our statistical tests and our optimization procedures. The statistics of experimental design and model estimation are linked in turn to some optimality criteria of minimum error or bias or a "best" fit. In turn, the rate of convergence and the precision of our optimization will be determined by our statistical estimates. These three areas of study and their interrelation will be considered in this chapter.

We consider first the role and ubiquity of *models* in our procedures. Our statistical procedures include assumptions about errors, for instance, the additivity of error, an error distribution, homogeniety of error variances, independence of observations, and so on. These form the statistical models upon which our results are based. For experimental designs models about error are used with additional elements about local approximations to the unknown system—itself a model of a real system—from which properties of design and the criteria for optimal designs are obtained. At times we even considered the effect of incorrect models in our errors or in our approximation by substituting alternative models. We may advance to explicit models about processes and even make statements about the relative merits of competing models.

Some of our models are indeed quite simple. We often assume unimodality for functions being optimized by direct search. We do this to design efficient optimization when this model about the function is true. If the model fails, we are not much worse off than before, and when it is true, we make rapid progress. Or, we can pick another procedure less sensitive to multimodality.

Our first point, therefore, is this: models underlie the statistical and optimization methods. Their choice has practical significance to the success of our work. This is true in general, but especially when our results are obscured by experimental uncertainties. Until some universal model appears that can isolate a model from statistics by optimization, our efforts in the several areas will be linked together.

We have in earlier chapters considered some of the implications of our statistical models. We have shown how experiments may be designed to minimize variances of results or in model interactions, and to test for significant deviations from our model assumptions. Tests on lack-of-fit and examinations of residuals were introduced to these model transformations and even regression methods based upon alternative distributions of errors beside the multivariate normal.

The success of our optimization depends upon the precision and the type of our statistical estimation. It is a basic point, noted in Box and Hunter [12] but worth repeating, that part of our effort should focus upon the implications of the representational model to design the experiment. If the data is poorly planned, then no amount of inspired effort will resurrect what has been lost. The purpose of the design is to contrast effects one from another as definitely as possible, to reduce the threshold of detectable effects while diminishing the possibility of mistaking one effect for another, or even random results for a real effect.

Our statistics also guide us in the methods of data analysis and in their weaknesses. Box [10] thus discusses the effect of errors in the independent variables, and Mezaki et al. [23] considers the biases inherent in the estimation based upon independent rather than dependent variables. In Box et al. [13], the implications of faulty estimation are considered and reduced, and "honest", estimation procedures are substituted instead.

We note again in passing that most statistical estimation problems are in fact optimization problems. Indeed, they are most often numerical problems as well. The implications of numerical procedures were touched upon in Chapter 3 with respect to linear estimation and problems arising from poor scaling and other types of poorly determined systems. The places for numerical difficulty are multiplied for nonlinear estimation, and many optimization procedures are advanced to answer them. The proper optimization procedure will therefore depend upon the structure of the estimation problem, the presence or absence of constraints, and so forth.

Besides choosing our optimization methodologies for a particular problem, we sometimes pick certain representations so that particular optimization methods can be used. For instance, with respect to plant designs,

Klimpel and Blau [20] divide the general scope of optimization into the "macro" and "micro" levels. The plant level or macro optimization concerns estimations of a necessarily aggregate nature, favoring simple linear models and linear or mixed linear and integer programming. It is likely that more sophisticated models and optimizations could be supported, although perhaps at too great an expense. Certain important subsystems are treated to further study in the micro level optimization. Here models and procedures to optimize them are likely to be more sophisticated. The greater sophistication in the model makes more powerful optimization necessary, whereas the greater precision in the model lends credence to the more precise estimate of the optimum.

What we see in experimental optimization is that there is a closed loop of interaction between model, statistical tratment, and optimization. The art and science of experimental optimization consists of the combination of these related items into a cyclic strategy of *hypothesis*, *estimation*, and *optimization*. We turn our attention to this general problem.

We pointed out earlier that experimental design presupposes some knowledge about the experimental system, since failing that a true design cannot be constructed. When our knowledge is minimal we turn to screening experiments. These produce estimates of the most qualitative sort, perhaps only groups of factors that contain at least one of importance or very rough approximations of the actual level of both factor responses and factor levels. Even so, it may be possible to estimate a very rough gradient and begin moving towards an optimum.

In later cycles information acquired from earlier cycles is incorporated into hypotheses about the system and the model, and design levels and the structure of the model is altered to reflect that. If adequate progress is made, or constraints are encountered, the modeling may be changed from first- to second-order approximation, or even to a theorectical model of some sort. As this occurs the optimization problem alters and more advanced algorithms selected.

At the close of the study experiments may also be run to confirm the optimum and to test the sensitivity of the solution to variations in the inputs. These *post-optimal* analyses may reveal the presence of false optima, moving optima, stationary ridges, or even relatively flat optima. In operating plants these analyses may be repeated at intervals to track moving optima.

Our general strategy of hypothesis, estimation, and optimization will further explored throughout this chapter. We distinguish three general situations:

1. Response surface methods: Unknown structural models with stationary or slowly varying optima.

2. Structural models with stationary or slowly varying optima.
3. Evolutionary systems whose optima tend to drift with time.

Strategies for the experimental optimization of these general structures will be considered in turn.

Response Surface Methods

The first experimental situation that we address is response optimization in the absence of theoretical models. Instead of theoretical models we fit interpolating polynomials as local approximations to the function in the region of interest and use these approximations as the basis of our optimization. In the region of interest we fit models of first or second order, depending upon the shape of the response surface and other experimental considerations. Our choice of optimization methodology will be dependent upon the order of the model and the objective of the optimization.

Most of the tools we use have been developed already in this text. Our purpose here is to clarify their application in the experimental realm.

Most of the techniques to be considered are not new. Many of them are loosely associated together as Response Surface Methodologies, the general title given to the optimization of responses via approximating functions. Perhaps the earliest synthesis of these techniques came with the paper of Box and Wilson [14], "On the Experimental Attainment of Optimum Conditions," wherein statistical tools of function estimation were derived that were especially suited to the methods of steepest ascent and the canonical characterization of second-order surfaces. In much of the work the emphasis is on the choice of suitable models and upon suitably efficient experimental designs for their estimation. The standard 2^k factorial designs, in particular, were augmented as Central Composite Designs to give designs that were rotatable and either uniform precision or orthogonal in the model estimates. Much of this development was assembled as *Response Surface Methodology* by Myers [28].

Response Surface Methodology has been primarily focused upon the unconstrained optimization problem. More recent work by Biles [5–7] and Swain [33] has incorporated standard nonlinear programming techniques into framework of response surface methods.

We consider two practical classifications among response surface methods in evaluating candidate tools in particular problems. With respect to experimentation we distinguish between sequential and simultaneous experimentation. In the former case later experiments are based upon the results of previous experiments. Simultaneous experiments are run without this effect. However, simultaneous experiments may represent a time constraint (i.e., several long-running experiments can be run in the time of a single experiment) or physical constraints, as for systems that run in batches. Many of the techniques we see contain elements of both approaches, such as the sequential block approach, a sequence of essentially simultaneous blocks of experiments.

Similarly, we will classify optimization techniques as being "on" or "off" line. The "on-line" optimizers are applied directly to the system in determining the optimum. Most such techniques are variations on steepest ascent where the optimization stage is a line search and progress can be observed directly from experimentation. In contrast, "off-line" techniques perform optimization upon the approximating model of the system. Because error is filtered out preliminary to optimization, a wider range of techniques can be applied to the "off-line" problem. The division in this classification is not hard, although in general on-line experiments will be first order, and those for the off-line case of second order.

Preliminary: Analysis and Screening. It is often the case that our knowledge about a system is limited. We begin, therefore, with a careful consideration of what our objective is and the resources we have available to meet these goals. In the preliminary analysis we attempt to identify the type of problem (constrained or unconstrained, single or multiple objective, etc.), the magnitude of error, the factors likely to contribute in determining the response, and perhaps even to estimate our payoff. At this stage it is still difficult to gauge our chances of success, although it is possible to estimate to an order of magnitude (at least) the likely amount of experimentation necessary to achieve our goal. We can do this based upon the choice of design and our "feeling" for how difficult the response surface will be to optimize.

These preliminary considerations are quite essential if there are severe time or budget constraints upon experimentation. To make the project feasible it may be necessary to simplify models, to eliminate marginal factors, or change the goals of the analysis.

Very likely there will remain uncertainties in the analysis that can be dealt with via a screening experiment. The purpose of a screening experiment is more than the generation of data, it is a testing of the assumptions of analysis and a clarification of matters only roughly estimated. Even early experimentation should have definite goals in mind so that they can be properly designed.

In many cases we want the screening experiment to shed light on several areas. First, we would like to determine which factors influence the response the most. To simplify experiments later, we attempt to identify the significant factors in the model so that those remaining can be eliminated. Second, we would like to know what the approximate level of the factor response is. This will aid us in scaling the model so that causal response can be distinguished from random error. A scaling that is too small or too big can significantly reduce the quality of model estimates. In addition, many optimization techniques depend upon scaling for their convergence rate. A third objective for experimentation is the characterization of the response surface. Is it first order, or are second (or higher) order terms significant?

We obtain from the screening experiment an estimate of experimental error, which can be refined with the replication of one or several points. This estimate can be used to estimate the precision of *proposed* experiments. This estimate will be important in the choice of an experimental design or in revising our expectation for convergence to a solution.

We may also use a screening experiment to choose an initial starting point or even to discover a feasible one if that is unknown.

The screening experiment can have important bearings on the planning of a response surface strategy and its ultimate success. The screening experiment is most likely to be useful when it is preceded by analysis and planning. There is no method of obtaining reliable information from a poorly planned experiment.

Steepest Ascent. The standard approach in unconstrained optimization far from an optimum is steepest ascent or one of its many variations. Its appeal derives from its simplicity and its general effectiveness in many situations.

Steepest ascent is a sequential block approach consisting of a sequence of gradient estimating and line search experimental blocks. The gradient estimate is essentially a simultaneous block since the estimates desired are averaged responses, dependent upon the aggregate rather than the individual effect of the responses. The line search can be either simultaneous or sequential in execution.

The choice of steepest ascent is not arbitrary; its utility lies in several features. First, convergence of gradient techniques is guaranteed. It is likely to be quite effective far from an optimum, although slow near a stationary point, where the technique is avoided. Second, the gradient is easy to estimate. A linear model is all that is required, and if provision for lack-of-fit is included, curvature and the likelihood of a stationary point can be detected. A third factor is that gradient search based on a good experimental design is more efficient than direct search techniques in generating the ascent direction. Many search techniques are error tolerant, but point for point none will be as effective unless the actual system is of high order and the linear model badly biased.

With respect to the gradient portion of a steepest ascent sequence, our only design consideration is in the choice of an experimental design. Now, this part of the sequence results in an estimate of the ascent direction, but the experiments themselves do not lead to an improved solution directly. The question becomes a tradeoff between precision of the estimate and the expected gain per experiment. However, extra experiments also buy us some protection from curvature and bias via lack-of-fit calculations.

Under the assumption of normally distributed error and no bias, Brooks and Mickey [15] show that the expected improvement per point is maximized with designs of minimum size, i.e., with orthogonal simplex designs. That is, the gain in precisely estimating the gradient is more than offset by the additional points needed to obtain the estimate. This indicates the relative robustness of gradient techniques, since the search will be effective even when the precise gradient cannot be obtained.

The choice of a first-order experimental design is therefore a matter of judgement. With respect to simplex designs, only indirect evidence can be obtained about the presence of curvature, since simplex designs are completely saturated. The most common sign of problems is the failure of a line search. The cautious designer will therefore consider slower, but less risky procedures, such as factorial or fractional factorial designs, or at least add a few center points. One may also choose to alternate saturated and unsaturated designs. Sometimes significant curvature will be indicated from the line search as well.

Line Searches. In sequential block approaches, including gradient searches, the line search is used to move from one base point to another one where the response is better. Our approach to this problem can be either sequential or simultaneous. As with the gradient estimating block of experiments, the overall goal is pointwise efficiency in estimation rather than an absolute precision of estimation.

The dichotomous, Fibonacci, and golden section searches are all candidate direct search algorithms for sequential experimentation along the line search direction. Use of these methods implies no penalty for sequential experimentation, unimodality in the response, and limited error in the observation of the response. It is possible that a large error could result in the elimination of the actual interval of uncertainty. This would soon become evident, but at an expenditure of unnecessary effort. These techniques are illustrated in the next chapter. Results from searches can be checked by fitting a polynomial to the responses observed during the course of the search. Many of the points will be clustered around the optimum so that such an estimate will be fairly accurate.

Simultaneous methods such as curve fitting responses to predetermined experimental points are most commonly employed for line searches. The response is not required to be unimodal. When constant intervals between points are used, orthogonal polynomials can be fitted. The prime attraction to orthogonal polynomials is their relative ease of calculation and the ease of statistically assessing the contribution of each factor. The polynomials themselves, with their implicit assumption of continuous derivatives, tend to act as an error filter.

When fitting polynomials to line search data, the emphasis is upon estimating the optimum step rather than providing an interpolating polynomial over the entire range of the data. As a result, lower order models, which maintain the essential features of the response, are preferred to those of higher order. It is only for obviously higher order or multimodal responses that a higher order polynomial is preferred.

Linear Techniques and Constraints. The basic devices of steepest ascent, gradient estimation, and "optimal step" solving line searches, are applicable to solving constrained optimization problems. These are multiple response problems in which a primary response and several secondary constraint responses are measured and gradients for each are calculated. In the interior of the feasible region the standard unconstrained search is conducted, although the optimal step may now terminate at a constraint boundary rather than at the optimum along the search direction. At the boundary the gradient of the primary response is altered to feasible direction via multiple gradient summation, gradient projection, or the best feasible direction algorithms of Zoutendijk. The line search is conducted along this deflected search direction thereafter.

It should be noted that penalty and barrier methods are not appropriate methods for direct application to experimental data as they tend to inflate the variance of random error. To illustrate, consider the simple exterior penalty function cy, $c>0$ for $y>0$. The measured response y includes random error of variance σ^2. The variance of the penalized variable is $c^2\sigma^2$, and as c increases, the quality of estimates based upon cy decreases quite rapidly. Penalty methods can only be used with uncontaminated responses or approximate responses.

Several general problems can occur with linear constraint procedures such as MGS, gradient projection, and best feasible directions. In the first place, linear approximation can be misleading. Consider the case in Figure 5.1(a). At the constrained point the two gradients appear to be colinear, indicating a constrained optimum. The tangent to both of these gradients suggests that no further improvement is possible. This difficulty may be detected by a suitable post optimal analysis containing terms of second

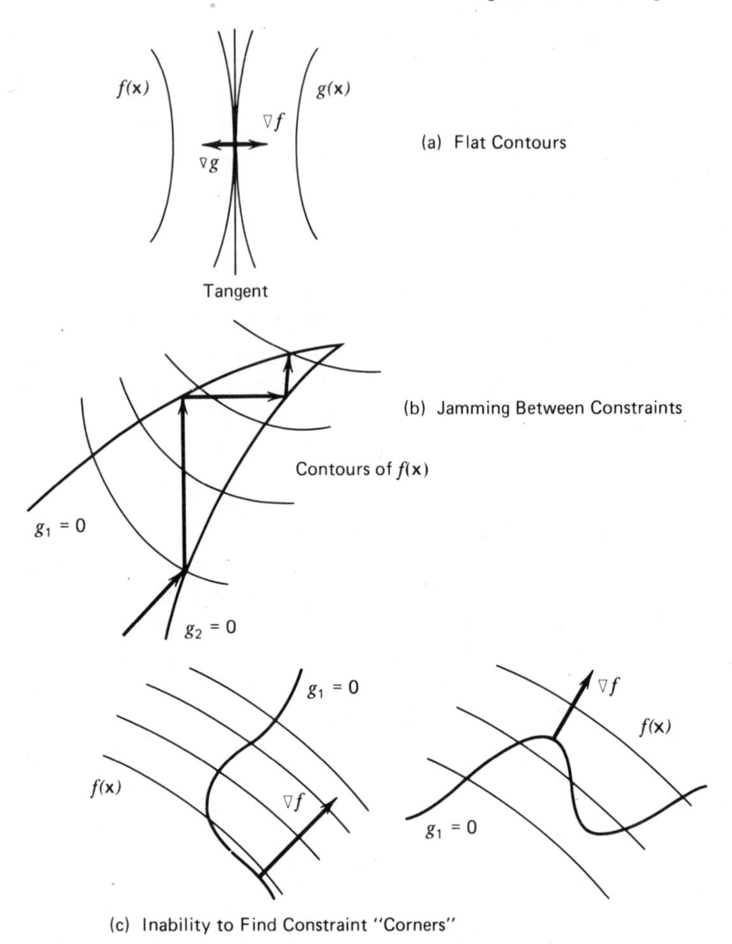

Figure 5.1. Situations where linear techniques fail.

order in it. In the second case, seen in Figure 5.1(b), convergence is slowed because of jamming between two constraints. Convergence in this situation might be speeded by a second-order model of both constraints and the response.

The last two figures, seen in Figure 5.1(c), show how linear approximations may fail through their inability to sense the "corners" in the constraint which promise increased improvement in the primary. Corners might be detected by second-order methods, especially in the first case. The second case has a cubic constraint that may or may not be observed, depending upon the size of the experiment.

Second-Order Designs. Near unconstrained stationary points and tightly constrained base points a second-order model can be necessary for proper analysis and efficient experimentation.

In the unconstrained case the base point is near enough to the stationary point for second-order terms to be significant. Their estimation provides both a better approximation of the response and of the optimal point. The same model can be used in the post optimal analysis using the canonical analysis of the response surface. This provides estimates of the shape and canonical sensitivities at the optimum.

At constrained points we use second-order models for off-line optimization, which can more thoroughly assess the possibility of multiple constrained optima. The additional experimentation is justified when the possibility exists for several jammed line searches in a tightly constrained region.

In this chapter we consider various approaches by which the experimental optimization problem can be formulated and solved. We consider the general case in which we have multiple independent variables x_i, $i=1,\dots,n$ and multiple experimental responses y_j, $j=1,\dots,m$. As discussed in previous chapters, the measured responses y_j, $j=1,\dots,m$ are observed values of the true but unknown response functions η_j, $j=1,\dots,m$. Thus in the response surface methodology context, we are dealing with m separate hypersurfaces each having its own characteristic variance structure. Our approach in this chapter is to outline several broad experimental optimization procedures that could be applied to a problem regardless of the formulation or precise experimental setting. In Chapter 6 we examine several realistic example problems in the realm of physical experimentation. Then in Chapter 7 we see the same general approaches extended to simulated systems.

In the remaining sections of this chapter we concentrate on describing various experimental optimization procedures and on comparing the performance of selected experimental design approaches as well as certain constrained optimization procedures. This chapter certainly does not exhaust the many possible approaches one might take in experimental optimization but attempts to offer a general methodology and discuss some of the most important issues involved in that methodology.

PROBLEM FORMULATIONS

We consider two basic approaches to formulating the problem of optimizing a multiple-variable, multiple-response experiment. One approach is the familiar constrained optimization formulation in which one of the system responses, say η_1, is to be maximized or minimized, subject to maintaining the remaining $m-1$ responses within prescribed bounds. The second approach is the multiple-objective formulation, in which the m responses are either weighted to form a single objective or treated in a manner akin to goal programming. Each of these two formulations is described in the following sections.

Constrained Optimization

Under the constrained optimization approach, the problem is stated as

$$\max \text{ (or min) } \eta_1 = g_1(x_1,\ldots,x_n) \tag{5.1}$$

subject to the constraints

$$a_i \leqslant x_i \leqslant c_i, \; i=1,\ldots,n \tag{5.2}$$

$$\eta_j = g_j(x_1,\ldots,x_n) \left\{ \begin{matrix} \geqslant \\ = \\ \leqslant \end{matrix} \right\} d_j, j=2,\ldots,m \tag{5.3}$$

Indeed, any given response η_j could be required to satisfy both upper and lower bounds, or

$$d_j^l \leqslant \eta_j \leqslant d_j^u \tag{5.4}$$

The constraints expressed in (5.2) are bounds on the controllable input variables $x_1,\ldots,x_n$ and are typically known *a priori*. These bounds generally form the known experimental region prior to conducting experimentation. In contrast to that, the response functions $g_j(x_1,\ldots,x_n)$ in (5.4) are not usually known *a priori* and hence the responses η_j must be estimated experimentally. Thus, experiments performed at points satisfying (5.2) may yield responses violating (5.4).

To complicate matters even more, the random error ε_j can lead to erroneous decisions relative to the constraints in (5.4), leading the experimenter to believe that a given response is feasible when it is not, or vice versa. The same is true relative to the objective function in (5.1). That is, one experiment can appear to represent an improvement over another when the true response at this particular set of values $x_1,\ldots,x_n$ does not.

Multiple-Objective Optimization

One approach to a multiple-objective formulation is to assign weights w_j, $j=1,\ldots,m$ to the m responses and form a single objective function

$$\max \text{ (or min) } W = \sum_{j=1}^{m} w_j g_j(x_1,\ldots,x_n) \tag{5.5}$$

The bounds (5.2) still apply, so that the problem remains one of constrained optimization, but one in which the entire feasible region is known *a priori*. The weights w_j, $j=1,\ldots,m$ are typically assigned through the subjective judgment of the experimenter. These weights are usually normalized, so

$$\sum_{j=1}^{m} w_j = 1 \tag{5.6}$$

One frequently encounters the situation in which certain of the responses η_j are to be maximized and others minimized. This case is handled by maximizing the negative of those functions that are to be minimized, so that the objective function in (5.5) is rearranged to the form

$$\max W = \sum_{j=1}^{s} w_j g_j(x_1,\ldots,x_n) - \sum_{j=s+1}^{m} w_j g_j(x_1,\ldots,x_n) \tag{5.7}$$

where s functions are maximized and $m-s$ functions are minimized.

A second approach to the multiple-objective formulation is one which casts the problems in the format

$$\max U[\, g_1(x_1,\ldots,x_n),\ldots,g_m(x_1,\ldots,x_n)] \tag{5.8}$$

subject to the bounds in (5.2). The formulation in (5.7) is a special case of that in (5.8), in which $U[g_j(x_1,\ldots,x_n)]$ is a linear additive function. Montgomery and Bettencourt [25] discuss various formulations of the multiple objective optimization problem, as well as several approaches to its solution, and demonstrate its application to multiple-response simulation.

Another multiple-objective optimization formulation is that called goal programming. This procedure was discussed in Chapter 4.

OPTIMIZATION TECHNIQUES

Various procedures have been applied in combining optimization and experimentation to seek the "optimum" solution to systems possessing a single response η. The multiple-response problem described here is complicated by the necessity to observe several responses at once, and to incorporate these values into the optimization technique. But many of the same techniques that have been applied successfully to the single-response problem can, with appropriate modifications, be extended to accommodate multiple responses. Moreover, these modified procedures are often applicable to more than one of the aforementioned formulations of the multiple-response problem.

The optimization procedures described below fall into three categories: (1) direct search techniques, (2) first-order response surface methods, and (3) second-order response surface procedures. Although numerous techniques are cited, only a few broadly stated procedures are outlined here. It should be remembered that, although we may refer to "optimization" techniques, the classical notion of an "optimum" solution is inapplicable due to the presence of the sampling error ε_j associated with each response variable η_j. Rather we shall seek a solution that hopefully lies close to the true solution. But the potential for having the estimated solution $(\hat{\mathbf{x}}, \hat{\mathbf{y}})$ differ somewhat from the true solution $(\mathbf{x}^*, \mathbf{y}^*)$ is not the only effect that arises due to the randomness inherent in experimentation. The entire progress of the optimization is affected. In a search procedure, for instance, both the estimated direction and the estimated step along this direction can be affected by random error. The combination of these errors causes the search path to differ from that which would be followed in the absence of error, thus reducing the efficiency of the search method. Sufficiently large error will, of course, render a search approach useless.

Direct Search Methods

Direct search methods are those that, applied in a purely computational manner, do not require the use of derivatives. These methods progress through a sequence of points according to some algorithm. Typical of this class of optimization techniques are the pattern search algorithm, sequential simplex search, and the so-called "complex" search method. These were discussed in Chapter 4. In general, these direct search procedures make rapid early progress toward an "optimum," but iterate laboriously as a solution is neared. This is particularly true in the presence of random error, as encountered in experimentation.

Among the direct search techniques, Box's "complex" method is most easily adapted to a multiple-response environment. It also performs better than any of the other direct search techniques in the face of random error and constraints. In fact, "complex" search is not at all complex, but derives its name from a contraction of the words "constrained simplex": it evolved from the sequential simplex method and the necessity to deal with constraints. Noh [29] has suggested a modification of Box's method, which makes it especially suitable for the multiple-response experimentation problem. The following procedure describes a generalized "complex" procedure as it might be applied to multiple-response experimentation.

1. Randomly generate a set of $N \geqslant n+2$ search points $\mathbf{x}^1, \ldots, \mathbf{x}^N$ satisfying the known bounds (5.2).

2. Perform an experimental trial at each of these N search points and record the mN estimated responses y_j^l, $j=1,\ldots,m$, $l=1,\ldots,N$.
3. Where a given point $\mathbf{x}^k$ is observed to violate one or more constraints (5.4), if such constraints apply with the particular problem formulation being employed, generate a replacement search point $\mathbf{x}^{k'}$, perform an experimental trial at $\mathbf{x}^{k'}$, and record the m estimated responses at $\mathbf{x}^{k'}$.
4. After N feasible search points have been established, evaluate the objective function for each of these N points. This "objective function" might be η_1 in (5.1), W in (5.7), U in (5.8), or the "achievement" function in a goal programming formulation. Among these N search points, find the worse point $\mathbf{x}^w$; that is, the search point giving the least desirable value of the objective function. Define $\mathbf{x}^c$ as the centroid of the $N-1$ remaining points. Project from $\mathbf{x}^w$ through $\mathbf{x}^c$ to the image $\mathbf{x}^{w'}$. If the known bounds (5.2) are violated by this move, shorten the step to $\mathbf{x}^{w'}$ until no violation occurs. Perform an experiment at $\mathbf{x}^{w'}$.
5. Repeat steps 3 and 4 until a solution $(\hat{\mathbf{x}},\hat{\mathbf{y}})$ is obtained that represents the best solution that can be achieved within the available experimental

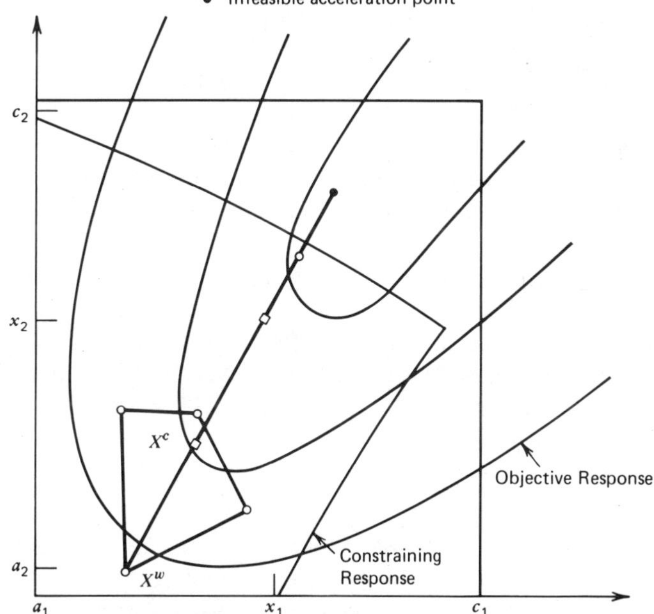

Figure 5.2. Box's "complex" search.

resources. Figure 5.2 illustrates complex search as applied to a constrained problem.

A significant advantage of complex search is that, once N feasible experiments (and perhaps several infeasible ones) have been performed, trials are conducted one at a time thereafter. The search can be continued as long as improved solutions are obtained. If several successive experiments are performed at scattered points around the known experimental region without achieving an improved solution, however, the search can be terminated and the best available solution adopted.

First-Order Response Surface Methods

First-order response surface methods attempt to accomplish experimentally what the "method of steepest ascent" accomplishes computationally. From a current point $\mathbf{x}^k$, a designed experiment is conducted (with a simulation trial at each design point) to estimate the gradient direction $\nabla g(\mathbf{x}^k)$. Experiments are then conducted at points along this direction to a new point $\mathbf{x}^{k+1}$ which represents the best solution obtained along the direction $\nabla g(\mathbf{x}^k)$. This process is an experimental approximation of

$$\mathbf{x}^{k+1} = \mathbf{x}^k + \lambda^k[\nabla g(\mathbf{x}^k)] \tag{5.9}$$

The step length λ^k can be estimated by a line search or by a regression procedure as described by Biles [6].

The gradient direction $\nabla g(\mathbf{x}^k)$ is estimated by placing an appropriate first-order experimental design, such as a 2^n factorial, 2^{n-p} fractional factorial, or n-dimensional simplex design [28] around the current point $\mathbf{x}^k$. An experiment is performed at each point in the selected experimental design. From these N observations the multiple linear regression model

$$\hat{y} = \hat{\beta}_0 + \sum_{i=1}^{n} \beta_i x_i \tag{5.10}$$

can be estimated. Since the gradient direction $\nabla g(\mathbf{x}^k)$ is mathematically defined as the n vector of first partial derivatives of $g(\mathbf{x})$ evaluated at $\mathbf{x}^k$, it is clear that $\nabla g(\mathbf{x}^k)$ is simply the n vector of regression coefficients, exclusive of the $\hat{\beta}_0$ term; that is,

$$\nabla g(\mathbf{x}^k) = (\hat{\beta}_1, \ldots, \hat{\beta}_n) \tag{5.11}$$

In the multiple-response experimental problem, an experiment is conducted at each design point in the selected first-order design and the m observations y_j^l, $j = 1, \ldots, m$ are recorded at each design point. Multiple linear regression is applied separately to each set of observations (assuming independence among the m responses), producing the m models

$$\hat{y}_j = \hat{\beta}_{j,0} + \sum_{i=1}^{n} \hat{\beta}_{j,i} x_i, \qquad j = 1,\ldots,m \tag{5.12}$$

and hence the m gradient vectors

$$\nabla g_j(X^k) = (\hat{\beta}_{j,1},\ldots,\hat{\beta}_{j,n}), \qquad j = 1,\ldots,m \tag{5.13}$$

These estimates can then be employed in any one of several optimization schemes to produce an improved solution $\mathbf{x}^{k+1}$.

Biles [6] has described a first-order response surface procedure for approaching the constrained formulation of the multiple-response process-optimization problem. This procedure involves performing a first-order design around a current point $\mathbf{x}^k$ to estimate the gradient direction $\nabla g(\mathbf{x}^k)$ according to relation (5.11). A line search is then performed along $\nabla g(\mathbf{x}^k)$ to estimate an optimal step λ in (5.9). As long as the search remains interior to the region bounded by the constraints (5.2) or (5.4), the procedure is basically the same as that proposed by Box and Wilson [14]. If one or more constraints (5.2) or (5.4) are encountered, however, Biles [6] proposes that the gradient projection direction be followed. The procedure for estimating the gradient projection direction is as follows.

Suppose that at an estimated boundary point $\mathbf{x}^k$, q constraints are satisfied as equalities. These can be either the (5.2) or (5.4) constraints, or both. Let B_q be the $n \times q$ matrix of first partial derivatives of these active constraints. Thus, B_q consists of the q gradient vectors $\nabla g_j(\mathbf{x}^k)$, $j = 1,\ldots,q$. That is

$$B_q = \begin{bmatrix} \dfrac{\partial g_1}{\partial x_1} & \cdots & \dfrac{\partial g_q}{\partial x_1} \\ \vdots & & \vdots \\ \dfrac{\partial g_1}{\partial x_n} & & \dfrac{\partial g_q}{\partial x_n} \end{bmatrix} \tag{5.14}$$

Since $g_j(\mathbf{x})$, $j = 1,\ldots,q$ denotes the set of binding constraint functions (a constraint (5.2) or (5.4) is said to be "binding" if it is satisfied at the equality), for the moment let $f(\mathbf{x})$ represent the objective function. Then $\nabla f(\mathbf{x}^k)$ and $\nabla g_j(\mathbf{x}^k)$, $j = 1,\ldots,q$ represent the gradient vectors of the objective and constraint functions, respectively, evaluated at the boundary point $\mathbf{x}^k$.

Performing a first-order response surface experiment about the boundary point $\mathbf{x}^k$ yields estimates of the gradient vectors $\nabla f(\mathbf{x}^k)$ and $\nabla g_j(\mathbf{x}^k)$, $j = 1,\ldots,q$ in the form of the vectors of regression coefficients. (If a constraint of type (5.2) is included in the set of binding constraints, the

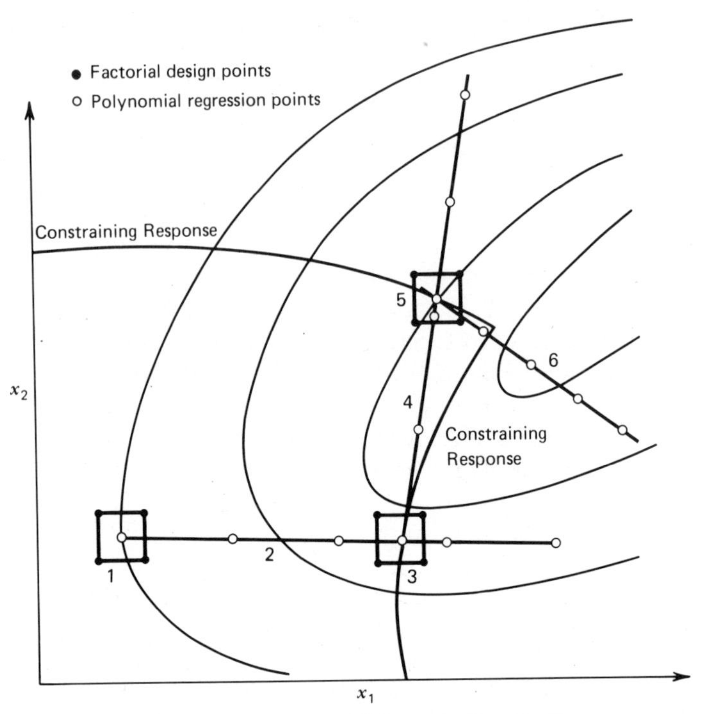

Figure 5.3. A first-order gradient projection search.

gradient vector has the form $(0, 0, \ldots, 1, \ldots, 0)'$, where all elements are zero except the ith element which is one). The gradient projection direction is given by

$$\mathbf{S}^k = \left[\nabla f(\mathbf{x}^k)\right]' - B_q\left(B_q' B_q\right)^{-1} B_q' \left[\nabla f(\mathbf{x}^k)\right]' \tag{5.15}$$

A line search is performed along direction $\mathbf{S}^k$ until either (a) a local "optimum" is found, or (b) other constraints are encountered. This new point is denoted $\mathbf{x}^{k+1}$. This procedure is repeated until the gradient projection direction $\mathbf{S}^k$ is approximately zero. This point $\mathbf{x}$ is taken as a "constrained optimal" solution. Figure 5.3 illustrates the application of the gradient projection procedure to a constrained optimization problem.

The following generalized procedure is followed in employing a first-order response surface approach to multiple-response experimentation. The particular problem formulation and optimization procedure will govern the precise sequence of steps in implementing this procedure.

1. Identify the known experimental region $a_i \leqslant x_i \leqslant c_i$, $i = 1, \ldots, n$. Select a starting point $\mathbf{x}^0$ within this region. With $\mathbf{x}^0$ as its center, array an orthogonal first-order response surface design within a selected design radius. Place $n_c = n/2 \geqslant 2$ points at the design center $\mathbf{x}^0$ (coded as the $\mathbf{0}$-vector).

2. Perform experiments at each of the N experimental design points and record the responses y_j^l, $j=1,\ldots,m$; $l=1,\ldots,N$. Using multiple linear regression, fit linear models of the form (5.12).
3. Apply an appropriate mathematical programming technique to locate the next center point in the search.
4. Repeat steps 1–3 until an "optimum" solution is located. It may be appropriate to add design points to complete a second-order response surface design to test this optimum solution. The procedure for accomplishing this is described in the next section.

Second-Order Response Surface Methods

A second-order response surface approach to the multiple-variable, multiple-response experimentation problem consists of one or more repetitions of a two-stage procedure: (1) the execution of an experimental trial at each point in a second-order response surface experimental design covering the known region given by (5.2), and the use of multiple linear regression to fit second-order regression models to the resulting data; and (2) the application of a suitable mathematical programming procedure to obtain a solution to the problem formulated in (5.1)–(5.4), in (5.7) together with (5.2) or in (5.8) together with (5.2). In contrast to the first-order methods, in which the optimization procedure was part and parcel with the experimental procedure, these procedures are distinct and sequential in the proposed second-order approaches.

The first step in the second-order approach is to identify the range of each input variable. A safe strategy is to cover the entire known region $a_i \leqslant x_i \leqslant c_i$, $i=1,\ldots,n$ with the first (and possibly only) experimental design. If we let α_i denote the radius of the n-dimensional hypersphere within which the design points are contained, then

$$\alpha_i = \frac{c_i - a_i}{2}, \qquad i=1,\ldots,n \tag{5.16}$$

is effectively the maximum radius we could construct. It is convenient to adopt the coding conventions described in Chapter 3, but choosing x_{iu} in such a way that α_i satisfied (5.16).

The second-order fitted response surface has the form

$$\hat{y} = \hat{\beta}_0 + \sum_{i=1}^{n} \hat{\beta}_i x_i + \sum_{i=1}^{n} \hat{\beta}_{ii} x_i^2 + \sum_{i=1}^{n} \sum_{j>i}^{n} \hat{\beta}_{ij} x_i x_j \tag{5.17}$$

where $\hat{y}$ is the estimate of the true response η at a given value $\mathbf{x}^T = (x_1,\ldots,x_n)$ and the β_i and β_{ij} are regression coefficients in the fitted model.

Since we must estimate m separate response relationships, (5.17) is modified to

$$\hat{y}_k=\hat{\beta}_{k,o}+\sum_{i=1}^{n}\hat{\beta}_{k,i}x_i+\sum_{i=1}^{n}\beta_{k,ii}x_i^2+\sum_{i=1}^{n}\sum_{j>i}^{n}\hat{\beta}_{k,ij}x_ix_j \qquad (5.18)$$

$$k=1,\dots,m$$

Given the independence of the m responses, these m regression equations can be estimated independently from a set of $N \geqslant (n+1)(n+2)/2$ data points obtained by performing an experiment at each point in a second-order response surface design.

Having estimated the m second-order regression equations (5.18) and formulated the appropriate optimization problem, it remains to apply mathematical programming to obtain a solution. For the constrained formulation, any of the following procedures could be employed: (1) Box's complex search, (2) Rosen's gradient projection method, or (3) one of Zoutendijk's methods of feasible directions. For the weighted objective function formulation and other multiple objective optimization formulations, these same three optimization procedures are applicable with only minor modifications.

EFFECT OF BIAS AND DESIGN ON OPTIMIZATION METHODOLOGY

This section illustrates the constrained experimental optimization methodologies. In this case the emphasis is on experimentation and the effects of various experimental designs on search progress and on the fitted parameter estimates.

Here the objective function contains a term of order three and one of order four, thus biasing both first- and second-order estimators. The errors are independently and normally distributed with zero mean and variance 0.25.

$$y_0=f(\mathbf{x})+\varepsilon_0=0.1x_1^2+0.3x_1^2x_2^2+0.2x_1x_2x_3+0.5x_3^2+\varepsilon_0(0,0.25) \qquad (5.19)$$

$$y_1=g_1(\mathbf{x})+\varepsilon_1=5x_1^2+5x_2^2+9x_3^2+2x_1x_2-6x_1x_3$$
$$-6x_2x_3+\varepsilon_1(0,0.25) \qquad (5.20)$$

$$y_2=g_2(\mathbf{x})+\varepsilon_2=4x_1^2+x_2^2+\varepsilon_2(0,0.25) \qquad (5.21)$$

The first constraint represents an ellipsoid whose major axis is along the $(1,1,1)'$ direction. The second constraint is an ellipsoidal cylinder. The constrained optimization problem to be solved is

$$\max f(\mathbf{x})=E(y_0) \qquad (5.22)$$

subject to

$$g_1(\mathbf{x}) = E(y_1) \leqslant 90 \tag{5.23}$$

$$g_2(\mathbf{x}) = E(y_2) \leqslant 50 \tag{5.24}$$

Two first-order designs will be considered: a full 2^3 factorial with center points and a 2^{3-1} fractional factorial used as an orthogonal simplex design, also with center points. The center points are included in the full factorial in order to make a formal lack-of-fit calculation. This calculation cannot be made with the simplex design due to cross-product terms, although one degree of freedom can be used for detecting curvature effects. The multiple gradient summation procedure is used to modify the gradient at constraint boundaries, as it is suitable for the convex constraints observed in this example.

Three second-order designs are also used to make an estimate of the constrained optima. Because of the biasing that could be expected from a higher order system, sufficient points have been included in the two uniform precision designs to permit lack-of-fit calculations. The augmented simplex, as with the first-order simplex, allows a limited, single degree of freedom calculation to test for error not due to pure experimental variation.

Steepest Ascent with a Full Factorial Design

All of the problems associated with experimental design come to the fore in approaching a surface that is completely unknown. Among these questions two are most insistent: "What is a feasible starting point?" and "What spacing will allow effects to be distinguished from experimental error without introducing significant higher order effect?" Here an arbitrary start point of $x^0 = (0.5, 0.5, 0.5)'$ and a spacing of $d = 0.20$ is used. The results are illustrated in Table 5.1.

The fitted objective function model from this experiment is

$$\hat{y}_0 = 0.50645 - 1.105x_1 + 0.6374x_2 - 0.11546x_3 \tag{5.25}$$

A quick look at the variance for this estimate, given in Table 5.2, shows that the most significant term belongs to pure error, and neither the regression nor lack-of-fit sum-of-squares is found to differ significantly from zero. Pure error was estimated from the four center points according to the formula.

$$SS_e = \sum_{i=1}^{n_2} \varepsilon_i^2 - \frac{1}{n_2}\left(\sum_{i=1}^{n_2} \varepsilon_i\right)^2 \tag{5.26}$$

Table 5.1. **First factorial direction determining block**

Design point	x_1	x_2	x_3	y_0	y_1	y_2
1	0.6	0.6	0.6	0.025999	3.3245	1.6857
2	0.6	0.6	0.4	0.12318	2.5126	1.6783
3	0.6	0.4	0.6	0.006789	2.5475	1.4851
4	0.6	0.4	0.4	0.2923	2.4076	1.6786
5	0.4	0.6	0.6	0.54249	2.7633	1.2842
6	0.4	0.6	0.4	0.45344	2.5087	0.53543
7	0.4	0.4	0.6	0.26868	2.3633	0.82399
8	0.4	0.4	0.4	0.06742	1.3666	0.88681
9	0.5	0.5	0.5	0.66869	2.4849	1.553
10	0.5	0.5	0.5	0.41072	2.2921	1.0679
11	0.5	0.5	0.5	0.015424	2.299	0.92738
12	0.5	0.5	0.5	−0.29445	2.247	1.3582

Table 5.2. **Analysis of variance for first direction determining block**

Source	Sum of squares	Degrees of freedom	Mean squares	F ratio
Regression	0.68619	4	0.1715	0.9482
Lack of Fit	0.15200	5	0.0304	0.1681
Error	0.54262	3	0.1809	
Total	1.3808	12		

Table 5.3. **A second factorial direction determining block**

Design point	x_1	x_2	x_3	y_0	y_1	y_3
1	1.2	1.2	1.2	1.8429	13.139	6.9689
2	1.2	1.2	0.8	1.4196	11.658	7.0149
3	1.2	0.8	1.2	1.5208	10.984	6.5502
4	1.2	0.8	0.8	0.75742	8.125	6.3548
5	0.8	1.2	1.2	0.95926	10.899	4.2231
6	0.8	1.2	0.8	0.73535	8.3576	4.3436
7	0.8	0.8	1.2	1.2217	8.7521	3.1591
8	0.8	0.8	0.8	0.71084	5.7974	3.1219
9	1.0	1.0	1.0	1.2599	9.081	4.9053
10	1.0	1.0	1.0	0.76956	9.0499	4.5988
11	1.0	1.0	1.0	0.78049	9.6411	4.3627
12	1.0	1.0	1.0	0.92409	8.8216	4.4854

The estimated models for the two constraint functions are both significant and indicate that we are well in the interior of the feasible region. A new start point $\mathbf{x}^1=(1,1,1)'$ is thus attempted. Table 5.3 gives the data for a second factorial design for estimating a gradient direction.

At this point a meaningful regression estimate occurs. The fitted objective function is

$$\hat{y}_0=-1.7882+1.196x_1+0.46653x_2+1.2009x_3 \tag{5.27}$$

The analysis of variance given in Table 5.4 is again included to illustrate a significant result. We are still well within the feasible region so only the steepest ascent direction (normalized) is used. The constraints are estimated to be 8 units away, so that four steps of 1.2 are used, underestimating this distance. The results of this step-determining block are given in Table 5.5.

Both constraints become active at about the fourth trial point, and the second constraint becomes the binding constraint. By polynomial regression using the y_2 data,

$$\hat{y}_2=4.5942+6.0989\lambda+1.9172\lambda^2 \tag{5.28}$$

Setting y_2 to 50 yields $\lambda=3.5293$. The new base point is therefore

$$\mathbf{x}^2=(3.4012, 1.9366, 3.411)' \tag{5.29}$$

The estimates on the three response functions at this point are

$$\hat{y}_0=25.56 \qquad \hat{y}_1=85.06 \qquad \hat{y}_2=50$$

Table 5.4. Analysis of variance for second direction determining block

Source	Sum of squares	Degrees of freedom	Mean squares	F ratio
Regression	14.86	4	3.7150	71.033
Lack of Fit	0.3822	5	0.0764	1.4608
Error	0.1569	3	0.0523	
	15.399	12		

Table 5.5. First factorial step determining block

Design point	λ	x_1	x_2	x_3	y_0	y_1	y_2
1	0	1.0	1.0	1.0	0.76915	8.9576	4.5807
2	1.2	1.8164	1.3185	1.8198	5.4032	25.695	14.577
3	2.4	2.6329	1.6369	2.6395	11.961	51.552	30.645
4	3.6	3.4493	1.9554	3.4593	25.356	87.387	51.0
5	4.8	4.2657	2.2739	4.2791	47.596	132.91	78.178

A third direction determining block is run at this point, giving the results in Table 5.6. Since y_2 is active it is fitted along with y_0. Both are significant,

$$\hat{y}_0 = -53.159 + 9.3812x_1 + 15.695x_2 + 4.4998x_3 \tag{5.30}$$

$$\hat{y}_2 = -50.734 + 27.828x_1 + 5.0488x_2 - 1.0232x_3 \tag{5.31}$$

The multiple gradient summation direction (5.15) is calculated from the gradients of these two functions and yields, after normalization, the direction

$$\mathbf{s}^2 = (-0.56381, 0.74988, 0.34612)'$$

This direction repels the search away from the second constraint into the interior of the feasible region and toward the first constraint. The first-order approximation

$$\lambda = \frac{90 - 85.418}{s^{2'} \nabla y_2'} = 0.81356$$

gives the maximum allowable step for the line search, the results of which are shown in Table 5.7. The first constraint does become active, at $\lambda = 0.63439$. The new base point is therefore

$$\mathbf{x}^3 = (3.0435, 2.4124, 3.6306)'$$

Another direction determining block (not shown) is run and the first two models fitted to give the new multiple gradient summation direction. This direction (normalized) is

$$\mathbf{s}^3 = (0.18871, 0.64773, -0.73813)'$$

It is used to estimate the maximum allowable step, calculated to be $\lambda = 0.018334$. This seems a bit short so a maximum step of $\lambda = 0.3$ is used instead. As Table 5.8 indicates, even this was too short and two additional experiments, at $\lambda = 0.6$ and $\lambda = 0.9$ were run before y_2 finally became active.

The estimate for y_2 is used to calculate the feasible step

$$\hat{y}_2 = 43.041 + 6.8223\lambda + 1.8433\lambda^2 \tag{5.32}$$

Equating this to 50 yields $\lambda = 0.83271$ so that

$$\mathbf{x}^4 = (3.2006, 2.9517, 3.0159)'$$

At this point 65 experiments have been run and a final sensitivity direction determining block is foregone. The function estimates from the step determining block reveal that at this last point

$$\hat{y}_0 = 38.468 \qquad \hat{y}_1 = 84.1871 \qquad \hat{y}_2 = 50.0$$

This solution is binding on the y_2 constraint.

Table 5.6. Third factorial direction determining block

Design point	x_1	x_2	x_3	y_0	y_1	y_2
1	3.5012	2.0366	3.511	27.845	91.345	53.284
2	3.5012	2.0366	3.311	26.804	84.776	53.587
3	3.5012	1.8366	3.511	23.97	89.077	52.456
4	3.5012	1.8366	3.311	23.176	83.832	52.614
5	3.3012	2.0366	3.511	25.415	87.168	47.781
6	3.3012	2.0366	3.311	24.256	81.521	48.177
7	3.3012	1.8366	3.511	22.613	86.307	46.88
8	3.3012	1.8366	3.311	22.006	80.003	46.84
9	3.4012	1.9366	3.411	24.732	85.201	50.648
10	3.4012	1.9366	3.411	24.31	84.92	50.066
11	3.4012	1.9366	3.411	24.332	85.377	50.366
12	3.4012	1.9366	3.411	24.433	85.479	49.747

Table 5.7. Second factorial step determining block

Design point	λ	x_1	x_2	x_3	y_0	y_1	y_2
1	0.0	3.4012	1.9366	3.411	24.469	85.142	50.009
2	0.3	3.232	2.1616	3.5149	26.729	87.045	46.757
3	0.6	3.0629	2.3866	3.6187	28.797	89.564	42.959
4	0.9	2.8938	2.6115	3.7225	30.296	93.199	40.106

Table 5.8. Final factorial step determining block

Design point	λ	x_1	x_2	x_3	y_0	y_1	y_2
1	0.0	3.0435	2.4124	3.6306	28.817	89.591	43.061
2	0.1	3.0624	2.4771	3.5568	30.065	88.145	43.728
3	0.2	3.0812	2.5419	3.483	31.325	87.087	44.465
4	0.3	3.1001	2.6067	3.4091	31.983	86.104	45.243
5*	0.6	3.1567	2.801	3.1877	35.164	84.15	47.825
6*	0.9	3.2133	2.9953	2.9663	39.508	84.411	50.664

*Run after the first four experiments in the block.

Steepest Ascent with a Simplex Design

In the last section an entire block of experiments had to be discarded at the first point since only the pure error proved to be significant. The gradient estimate discarded was quite poor and the rejection saved us a fruitless search. The same risk occurs with the simplex design except that now it will not be possible to detect this lack of real significance. On the other hand less than half the points are now needed for the estimate as were required to the full factorial, as shown in Table 5.9.

The first model estimates are, fortunately, quite good, and the first line search will actually be along an ascent direction.

$$\hat{y}_0 = -0.64785 + 0.76977x_1 + 0.5994x_2 + 0.17471x_3 \tag{5.33}$$

The results of the line search in terms of λ are given in Table 5.10. The maximal step calculations yield steps of 14 and 32, which are guessed to be overestimations. Instead, four steps every $\lambda = 1.2$ are used.

The two constraint functions are fitted to second-order polynomials

$$\hat{y}_1 = 1.9367 + 4.8829\lambda + 4.4737\lambda^2 \tag{5.34}$$

$$\hat{y}_2 = 1.2668 + 3.4715\lambda + 2.7765\lambda^2 \tag{5.35}$$

and yield feasible steps of $\lambda_1 = 3.9245$ and $\lambda_2 = 3.6108$. The estimate for y_0

Table 5.9. First simplex direction determining block

Design point	x_1	x_2	x_3	y_0	y_1	y_2
1	0.7	0.7	0.7	0.42697	4.2773	2.4039
2	0.7	0.3	0.3	0.11733	2.7688	2.0401
3	0.3	0.7	0.3	0.049179	2.4845	1.0426
4	0.3	0.3	0.7	−.1207	2.9353	0.40451
5	0.5	0.5	0.5	0.1477	2.7062	1.0789

Table 5.10. First simplex step determining block

Design point	λ	x_1	x_2	x_3	y_0	y_1	y_2
1	0.0	0.5	0.5	0.5	0.50096	1.9832	1.3843
2	1.2	1.4237	1.2193	0.70965	1.7145	14.028	9.5946
3	2.4	2.3474	1.9386	0.9193	8.5636	39.775	25.8
4	3.6	3.2712	2.6578	1.129	26.709	77.236	49.866
5	4.8	4.1949	3.3771	1.3386	66.77	128.52	81.794

Table 5.11. Second simplex direction determining block

Design point	x_1	x_2	x_3	y_0	y_1	y_2
1	3.3795	2.7643	1.2308	30.397	82.51	53.224
2	3.3795	2.5643	1.0308	26.179	79.906	52.301
3	3.1795	2.7643	1.0308	26.496	79.637	47.926
4	3.1795	2.5643	1.2308	23.688	71.605	47.165
5	3.2795	2.6643	1.1308	26.348	78.136	49.95

is not a very good fit at second order so a third term is added

$$\hat{y}_0 = 0.43012 + 1.6623\lambda - 1.2386\lambda^2 + 0.78512\lambda^3 \tag{5.36}$$

so that at the new base point

$$\mathbf{x}^1 = (3.2795, 2.6643, 1.1308)'$$

y_0 is estimated to be

$$\hat{y}_0 = 27.245$$

This estimate is a trifle high, as the next block given in Table 5.11 shows, but it is close enough.

The fitted models here yield the following (normalized) multiple gradient summation direction,

$$\mathbf{s}^1 = (-0.48022, 0.85487, 0.19643)'$$

and a maximum step of $\lambda = 1.7506$.

The next two step determining blocks and one direction determining block are omitted. The first step leaves the second constraint and encounters the first at

$$\mathbf{x}^2 = (2.6741, 3.742, 1.3785)'$$

The direction determining block then uses the first constraint gradient to calculate the new search direction

$$\mathbf{s}^2 = (0.46611, -0.48273, 0.74143)'$$

This is used in a final step determining block where the second constraint is again encountered at

$$\mathbf{x}^3 = (3.1402, 3.2593, 2.1198)'$$

At this point only 30 experiments have been run (compared to 65 for the factorial) so a final sensitivity experiment is run to confirm the estimated results from the last line search. These data are given in Table 5.12.

The final block reveals that the last point is right at the y_2 constraint and that (presumably) further increases could be made using a multiple

Table 5.12. **Final simplex direction determining block**

Design point	x_1	x_2	x_3	y_0	y_1	y_2
1	3.2402	3.3593	2.2198	43.648	87.351	53.224
2	3.2402	3.1593	2.0198	39.241	81.812	52.485
3	3.0402	3.3593	2.0198	38.502	82.353	47.687
4	3.0402	3.1593	2.2198	35.51	77.139	46.855
5	3.1402	3.2593	2.1198	38.902	81.739	50.096

gradient summation direction. The estimated functions at this point

$$\hat{y}_0 = 39.16 \qquad \hat{y}_1 = 82.078 \qquad \hat{y}_2 = 50.068$$

suggest that the results obtained here are slightly better than the extra 30 experiments obtained with the factorial steepest ascent in the last section.

Three Second-Order Experiments The *modus operandi* with second order designs is to spend all your experiments at once and estimate the optimum point directly. Here, three designs are used, using $\mathbf{x}^0 = (2,2,2)'$ and a spacing of $(3,3,3)'$. The augmented simplex used has ten points plus four center points. The Icosahedron [28] is uniform precision with 5 center points, for a total of 17. The full uniform precision central composite is also used with a total of 20 points.

The simplex used is the same 2^{3-1} fractional factorial (defining contrast $ABC = I$) used in the previous section. The augmenting points for this design are axial and are the points which will be added to the factorial to produce the central composite design. An $\alpha = \sqrt{2}$ is chosen to make all points equiradial and the design rotatable, since for the coded design

$$\lambda_4 = \sum_{u=1}^{14} x_{iu}^2 x_{ju}^2 = 4$$

$$3\lambda_4 = \sum_{u=1}^{14} x_{iu}^4 = 12$$

The experimental results for this design are given in Table 5.13. A quick look at the variance for the three fitted models shows a good fit for both of the constraint functions but a lack-of-fit indication for the fitted objective function, as seen in Table 5.14. Both of these F terms are significant, so that the regression and lack-of-fit contributions are found to differ significantly from zero. A quick look at the fitted objective function

$$\begin{aligned}\hat{y}_0 = {} & 33.191 - 15.192x_1 - 15.143x_2 - 8.8871x_3 + 1.6057x_1^2 \\ & + 1.4989x_2^2 + 0.69911x_3^2 + 5.2255x_1x_2 + 2.2068x_1x_3 \\ & + 2.1738x_2x_3 \end{aligned} \tag{5.37}$$

Table 5.13. Augmented simplex second-order design with four center points

Design point	x_1	x_2	x_3	y_0	y_1	y_2
1	3.5	3.5	3.5	60.983	110.37	61.535
2	3.5	0.5	0.5	2.5737	56.226	49.504
3	0.5	3.5	0.5	1.3898	56.524	13.299
4	0.5	0.5	3.5	6.0332	91.946	1.03
5	4.1213	2.0	2.0	27.737	84.138	71.925
6	2.0	4.1213	2.0	26.315	84.07	33.014
7	2.0	2.0	4.1213	16.394	102.28	19.863
8	2.0	2.0	−0.1213	5.0639	51.252	20.061
9	2.0	−0.1213	2.0	2.3409	33.236	15.82
10	−0.1213	2.0	2.0	1.8791	33.259	3.9592
11	2.0	2.0	2.0	8.6223	36.007	20.239
12	2.0	2.0	2.0	8.2913	36.077	20.314
13	2.0	2.0	2.0	9.0403	35.911	20.035
14	2.0	2.0	2.0	9.1891	35.758	20.779

is hardly recognizable as the original function. The x_1^2 and x_3^2 terms are biased (they also occur in the original) as are all the other terms. Nevertheless, the F ratio for the regression indicates that in some sense the data is well explained by the fitted model.

The fitted models for the two constraint functions are recognizable as the original functions and give no significance in the lack-of-fit tests.

$$\begin{aligned}\hat{y}_1 = {} & 0.41305 - 0.13248x_1 + 0.044107x_2 - 0.32876x_3 \\ & + 5.0186x_1^2 + 5.0084x_2^2 + 9.0336x_3^2 + 1.9332x_1x_2 \\ & - 5.908x_1x_3 - 5.9812x_2x_3 \end{aligned} \tag{5.38}$$

and

$$\begin{aligned}\hat{y}_2 = {} & -0.058896 + 0.19864x_1 + 0.11922x_2 - 0.007056x_3 \\ & + 3.9539x_1^2 + 0.98427x_2^2 - 0.04177x_3^2 + 0.004587x_1x_2 \\ & - 0.001843x_1x_3 + 0.06562x_2x_3 \end{aligned} \tag{5.39}$$

Table 5.14. Analysis of variance for fitted objective function

Source	Sum of squares	Degrees of freedom	Mean squares	F ratio
Regression	5830.8	10	583.08	3507.6
Lack of Fit	7.0725	1	7.0725	42.5456
Error	0.4987	3	0.1662	
	5838.4	14		

The actual terms are well estimated and contribute significantly to the regression sum of squares. Since x_1, x_2, x_3 are all greater than one in the region of interest the quadratic terms predominate to the sum of squares and in predicting the behavior of the response surface. Many of the less important terms would be found to "not differ significantly from zero" if tested, but the values given are the best estimates and should therefore be retained in the model.

An optimization based on these fitted models fared quite well. The major reason for this is that the estimated objective function was a good enough estimate to take the optimization to the constraints, which were estimated with higher precision. The estimated optimum, both constraints active, was estimated to be

$$\hat{\mathbf{x}}^* = (3.077, 3.44, 2.897)'$$

with a functional estimate at that point of

$$\hat{y}_0 = 44.06$$

The actual optimum was closer to

$$\mathbf{x}^* = (3.05, 3.519, 2.764)'$$

and

$$y_0 = 45.809$$

The next design is a twelve point figure called an icosahedron. With the addition of 5 center points this rotatable (and singular) design is made rotatable uniform precision and nonsingular. With the extra degrees of freedom a more meaningful lack-of-fit calculation can be made. The experimental results from this designed experiment are shown in Table 5.15. The results obtained are similar to those of the augmented simplex: the y_0 estimates are heavily biased, and the constraint estimates are quite good.

$$\begin{aligned}\hat{y}_0 = {} & 13.47 + 10.744x_1 - 9.3217x_2 - 0.24508x_3 + 1.5734x_1^2 \\ & + 1.215x_2^2 + 0.39291x_3^2 + 5.1219x_1x_2 + 0.43946x_1x_3 \\ & + 0.26138x_2x_3 \end{aligned} \tag{5.40}$$

The analysis of variance results, shown in Table 5.16 for the objective response y_0, are also similar, except that the lack-of-fit is now more marked, although much less, with respect to the actual sum of squares involved, than what is explained by the regression. With this type of agreement subtle behavior of the response function will not be detected (two orders are missing, after all) but directions and locations of improvement are still possible. This approximating problem was also solved, in spite of the significant lack-of-fit, and also yielded a good estimate of the

Table 5.15. A uniform precision icosahedron second order design

Design point	x_1	x_2	x_3	y_0	y_1	y_2
1	2.0	3.473	2.911	2.683	74.943	27.988
2	2.0	3.473	1.089	17.296	69.294	28.103
3	2.0	0.527	2.911	5.6547	55.868	15.994
4	2.0	0.527	1.089	1.671	17.365	16.235
5	2.911	2.0	3.473	21.255	80.529	37.904
6	1.089	2.0	3.473	9.0035	74.279	8.2768
7	1.089	2.0	0.527	1.9688	22.729	8.4693
8	2.911	2.0	0.527	11.862	60.952	37.935
9	3.473	2.911	2.0	37.883	82.508	56.521
10	3.473	1.089	2.0	9.0157	55.576	49.228
11	0.527	2.911	2.0	3.8101	41.405	9.8
12	0.527	1.089	2.0	2.4352	25.496	2.0873
13	2.0	2.0	2.0	9.005	36.083	19.885
14	2.0	2.0	2.0	8.7863	36.433	20.313
15	2.0	2.0	2.0	8.8791	36.075	20.093
16	2.0	2.0	2.0	8.7447	36.077	19.744
17	2.0	2.0	2.0	8.9044	35.503	20.191

constrained optimum

$$\hat{\mathbf{x}}^* = (3.048, 3.596, 2.583)'$$

where $\hat{y}_0 = 41.555$.

A sensitivity experiment at that point would reveal that the objective function had been underestimated somewhat, but the estimate of the point itself is quite good. This is in part due to the precision of the constraint function estimates which allow the constrained point to be estimated precisely.

The full uniform precision central composite design requires 20 points, including 6 center points. This will allow the most refined test for lack-of-fit of all, and will give the same conclusions as before. The experimental data are arrayed in Table 5.17.

Table 5.16. Analysis of variance for icosahedron design

Source	Sum of squares	Degrees of freedom	Mean squares	*F* ratios
Regression	3435.6	10	343.56	31232.7
Lack-of-Fit	19.96	3	6.653	633.59
Error	0.042	4	0.011	
	3455.6	17		

Table 5.17. A uniform precision central composite design

Design point	x_1	x_2	x_3	y_0	y_1	y_2
1	3.5	3.5	3.5	61.128	110.15	61.482
2	3.5	3.5	0.5	47.214	127.87	61.579
3	3.5	0.5	3.5	9.5837	92.386	48.923
4	3.5	0.5	0.5	2.7296	56.255	49.341
5	0.5	3.5	3.5	8.5366	93.05	13.201
6	0.5	3.5	0.5	1.1551	56.184	13.396
7	0.5	0.5	3.5	6.3088	92.323	0.91394
8	0.5	0.5	0.5	0.26329	2.1322	0.7487
9	4.526	2.0	2.0	32.229	97.951	85.973
10	−0.521	2.0	2.0	1.6832	37.102	5.1818
11	2.0	4.525	2.0	30.555	97.868	36.636
12	2.0	−0.526	2.0	2.213	37.563	16.404
13	2.0	2.0	4.526	18.945	123.81	19.764
14	2.0	2.0	−0.526	5.461	63.441	19.886
15	2.0	2.0	2.0	9.2361	36.339	20.205
16	2.0	2.0	2.0	8.9438	36.333	20.057
17	2.0	2.0	2.0	8.8052	35.728	20.086
18	2.0	2.0	2.0	8.7536	35.999	19.823
19	2.0	2.0	2.0	8.8041	36.258	19.909
20	2.0	2.0	2.0	8.8612	36.014	19.945

The estimate of the objective function y_0 is similar to the two previous

$$\begin{aligned}\hat{y}_0 = {} & 11.496 - 9.1768x_1 - 9.3719x_2 - 1.6306x_3 + 1.4092x_1^2 \\ & + 1.3194x_2^2 + 0.66417x_3^2 + 5.1616x_1x_2 + 0.40785x_1x_3 \\ & + 0.46646x_2x_3 \end{aligned} \tag{5.41}$$

Notice that while the general results in Table 5.18 are similar to the analyses of variance that have gone on earlier, the sum of squares and mean squares lack-of-fit error has been increasing with its degrees of

Table 5.18. Analysis of variance for uniform precision central composite design

Source	Sum of squares	Degrees of freedom	Mean squares	F ratio
Regression	8905.9	10	890.59	28544.
Lack of Fit	117.34	5	23.47	752.18
Error	0.156	5	0.0312	
	9023.4	20		

Table 5.19. A factorial sensitivity experimental block

Design point	x_1	x_2	x_3	y_0	y_1	y_2
1	3.134	3.7	2.71	51.209	96.563	53.244
2	3.134	3.7	2.51	50.247	94.323	53.419
3	3.134	3.5	2.71	46.702	90.516	51.609
4	3.134	3.5	2.51	45.412	89.332	51.485
5	2.934	3.7	2.71	45.935	90.992	47.67
6	2.934	3.7	2.51	45.237	89.35	48.396
7	2.934	3.5	2.71	41.809	86.33	46.977
8	2.934	3.5	2.51	40.76	84.665	46.828
9	3.034	3.6	2.61	45.821	90.221	49.306
10	3.034	3.6	2.61	45.82	90.029	49.573
11	3.034	3.6	2.61	44.983	90.039	49.856
12	3.034	3.6	2.61	45.836	90.079	50.029

freedom. The overall regression is still predicting, according to these calculations, more than it is unable to predict. Nevertheless the true extent of its adequacy (or lack thereof) does require more than one or two degrees of freedom.

This approximating optimization result is once again good.

$$\hat{\mathbf{x}}^* = (3.0344, 3160, 2.6)$$

with

$$\hat{y}_0 = 44.297$$

For confirmation a final experiment is executed, a sensitivity block of first order. A full 2^3 factorial with 4 center points is included. The results, especially because of the good constraint estimates, are considered good and a small design of spacing $d = 0.20$ is used. The experimental results for the sensitivity design are presented in Table 5.19.

The newly fitted models confirm our earlier predictions,

$$\hat{y}_0 = 45.814 \qquad \hat{y}_1 = 90.12 \qquad \hat{y}_2 = 49.866$$

Table 5.20. Analysis of variance for the sensitivity block

Source	Sum of squares	Degrees of freedom	Mean squares	F ratio
Regression	25279.0	4	6319.7	35504.0
Lack of Fit	0.3939	5	0.0788	0.4426
Error	0.5340	3	0.1780	
	25280.	12		

showing that our estimate of the objective function was a little low, and that we are very close to the constraint boundaries. The regression is seen in Table 5.20 to be significant and without significant lack-of-fit. A good estimate of the optimum has been found.

APPLICATION OF CONSTRAINED EXPERIMENTAL OPTIMIZATION METHODOLOGIES

In this section the experimental design fundamentals and mathematical programming techniques outlined in the previous two chapters are applied to a two-dimensional example problem, illustrating their use in a constrained experimental problem. The essential features of the techniques can be seen with two dimensions and easily extended to problems of higher dimensionality.

The problem to be used was given by Biles [6] and contains a primary response and two constraints.

$$y_0 = f(\mathbf{x}) + \varepsilon_0 = 3x_1^2 + 2x_2^2 + \varepsilon_0(0, 0.25) \tag{5.42}$$

$$y_1 = g_1(\mathbf{x}) + \varepsilon_1 = x_1^2 + x_2^2 + \varepsilon_1(0, 0.25) \tag{5.43}$$

$$y_2 = g_2(\mathbf{x}) + \varepsilon_2 = 9x_1 - x_2^2 + \varepsilon_2(0, 0.25) \tag{5.44}$$

The constrained optimization problem to be solved is

$$\max f(\mathbf{x}) = E(y_0) \tag{5.45}$$

subject to

$$g_1(\mathbf{x}) = E(y_1) \leqslant 25 \tag{5.46}$$

$$g_2(\mathbf{x}) = E(y_2) \leqslant 27 \tag{5.47}$$

$$x_1 \geqslant 0 \qquad x_2 \geqslant 0 \tag{5.48}$$

The notation for the error terms $\varepsilon_j(0, 0.25)$ will indicate a process error which is normally distributed with a mean of zero and a variance of $\sigma^2 = 0.25$.

In this section the gradient projection, multiple gradient summation, constrained gradient and two formulations of Zoutendijk's best feasible directions method will be used to illustrate the first-order methods. The gradient projection material is taken from Biles, except for the concluding sensitivity analysis, and as in that paper each technique will begin at the point $(1, 1)'$. The direction estimating blocks will consist of 2^2 factorial designs of 4 points, and the line searches will consist of 4 equally spaced points in the search direction given by the gradient or a modified gradient. The initial step length for the line search is generally taken as the distance to the nearest active constraint, as estimated from the gradients of the constraint functions along the proposed search direction.

This problem will also be investigated using a second-order design and an off-line optimization of the resultant approximating optimization problem.

Gradient Projection Method

The results of the gradient projection method are given in the paper by Biles and summarized here for comparison with the methods that follow. The first direction determining experimental block shows that the initial point $\mathbf{x}_0=(1,1)'$ is an interior point, so only the gradient for the primary response function need be used for the first line search. The regression relationship for the primary response is shown to be

$$\hat{y}_0=-2.92+5.5x_1+2.675x_2 \tag{5.49}$$

for which the gradient is

$$\nabla\hat{y}_0=(5.5,2.675)$$

We note that constraint violations would occur if x_1 exceeded 5 or x_2 exceeded 3, so a step length of $\lambda=1.2$ is chosen, and four steps along this direction are taken. We calculate the fitted equations from these observations to be

$$\hat{y}_0=5.51+41.4\lambda+106.7\lambda^2 \tag{5.50}$$

$$\hat{y}_1=1.82+18.3\lambda+35.9\lambda^2 \tag{5.51}$$

$$\hat{y}_2=6.95+46.0\lambda+7.83\lambda^2 \tag{5.52}$$

Equation (5.50) suggests that λ be chosen as large as possible, ruling out the possibility of an interior stationary point. Solving (5.51) equal to 25 for λ yields $\lambda_1=0.588$ and solving (5.52) equal to 27 yields $\lambda_2=0.472$. The second constraint is thus the binding constraint, $\lambda^0=0.472$

$$\begin{aligned}\mathbf{x}^0&=\mathbf{x}^0+\lambda^0\nabla y_0(\mathbf{x}^0)\\&=(1,1)'+0.472(5.5,2.675)'\\&=(3.70,2.26)'\end{aligned}$$

The point $\mathbf{x}^1$ lies at an estimate of the boundary $g_2(\mathbf{x}^1)=27$.

The next direction determining block requires estimates of the gradients of y_0 and y_2 since the second constraint is active. The regression model for y_0 is given by

$$\hat{y}_0=-45.8+21.3x_1+8.25x_2 \tag{5.53}$$

whose gradient is seen to be

$$\nabla \hat{y}_0 = (21.3, 8.25)$$

The gradient projection direction is estimated using the gradient of y_2 since it is the only constraint active

$$\hat{y}_2 = 6.57 + 7.63x_1 - 2.88x_2 \tag{5.54}$$

and

$$\nabla \hat{y}_2 = (7.63, -2.88)$$

Therefore, the gradient projection direction as given by (5.15) is

$$\mathbf{s}^1 = \begin{pmatrix} 21.3 \\ 8.25 \end{pmatrix} - \begin{pmatrix} 7.63 \\ -2.88 \end{pmatrix} \left(\begin{pmatrix} 7.63 \\ -2.88 \end{pmatrix}' \begin{pmatrix} 7.63 \\ -2.88 \end{pmatrix} \right)^{-1} \begin{pmatrix} 7.63 \\ -2.88 \end{pmatrix}' \begin{pmatrix} 21.3 \\ 8.25 \end{pmatrix}$$

$$\mathbf{s}^1 = (5.35, 14.3)'$$

The constraint $y_1 = 25$ is quite close to the present point, so λ will only vary between zero and 0.15.

The fitted curvilinear relations from this fourth experimental block are

$$\hat{y}_0 = 49.0 + 242\lambda + 535\lambda^2 \tag{5.55}$$

$$\hat{y}_1 = 18.3 + 70.3\lambda + 475\lambda^2 \tag{5.56}$$

$$\hat{y}_2 = 27.8 - 13.1\lambda - 258\lambda^2 \tag{5.57}$$

Equating (5.56) to 25 and (5.57) to 27 and solving for λ yields $\lambda_1 = 0.066$ and $\lambda_2 = 0.0345$. However, y_2 is not active after $\lambda = 0.0345$, and what has occurred is that the base point $\mathbf{x}^1$ was slightly infeasible and the feasible region was reentered at $\lambda = 0.0345$. Using $\lambda = 0.066$ as the optimal step, the new base $\mathbf{x}^2$ becomes

$$\mathbf{x}^2 = (3.95, 3.20)'$$

and the estimated functions at that point

$$\hat{y}_0 = 67.2 \qquad \hat{y}_1 = 25 \qquad \hat{y}_2 = 25.8$$

A third direction determining block (not shown) is taken at this new base point. Since both constraints are nearby the design spacing is reduced from 0.20 to 0.10. The new regression relations yield gradients for y_0 and y_1, since y_1 is now active.

$$\hat{y}_0 = -65.2 + 25.2x_1 + 10.4x_2 \tag{5.58}$$

$$\hat{y}_1 = -28.5 + 9.1x_1 + 5.75x_2 \tag{5.59}$$

The gradient projection direction is calculated as before using y_1 as the active constraint and a line search conducted along this direction. The results of the line search are given in Table 5.21.

Table 5.21. Third step-determining experimental block

Design point	λ	x_1	x_2	y_0	y_1	y_2
1	0	3.95	3.20	67.05	25.71	24.63
2	0.02	4.00	3.12	67.08	24.98	26.99
3	0.04	4.05	3.04	67.23	25.35	26.93
4	0.06	4.10	2.96	68.03	25.85	28.95

At the second experimental point both constraints are satisfied and this is taken to be the optimal step, with $\mathbf{x}^3=(4.00, 3.12)'$ and

$$\hat{y}_0=67 \qquad \hat{y}_1=25 \qquad \hat{y}_2=27$$

Recalling our experience with the point $\mathbf{x}^1$, it is often useful to place a final experiment at the estimated optimum. This final experiment takes the form of a direction determining block, although its use is to verify whether the point is feasible or to make a small refinement in the estimate of that optimum. Since information gained at such a block is sometimes used to estimate the dual variables the final experiment is also known as a sensitivity experiment or sensitivity analysis. These results are arrayed in Table 5.22.

Table 5.22. Final direction determining block at the estimated optimum

Design point	x_1	x_2	y_0	y_1	y_2
1	3.90	3.02	63.674	23.908	26.112
2	3.90	3.22	66.152	25.553	24.439
3	4.10	3.02	68.246	25.641	27.685
4	4.10	3.22	71.289	27.614	26.417

The three fitted relations from this data are

$$\hat{y}_0=-72.815+24.273x_1+13.803x_2 \tag{5.60}$$

$$\hat{y}_1=-40.472+9.4835x_1+9.0438x_2 \tag{5.61}$$

$$\hat{y}_2=13.597+8.8773x_1-7.3535x_2 \tag{5.62}$$

Solving for the point at which $\hat{y}_1=25$ and $\hat{y}_2=27$ yields a new estimate for the optimum point

$$\mathbf{x}^4=(4.017, 3.0271)'$$

which is slightly closer to the actual optimum at $(4,3)'$. The estimated function values at this point are

$$\hat{y}_0=66.47 \qquad \hat{y}_1=25 \qquad \hat{y}_2=27$$

This final experiment showed that the estimated optimum was not quite at the boundary. It also gives the experimenter a slightly better estimate of the functions at the optimum than is available from the line search estimate.

Multiple Gradient Summation Technique

The multiple gradient summation technique, and all the first-order techniques to follow, does not modify the gradient at an interior point so that the illustration begins at the point $\mathbf{x}^1=(3.6,2.26)'$. At this point y_2 is active so that its estimated gradient is used to modify the unconstrained gradient.

$$\begin{aligned}\mathbf{s}^1&=(22.84)^{-1}(21.3,8.25)'-(8.16)^{-1}(7.63,-2.88)'\\&=(0.932,0.361)'-(0.936,-0.353)'\\&=(-0.004,0.714)'\end{aligned}$$

Normalized, the new search direction is

$$\mathbf{s}^1=(-0.006,1.000)'$$

The first constraint becomes active at $\lambda=2.228$ so λ is varied between zero and 2.4 in the line search, as shown in Table 5.23.

In this particular experiment $\mathbf{x}^1$ appears to have been feasible and the estimate on the y_2 boundary seems to have been quite good. The regression relations

$$\hat{y}_0=49.30+8.4256\lambda+2.1992\lambda^2 \tag{5.63}$$

$$\hat{y}_1=18.18+3.898\lambda+1.1836\lambda^2 \tag{5.64}$$

$$\hat{y}_2=26.871-4.4481\lambda-.94922\lambda^2 \tag{5.65}$$

show that λ should be increased indefinitely to minimize y_0. The constraint on y_1 becomes active at $\lambda=1.264$, so the new trial point $\mathbf{x}^2$ is

$$\begin{aligned}\mathbf{x}^2&=(3.6,2.26)'+1.264(-0.006,1.0)'\\&=(3.6,3.52)'\end{aligned}$$

Table 5.23. First multiple gradient summation step determining block

Design point	λ	x_1	x_2	y_0	y_1	y_2
1	0	3.6	2.26	49.32	18.22	26.92
2	0.8	3.6	3.06	57.39	21.95	22.53
3	1.6	3.59	3.86	68.47	27.56	17.50
4	2.4	3.59	4.66	82.17	34.32	10.67

Table 5.24. First multiple gradient summation direction determining block

Design point	x_1	x_2	y_0	y_1	y_2
1	3.5	3.42	59.967	24.251	20.129
2	3.5	3.62	62.825	24.881	18.805
3	3.7	3.42	64.798	25.399	21.548
4	3.7	3.62	67.645	26.728	20.185

At this point the estimated function values are

$$\hat{y}_0=63.46 \qquad \hat{y}_1=25 \qquad \hat{y}_2=19.73$$

We have left one constraint for another and we are approaching their intersection. Accordingly we reduce the spacing in the factorial experiment from 0.20 to 0.10. A new direction determining step is run as summarized in Table 5.24.

At this point y_1 is active so the regressions on y_0 and y_1 are necessary.

$$\hat{y}_0=-73.249+24.126x_1+14.262x_2 \tag{5.66}$$

$$\hat{y}_1=-18.882+7.4806x_1+4.9056x_2 \tag{5.67}$$

Using the gradients from these two equations,

$$\mathbf{s}^3=(0.861,0.509)'-(0.836,0.548)'=(0.025,-0.039)'$$

This is normalized to yield

$$\mathbf{s}^3=(0.536,-0.836)'$$

A step-determining experimental block shown in Table 5.25 is conducted. The estimated maximum allowable step to the y_2 constraint is given by $\lambda=0.73$.

In this experiment it appears that y_1 nears its constraint boundary and then begins to depart it somewhat. It is clear that the y_2 constraint is encountered, and this will be used to determine the optimal step

$$\hat{y}_2=20.312+8.5531\lambda+2.3088\lambda^2 \tag{5.68}$$

Solving for y_2 set equal to 27 yields $\lambda=0.66317$. In the previous procedure

Table 5.25. Second multiple gradient summation step determining block

Design point	λ	x_1	x_2	y_0	y_1	y_2
1	0	3.6	3.52	63.752	25.301	20.317
2	0.25	3.73	3.311	63.626	24.418	22.580
3	0.50	3.87	3.102	64.659	24.892	25.181
4	0.75	4.002	2.893	64.751	24.318	28.021

Table 5.26. Second multiple gradient summation direction determining block

Design point	x_1	x_2	y_0	y_1	y_2
1	3.855	2.866	60.861	23.121	26.649
2	3.855	3.066	63.173	24.359	25.488
3	4.055	2.866	65.562	24.704	28.079
4	4.055	3.066	67.447	25.863	26.841

both constraints were estimated to be active and this was felt to be the optimum. Here the function at the new point may be capable of improvement as only one constraint is active. Another direction determining block is therefore run with the experimental data as seen in Table 5.26. From these results we obtain

$$\mathbf{x}^3 = (3.6, 3.52)' + 0.66317(0.536, -0.836)'$$
$$= (3.9555, 2.9656)'$$

The fitted regression relations are all considered since both constraints are actually or nearly active at this point after all.

$$\hat{y}_0 = -55.604 + 22.439x_1 + 10.491x_2 \tag{5.69}$$
$$\hat{y}_1 = -23.776 + 7.7149x_1 + 5.9931x_2 \tag{5.70}$$
$$\hat{y}_2 = 17.046 + 6.956x_1 - 5.9988x_2 \tag{5.71}$$

It is noted from these relations that the refined estimates at the design center $\mathbf{x}^3$ yield

$$\hat{y}_0 = 64.268 \qquad \hat{y}_1 = 24.513 \qquad \hat{y}_2 = 26.770$$

It would appear that this point is very nearly at the intersection of the two constraints. This point is calculated and is found to be

$$\mathbf{x}^4 = (4.0043, 2.9839)'$$

which is inside the design region. This becomes the final estimate of the optimum, as a line search is not warranted for such a short move and the meager functional improvement to be anticipated. At this estimated optimum

$$\hat{y}_0 = 65.55 \qquad \hat{y}_1 = 25 \qquad \hat{y}_2 = 27$$

Constrained Gradient Approach

The constrained gradient approach is similar in motivation to the multiple gradient summation technique, although not quite as simple to calculate. The base point $\mathbf{x}^1$ from the gradient projection example is again the starting point, and the gradients are calculated from the direction determining block at that point. Here

$$\nabla\hat{y}_0=(21.3,8.25)$$
$$\nabla\hat{y}_2=(7.63,-2.88)$$

Therefore

$$m_{00}=\nabla\hat{y}_0'\nabla\hat{y}_0=521.75$$
$$m_{02}=\nabla\hat{y}_0'\nabla\hat{y}_2=138.76$$
$$m_{22}=\nabla\hat{y}_2'\nabla\hat{y}_2=66.51$$
$$M=\begin{vmatrix} m_{00} & m_{02} \\ m_{20} & m_{22} \end{vmatrix}=15447.93$$

The new search direction is calculated to be

$$\mathbf{s}^1=\frac{y}{\rho}=(2839.01)^{-1}\left(66.51\begin{pmatrix}21.3\\8.25\end{pmatrix}-138.76\begin{pmatrix}7.63\\-2.88\end{pmatrix}\right)$$
$$=2839.01^{-1}(357.92,948.34)'$$
$$=(0.1261,0.3340)'$$

This is normalized to yield the new search direction

$$\mathbf{s}^1=(0.3532,0.9355)'$$

The estimated maximum allowable step for the line search using the experimental data in Table 5.27 is calculated to be $\lambda=1.1890$.
Here the situation is similar to the second line search in the gradient projection procedure; between the first and second design points y_2 reenters the feasible region until the y_1 boundary is encountered past the third design point. Using the regression relation for y_1

$$\hat{y}_1=17.958+6.2688\lambda+1.6986\lambda^2 \tag{5.72}$$

the optimal step is calculated to be $\lambda=0.90258$. The new base point becomes

$$\mathbf{x}^2=(3.9188,3.1044)'$$

A gradient direction block is conducted here, as shown in Table 5.28.

Table 5.27. First constrained gradient step determining block

Design point	λ	x_1	x_2	y_0	y_1	y_2
1	0	3.6	2.26	48.479	17.954	27.188
2	0.4	3.7413	2.6342	56.068	20.750	26.774
3	0.8	3.8826	3.0084	63.820	24.048	25.941
4	1.2	4.0238	3.3826	71.459	27.931	24.685

Table 5.28. First constrained gradient direction determining block

Design point	x_1	x_2	y_0	y_1	y_2
1	3.82	3.00	62.084	23.326	25.514
2	3.82	3.20	64.367	24.542	23.639
3	4.02	3.00	67.022	25.526	27.276
4	4.02	3.20	69.323	26.115	25.757

It appears that we are already near a constrained optimal point. Solving the two regression relations for the constraint functions simultaneously at their constraint values from

$$\hat{y}_1 = -26.092 + 9.4344x_1 + 4.5116x_2 \tag{5.73}$$

$$\hat{y}_2 = 13.94 + 9.6822x_1 - 8.4998x_2 \tag{5.74}$$

yields an estimate of the optimal point $\mathbf{x}^3 = (3.9815, 2.9988)'$, at which point the regression relation for y_0

$$\hat{y}_0 = -66.755 + 24.733x_1 + 11.458x_2 \tag{5.75}$$

takes on the value of 66.061.

It will be noted that the gradients of $\hat{y}_0$ and $\hat{y}_1$ are nearly parallel, which might normally suggest that perhaps only y_1 will be active at the optimum. However, since $\nabla\hat{y}_0\nabla\hat{y}_1'$ is positive, y_0 can be increased indefinitely by increasing x_1 indefinitely and letting x_2 satisfy the constraint $\hat{y}_1$

$$-26.092 + 9.4344x_1 + 4.5116x_2 = 25$$

or

$$x_2 = \frac{51.092 - 9.4344x_1}{415116}$$

By the nature of gradients of linear functions it will take at least as many active constraints as variables to specify a unique optimum from a direction determining experimental block. When the number of constraints is less than the number of variables some of the variables are "free". Since in practice extrapolation is a risky procedure, we can add extra constraints limiting the solution to the experimental region, allowing the calculation of a unique point. Here, for example, let x_2 take on its maximum value (from the experimental block) of 3.20, which means an x_1 equal to 3.885. The calculated y_0 at that point is 65.98.

In the first-order situation then, the sensitivity block will provide a refined estimate of the function and its behavior in the vicinity of the optimum. An improved estimate of the constrained optimum can only be made with any confidence when the system is exactly determined, that is, when the number of active constraints equals the number of variables.

A Simplified Zoutendijk Approach

The simplified Zoutendijk approach is a simplified best feasible directions approach. As shown in the previous chapter, the direction sought is to maximize the expected gain in the objective function while not allowing any active constraint to grow larger. The measure in each case is the inner product of a gradient and a directional; these are also known as direction derivatives.

Starting at the base point $\mathbf{x}^1 = (3.6, 2.26)'$ we need to solve the following constrained maximization problem

$$\max \nabla \hat{y}_0(\mathbf{x}^1)\mathbf{d} \tag{5.76}$$

such that

$$\nabla \hat{y}_2(\mathbf{x}^1)\mathbf{d} \leqslant 0 \tag{5.77}$$

$$d_1^2 + d_2^2 = 1 \tag{5.78}$$

Using the values calculated from the gradient estimating block at $\mathbf{x}^1$, this problem is

$$\max 21.3d_1 + 8.25d_2 \tag{5.79}$$

such that

$$7.63d_1 - 2.88d_2 \leqslant 0 \tag{5.80}$$

$$d_1^2 + d_2^2 = 1 \tag{5.81}$$

In this case the gradient $\nabla \hat{y}_0$ points outside the feasible region so the optimum occurs at the intersection of the two constraints. The first constraint is substituted into the second, yielding

$$\mathbf{s}^1 = \mathbf{d} = (0.3531, 0.9356)'$$

Solving for the estimated λ which will take us to the first constraint yields $\lambda = 1.184$.

Table 5.29. First best feasible direction step determining block

Design point	λ	x_1	x_2	y_0	y_1	y_2
1	0	3.6	2.26	48.756	18.011	27.378
2	0.4	3.7412	2.6342	55.579	21.180	26.753
3	0.8	3.8825	3.0085	63.211	23.997	25.881
4	1.2	4.0237	3.3827	71.545	27.799	24.897

The curvilinear equations from the data in Table 5.29 for y_0 and y_1 are

$$\hat{y}_0 = 48.751 + 16.168\lambda + 2.3597\lambda^2 \tag{5.82}$$

$$\hat{y}_1 = 18.078 + 6.8584\lambda + 0.98956\lambda^2 \tag{5.83}$$

Equation (5.82) shows that y_0 increases indefinitely along the search direction. Setting $\hat{y}_1$ equal to 25 and solving for λ yields $\lambda = 0.89402$. As in earlier examples, $\hat{y}_2 - 27$ also has a root along the search length at $\lambda = 0.24887$, which represents an estimated reentering of the feasible region from the base point. The new base point $\mathbf{x}^2$ is calculated to be

$$\mathbf{x}^2 = (3.9157, 3.0964)'$$

The next gradient estimating block shows that we are already in the vicinity of the optimum, and becomes a sensitivity block instead. All three functions are fitted to the data at this point.

$$\hat{y}_0 = -58.106 + 23.597x_1 + 9.9854x_2 \tag{5.84}$$

$$\hat{y}_1 = -18.995 + 6.8752x_1 + 5.5158x_2 \tag{5.85}$$

$$\hat{y}_2 = 10.425 + 8.8504x_1 - 6.3329x_2 \tag{5.86}$$

Judging from the data in Table 5.30, the optimal point occurs in the immediate vicinity of design point 3 with both constraints active. The point solving that problem is given by

$$\mathbf{x}^3 = (4.0066, 2.9821)'$$

where

$$\hat{y}_0 = 66.2151 \qquad \hat{y}_1 = 25 \qquad \hat{y}_2 = 27$$

What has occurred is that the best feasible direction originally calculated has moved almost parallel to the active second constraint, rather than being directed inward to the first constraint as with the previous methods.

Table 5.30. First best feasible direction determining block

Design point	x_1	x_2	y_0	y_1	y_2
1	3.816	3.00	62.199	23.822	25.171
2	3.816	3.20	63.594	24.858	23.961
3	4.016	3.00	66.316	25.130	26.997
4	4.016	3.20	68.915	26.300	25.674

A Nonlinear Zoutendijk Approach

This version of the best feasible directions problem of Zoutendijk is more conservative than the previous one, since the direction sought is repelled from active constraints and, in addition, nonactive constraints now play a role. These changes essentially have to do with the difference between linear and nonlinear constraints. The previous problem is exact in the linear constraint case (if the gradient estimates are good) and even in the nonlinear case if the constraint set is concave. If the constraint set is concave the gradient direction leads to the interior of the constraint set and all is well. The experimenter usually has no a priori way of knowing whether his constraints are linear or nonlinear, let alone whether they are nonlinear concave or convex. Therefore a more conservative approach may be chosen: as a constraint is approached or as a constraint becomes active the experimenter will force the best feasible direction to veer away from these constraints.

The problem at the base point $\mathbf{x}^1$, where

$$\hat{y}_0 = 49.53 \qquad \hat{y}_1 = 17.79 \qquad \hat{y}_2 = 27.53$$

becomes

$$\max \eta \tag{5.87}$$

such that

$$-21.3d_1 - 8.25d_2 + \eta \leqslant 0 \tag{5.88}$$

$$17.79 + 8.50d_1 + 3.30d_2 + \eta \leqslant 25 \tag{5.89}$$

$$27.53 + 7.63d_1 - 2.88d_2 + \eta \leqslant 27 \tag{5.90}$$

$$|d_1| \leqslant 1 \tag{5.91}$$

$$|d_2| \leqslant 1 \tag{5.92}$$

The last two inequalities are used instead of the nonlinear normalization

$$d_1^2 + d_2^2 = 1 \tag{5.93}$$

so that the entire problem can be now formulated as a linear programming problem.

The linear programming problem must be modified somewhat to be put into standard form. The component directions d_1 and d_2 have a nonnegativity requirement and so are redefined as the combination of two variables, each of which must be nonnegative.

$$d_1 = p_1 - p_2 \tag{5.94}$$

$$d_2 = p_3 - p_4 \tag{5.95}$$

where the p_i satisfy

$$p_i \geqslant 0 \qquad i=1,2,3,4 \tag{5.96}$$

The constraints on the magnitudes of d_1, d_2, (5.91) and (5.92) can be recast in these new variables

$$p_i \leqslant 1 \qquad i=1,2,3,4 \tag{5.97}$$

d_1 and d_2 are replaced in the other equations by the p_i according to (5.94) and (5.95). Since it is known that at most one of p_1 and p_2, and at most one of p_3 and p_4 will be nonzero, these four constraints will be equivalent to the magnitude constraints on d_1 and d_2, and all the other relations will be preserved as well.

Solving the modified problem by linear programming yields the solution

$$\eta = \nabla y_0 \mathbf{d} = 3.904$$
$$\mathbf{d} = (-0.204, 1)'$$

which becomes, after normalization,

$$\mathbf{d} = (-0.1999, 0.9798)'$$

At this point y_2 is active so the step to where y_1 is active is solved, which yields $\lambda = 4.70$. Since **d** has been normalized, this λ represents the distance to the estimated intercept. Here, $\lambda = 4.70$ is a large step, since it represents a change in x_2 of 4.2, and is far beyond the range of the local gradient estimating experimental points. A maximum step of 1.50 is used instead of 4.70. The smaller search length permits more accurate interpolation between experimental points should the actual step turn out to be less than the 4.70 estimated. The data for this line search are arrayed in Table 5.31. If the function remains feasible over the entire length of the reduced line search with the objective function increasing monotonically, extra experiments can easily be run until a peak in the objective is noted or an infeasible region is encountered. Such experiments would be justified as further functional improvement could then be anticipated.

Table 5.31. First nonlinear Zoutendijk step determining block

Design point	λ	x_1	x_2	y_0	y_1	y_2
1	0	3.60	2.26	48.742	18.324	27.494
2	0.5	3.50	2.75	51.940	19.630	23.833
3	1.0	3.40	3.24	55.810	21.904	20.209
4	1.5	3.30	3.73	60.067	24.947	15.922

Table 5.32. First nonlinear Zoutendijk direction-determining block

Design point	x_1	x_2	y_0	y_1	y_2
1	3.2	3.636	57.353	23.269	15.689
2	3.2	3.836	60.095	25.612	14.251
3	3.4	3.636	61.079	24.760	17.269
4	3.4	3.836	63.881	25.937	15.664

Fitting the three functions to the data yields

$$\hat{y}_0 = 48.728 + 5.979\lambda + 1.0595\lambda^2 \tag{5.98}$$

$$\hat{y}_1 = 18.314 + 1.823\lambda + 17373\lambda^2 \tag{5.99}$$

$$\hat{y}_2 = 27.459 - 6.7279\lambda - 0.62657\lambda^2 \tag{5.100}$$

Equating (5.99) to 25 and solving for λ yields a value of 1.5061, a slight extrapolation (0.4% beyond the experimental region). Equation (5.100) becomes feasible for $\lambda \geqslant 0.0677$. The new base point is $\mathbf{x}^2 = (3.3, 3.736)'$. Table 5.32 gives the experimental data for a Zoutendijk direction-determining block.

At this point y_0 is estimated to be

$$\hat{y}_0 = -53.152 + 18.779x_1 + 13.86x_2 \tag{5.101}$$

y_1 is active at this point, but both y_1 and y_2 must be estimated to compute the best feasible direction.

$$\hat{y}_1 = -22.956 + 4.5406x_1 + 8.7972x_2 \tag{5.102}$$

$$\hat{y}_2 = 19.438 + 7.4837x_1 - 7.6059x_2 \tag{5.103}$$

Extracting the gradients from these relations, the new linear programming problem (before converting, as previously, to standard form) becomes

$$\max \eta \tag{5.104}$$

such that

$$-18.779d_1 - 13.860d_2 + \eta \leqslant 0 \tag{5.105}$$

$$24.887 + 4.5406d_1 + 8.7972d_2 + \eta \leqslant 25 \tag{5.106}$$

$$15.712 + 7.4837d_1 - 7.6059d_2 + \eta \leqslant 27 \tag{5.107}$$

$$|d_1| \leqslant 1 \qquad |d_2| \leqslant 1 \tag{5.108}$$

This problem can then be put into standard form and solved as a linear

Table 5.33. Second nonlinear Zoutendijk step determining block

Design point	λ	x_1	x_2	y_0	y_1	y_2
1	0	3.2989	3.7357	60.565	24.791	15.993
2	0.35	3.5439	3.4857	62.245	24.729	19.480
3	0.70	3.7888	3.2358	63.841	24.603	23.652
4	1.05	4.0338	2.9858	66.687	24.710	27.919

programming problem with the result

$$\eta = \nabla y_0 \mathbf{d} = 2.632$$

$$\mathbf{d} = (0.56779, -0.5794)'$$

This direction is normalized to be

$$\mathbf{d} = (0.6999, -0.7142)'$$

Using the normalized direction we calculate that y_2 will become active at $\lambda = 1.06$. Thus we array a line search as shown in Table 5.33.
y_1 is clearly feasible and y_0 increasing along the search direction, so the optimal step will be determined by the y_2 constraint

$$\hat{y}_2 = 15.963 + 9.7438\lambda + 1.5909\lambda^2 \tag{5.109}$$

Equating this to 27 yields a calculated value of $\lambda = 0.97687$. The new base point is $\mathbf{x}^3 = (3.9826, 3.038)'$, where y_2 is estimated to be active and y_1 near its constraint value. A 2^2 factorial experiment at $\mathbf{x}^3$ gives the data in Table 5.34.
Judging from the data a constrained optima has once again been approached. Using

$$\hat{y}_1 = -27.273 + 7.8509x_1 + 7.0229x_2 \tag{5.110}$$

$$\hat{y}_2 = 8.8279 + 7.4007x_1 - 3.8643x_2 \tag{5.111}$$

and setting them equal to their constraint values allows $\mathbf{x}^4$ to be calculated,

$$\mathbf{x}^4 = (4.0045, 2.9666)'$$

at which point y_0 is estimated to be 65.57.

Table 5.34. Second nonlinear Zoutendijk direction determining block

Design point	x_1	x_2	y_0	y_1	y_2
1	3.8826	2.938	62.196	23.773	25.960
2	3.8826	3.138	64.689	25.316	25.684
3	4.0826	2.938	67.062	25.482	27.937
4	4.0826	3.138	69.964	26.747	26.668

Second-Order Approach

The second-order approach does not utilize the first gradient estimating block of the gradient projection section since a larger portion of the feasible region is to be covered by this first block. The point $(2,2)'$ becomes the base $\mathbf{x}^0$ instead. The spacing d will be 1.50 and a rotatable design ($\alpha=\sqrt{2}$) is desired. This parameterization gives rise to two points that are infeasible, violating the explicit bounds, since $2-\sqrt{2}\,(1.5)=-0.1213$. The experimenter has several alternatives in this case. He can move the base point, or reduce the spacing, or alter the properties of the design by altering those two points by moving them into the feasible region, or use the two infeasible points. The alternative chosen is dependent upon whether or not an infeasible point can be run at all. If a variable represents a volume or a mass, for instance, specification of a negative quantity is impossible. On the other hand, the nonnegativity constraints may have been arbitrary, that is negative values are possible but undesired. In this section a negative value is accepted, although it is expected that the solution vector will contain only nonnegative components. The data for this experiment are presented in Table 5.35.

The fitted second-order equation for the objective function y_0 is

$$\hat{y}_0=-0.15815-0.55189x_1+0.88349x_2+0.22203x_1x_2$$
$$+3.0236x_1^2+1.7075x_2^2 \qquad (5.112)$$

The system is positive definite and has an unconstrained minimum at $\mathbf{x}^*=(0.11133,-0.27318)$. At least one of the two constraints will therefore

Table 5.35. A second-order experimental block

Design point	x_1	x_2	y_0	y_1	y_2
1	0.500	0.500	1.2608	−0.0466	4.1628
2	0.500	3.5	25.347	12.263	−7.709
3	3.500	0.500	27.2081	12.609	31.425
4	3.500	3.500	61.7011	24.6618	19.4591
5	4.1213	0.0	48.3590	16.8313	37.0458
6	−0.1213	0.0	0.0759	0.4824	−0.7166
7	0.0	4.1213	32.1857	16.5030	16.7874
8	0.0	−0.1213	−0.3822	−0.0291	0.0096
9	2.0	2.0	19.6481	7.4969	13.7534
10	2.0	2.0	20.1614	7.7606	14.1889
11	2.0	2.0	20.3581	8.0276	14.3219

Table 5.36. A first-order direction determining block

Design point	x_1	x_2	y_0	y_1	y_2
1	3.69	2.75	56.625	20.847	25.833
2	3.69	3.15	60.484	23.258	23.411
3	4.09	2.75	65.499	24.08	28.954
4	4.09	3.15	70.17	26.604	27.151

be active at the optimum.

$$\hat{y}_1 = 0.05692 - 1.5872x_1 + 1.1563x_2 + .065094x_1x_2 + 1.3833x_1^2 + 0.68833x_2^2 \quad (5.113)$$

$$\hat{y}_2 = 0.13853 + 8.3192x_1 + 0.55875x_2 + 0.018706x_1x_2 + 0.16104x_1^2 - 1.1369x_2^2 \quad (5.114)$$

Since the number of constraints is equal to the number of variables and not less, the primal problem was solved using an adaptive random search at several start points. Each start point converged to $\mathbf{x}^2 = (3.89, 2.95)'$; this is already a good approximation to the optimum. At that point y_0 is estimated to be 63.462. A first-order experiment is now run, with the results shown in Table 5.36, to determine whether this is a good estimate of the optimum and whether any substantial improvement in the function can be expected. A 2^2 factorial design is chosen as the first-order design, since it can be expanded into a second-order design if the expected improvement warrants extra experimentation.

The data confirms that we are near both constraint boundaries. The two constraint functions are fitted

$$\hat{y}_1 = -26.491 + 8.223x_1 + 6.1697x_2 \quad (5.115)$$

$$\hat{y}_2 = 8.5543 + 8.5756x_1 - 5.28x_2 \quad (5.116)$$

Solving for the intersection of the two constraints yields a refined estimate of the optimal point, $\mathbf{x}^2 = (4.0039, 3.0094)'$. Since

$$\hat{y}_0 = -58.51 + 23.2x_1 + 10.662x_2 \quad (5.117)$$

the estimate of the function at this point is $\hat{y}_0 = 66.47$.

SUMMARY

The examples set forth in this chapter were chosen to illustrate the methodologies proposed herein. Although not set in any particular physical experimental setting, these example problems enable a clear concept of how the proposed first-order experimental methodologies operate. Direct search and second-order response surface approaches are less well illustrated here, but they are stressed in the chapters to follow.

REFERENCES

1. Ames, W. F., "Canonical Forms for Nonlinear Kinetic Differential Equations," *I & EC Fundamentals*, vol. 1, no. 3, August 1962, pp. 214–218.
2. Anderson, T. F., D. S. Abrams, and E. A. Grens, II, "Evaluation of Parameters for Nonlinear Thermodynamic Models," *American Institute of Chemical Engineering Journal*, vol. 24, no. 1, January 1978, pp. 20–29.
3. Bard, Y., *Nonlinear Parameter Estimation*, Academic Press, New York, 1974.
4. Bard, Y., and L. Lapidus, "Kinetics Analysis by Digital Parameter Estimation," *Catalysis Reviews*, vol. 2, no. 1, 1968, pp. 67–112.
5. Biles, W. E., "Optimization of Multiple-Resonse Simulation Models," Final Report to the Office of Naval Research, Notre Dame, October 1978.
6. Biles, W. E., "A Response Surface Method for Experimental Optimization of Multi-Response Processes," *I & EC Process Design and Development*, vol. 14, no. 2, 1975, pp. 152–158.
7. Biles, W. E., and J. J. Swain, "Strategies for Optimization of Multiple-Response Simulation Models," Proceedings of the 1977 Winter Simulation Conference, Gaithersburg, MD, December 1977.
8. Blau, G. E., R. R. Klimpel, and E. C. Steiner, "Equilibrium Constant and Model Distinguishability," *I & EC Fundamentals*, vol. 11, no. 3, March 1972, pp. 324–332.
9. Boag, I. F., D. W. Bacon, and J. Downie, "Using a Statistical Multiresponse Method of Experimental Design in a Reaction Network Study," *Canadian Journal of Chemical Engineering*, vol. 56, June 1978, pp. 389–395.
10. Box, G. E. P., "The Effects of Errors in the Factor Levels and Experimental Design," *Technometrics*, vol. 5, no. 2, May 1963, pp. 247–262.
11. Box, G. E. P., and N. R. Draper, "The Bayesian Estimation of Common Parameters from Several Responses," *Biometrika*, vol. 52, 1965, pp. 355–365.
12. Box, G. E. P., and W. G. Hunter, "The Experimental Study of Physical Mechanisms," *Technometrics*, vol. 7, no. 1, Feb. 1965, pp. 23–42.
13. Box, G. E. P., W. G. Hunter, J. F. MacGregor, and J. Erjavec, "Some Problems Associated with the Analysis of Multiresponse Data," *Technometrics*, vol. 15, no. 1, February 1973, pp. 33–51.
14. Box, G. E. P., and K. B. Wilson, "On the Experimental Attainment of Optimum Conditions," *Journal of the Royal Statistical Society*, ser. B, vol. 13, 1951, pp. 1–45.
15. Brooks, S. H., and M. R. Mickey, "Optimum Estimation of Gradient Direction in Steepest Ascent Experiments," *Biometrics*, vol. 17, no. 1, March 1961, pp. 48–56.
16. Eakman, J. M., "Strategy for Estimation of Rate Constants from Isothermal Reaction Data," *I & EC Fundamentals*, vol. 8, no. 1, February 1969, pp. 53–58.
17. Eldredge, D. L., "Screening Designs for Simulation," Presented at the ORSA/TIMS Joint National Meeting, Miami, FL, Nov. 1976.
18. Hahn, G. J., W. Q. Meeker, Jr., and P. I. Feder, "The Evaluation and Comparison of

Experimental Designs for Fitting Regression Relationships," *Journal of Quality Technology*, vol. 8, no. 3, July 1976, pp. 140–157.

19. Himmelblau, D. M., C. R. Jones, and K. B. Bischoff, "Determination of Rate Constants for Complex Kinetics Models," *I & EC Fundamentals*, vol. 6, no. 4, November 1967, pp. 539–543.
20. Klimpel, R. R., and G. E. Blau, "The Role of Optimization in Processs Plant Design," Presented at the 1975 ORSA Meeting, Chicago, IL, 1975.
21. Lapidus, L., and T. I. Peterson, "Analysis of Heterogeneous Catalytic Reactions by Nonlinear Estimation," *American Institute of Chemical Engineering Journal*, vol. 11, no. 5, September 1965, pp. 891–897.
22. Mandelbaum, J., "HRB Singer Optimizer Performance on an Evaluation Test Problem," Naval Ship Research and Development Center, Computation and Mathematics Dept. Technical Note CMD-2-73, January 1973.
23. Mezaki, R., N. R. Draper, and R. A. Johnson, "On the Violation of Assumptions in Nonlinear Least Squares by Interchange of Response and Predictor Variables," *I & EC Fundamentals*, vol. 12, no. 2, 1973, pp. 251–254.
24. Montgomery, D. C., *Design and Analysis of Experiments*, John Wiley and Sons, New York, 1976.
25. Montgomery, D. C., and V. M. Bettencourt, "Multiple Response Surface Methods in Computer Simulation," *Simulation*, vol. 29, no. 4, October 1977, pp. 113–121.
26. Montgomery, D. C., and D. M. Evans, Jr., "Second Order Response Surface Designs in Digital Simulation," Presented at the 41st National ORSA Meeting, New Orleans, LA, April 1972.
27. Morshedi, A. M., and R. H. Luecke, "On-Line Optimization of Stochastic Processes," *I & EC Process Design Development*, vol. 16, no. 4, 1977, pp. 473–478.
28. Myers, R. H., *Response Surface Methodology*, distributed by Edwards Brothers, Inc., Ann Arbor, MI, 1976.
29. Noh, J. C., "A Two Phase Complex Method for Nonlinear Process Optimization," Presented to the 45 National ORSA/TIMS Meeting, Boston, MA, 1974.
30. Sane, P. P., R. E. Eckert and J. M. Woods, "On Fitting Combined Integral and Differential Reaction Kinetic Data. Rate Modeling for Catalytic Hydrogenation of Butadiene," *I & EC Fundamentals*, vol. 13, no. 1, 1974, pp. 52–56.
31. Seinfeld, J. H., and L. Lapidus, *Mathematical Methods in Chemical Engineering, Vol. 3, Process Modeling, Estimation and Identification*, Prentice-Hall, Englewood Cliffs, NJ, 1974.
32. Spendley, W., G. R. Hext, and R. F. Himsworth, "Sequential Application of Simplex Designs in Optimization and Evolutionary Operations," *Technometrics*, vol. 4, 1962.
33. Swain, J. J., "Efficient Strategies for Constrained Experimental Optimization," unpublished Master's Thesis, Dept. of Aerospace and Mechanical Engineering, University of Notre Dame, 1977.
34. Tanner, R., "Estimating Kinetic Rate Constants Using Orthogonal Polynomials and Picard's Iteration Method," *I & EC Fundamentals*, vol. 11, no. 1, 1972, pp. 1–8.
35. Ulrichson, D. L., and F. D. Stevenson, "Effects of Experimental Errors on Thermodynamic Consistency and on Representation of Vapor-Liquid Equilibrium Data," *I & EC Fundamentals*, vol. 11, no. 3, 1972, pp. 287–291.
36. Wilde, D. J., and C. S. Beightler, *Foundations of Optimization*, Prentice-Hall, Englewood Cliffs, NJ, 1967.

Chapter 6

Optimization and Experimentation with Physical Processes

In the previous chapters of this book, we have described various methodologies by which one can approach the problem of optimizing a system or process through experimentation. In this chapter, we examine some of the ways in which *physical* systems or processes can be optimized experimentally. In the next chapter, we explore the application of experimental optimization to *computer simulation* models. The unique characteristics of computer simulation deserve special treatment.

This chapter illustrates various experimental optimization procedures as applied to realistic systems and processes found in the literature. The examples focus on problems in metals machining and processing and in chemical process optimization, simply because these areas comprise two of the main endeavors in industrial experimentation. Although no examples are cited for the life sciences or social sciences, the methodologies described here apply to those areas, too.

EXAMPLE 1. OPTIMIZING CUTTING FLUID PRESSURE IN MACHINING

Nagpal and Sharma [9] described experiments to characterize the performance of cutting fluids in machining operations. Citing the work of Pigott and Colwell [10] in optimizing the jet pressure (psi) of cutting fluid, which is applied at high pressure directly on the clearance crevice of the tool, Nagpal and Sharma conclude that tool life is maximized at an optimum jet pressure due to a combination of a lubricating effect, a chip-removal effect, and a heat-transfer effect that the cutting fluid provides. This section describes two approaches to optimizing the tool life y (minutes) as a function of cutting fluid jet pressure x (psi). The results shown here are not exactly as reported in Nagpal and Sharma [9], but are indicative of the results that Pigott and Colwell [10] might have obtained had they employed the experimental approaches described herein. This example illustrates the application of several different approaches to a physical problem involving a *single* control variable x and a *single* response y.

Golden Section Search

Employing the golden section technique described in Chapter 4, we search the interval $0 \leqslant x \leqslant 600$ psi. It is desired to find an estimated optimum $(\hat{x}, \hat{y})$ within a final interval of uncertainty of 50 psi for x. Recall that golden section search requires that the function $y = g(x)$ be unimodal in the search interval (a, b). Here, $a = 0$ and $b = 600$.

The initial pair of search points are placed at

$$x_1 = b - 0.618(b - a) = 600 - 0.618(600 - 0) = 229 \text{ psi}$$

$$x_2 = a + 0.618(b - a) = 0 + 0.618(600 - 0) = 371 \text{ psi}$$

The observed values of y at these search points are

$$y(x_1) = 39 \text{ min}$$

$$y(x_2) = 81 \text{ min}$$

Thus we eliminate the undesired interval $0 \leqslant x < 229$ psi. Placing a second point symmetrically in the remaining interval $229 \leqslant x \leqslant 600$ psi, we get

$$x_3 = 229 + 0.618(600 - 229) = 229 + 0.618(371) = 458 \text{ psi}$$

$$y(x_3) = 82 \text{ min.}$$

Comparing $y(x_2) = 81$ min and $y(x_3) = 82$ min, we now eliminate $x < 371$ psi. Placing x_4 symmetrically in the remaining interval $371 \leqslant x \leqslant 600$ psi, we obtain

$$x_4 = 371 + 0.618(600 - 371) = 371 + 0.618(229) = 513 \text{ psi}$$

$$y(x_4) = 79 \text{ min}$$

Thus we eliminate the interval $x > 513$ psi. Placing x_5 in the remaining interval $371 \leqslant x \leqslant 513$ psi, we find

$$x_5 = 513 - 0.618(513 - 371) = 513 - 0.618(142) = 425 \text{ psi}$$

$$y(x_5) = 84 \text{ min.}$$

We can now eliminate $x > 458$ psi. Placing x_6 in the remaining interval $371 \leqslant x \leqslant 458$ psi, we get

$$x_6 = 458 - 0.618(458 - 371) - 458 - 0.618(87) = 404 \text{ psi}$$

$$y(x_6) = 85 \text{ min}$$

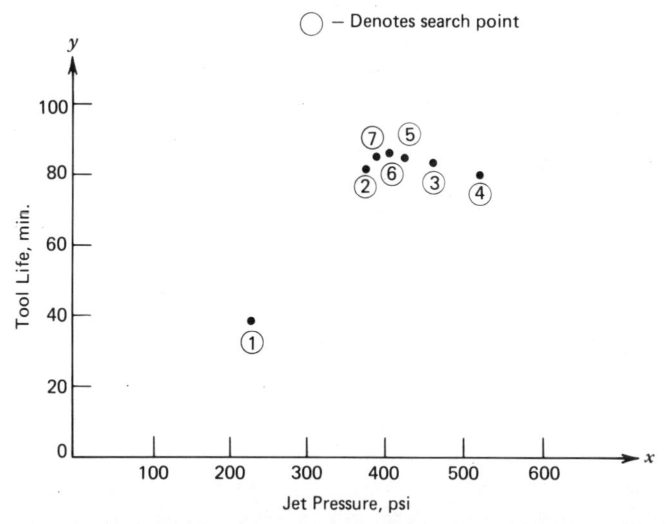

Figure 6.1. Golden section search to optimize cutting fluid pressure.

Eliminating $x > 425$ psi, our remaining interval of uncertainty is $371 \leqslant x \leqslant 425$ psi. Since we desired a final interval of uncertainty no greater than 50 psi, we place a final experiment

$$x_7 = 425 - 0.618(425 - 371) = 425 - 0.618(54) = 392 \text{ psi}$$

$$y(x_7) = 84 \text{ min}$$

Our final interval of uncertainty is therefore $392 \leqslant x \leqslant 425$ psi, with an estimated optimum $\hat{x} = 404$ psi, $\hat{y} = 85$ min. The tool life changes remarkably little over this final range of values of cutting fluid jet pressure, suggesting that some "lack of control" can be tolerated in jet pressure without sustaining significant reduction in tool life. Figure 6.1 illustrates the progress of this golden section search approach to optimizing the cutting fluid jet pressure.

Polynomial Regression

Suppose that we take a polynomial regression approach to seeking the optimum cutting fluid jet pressure, performing seven experiments at uniform intervals over the range $0 \leqslant x \leqslant 600$ psi. Table 6.1 presents the results of this experiment.

Figure 6.2 shows the results of this polynomial regression experiment with the resulting polynomial model

$$y = 6.41 + 0.01765x + 0.00104x^2 - 0.00000153x^3$$

Table 6.1. Polynomial regression experiment: Tool life as a function of jet pressure

Experiment j	Jet pressure (psi) x_j	Tool life (min.) y_j
1	0	6
2	100	20
3	200	33
4	300	68
5	400	85
6	500	80
7	600	63

The polynomial regression experiment would lead us to choose estimated optimum solution $\hat{x}=462$ psi, $\hat{y}=85.7$ min. Although not significantly different from the y obtained through golden section search, the corresponding "interval of uncertainty" after seven uniformly spaced polynomial regression experiments is $400 \leqslant x \leqslant 500$ psi.

There is no "rule" that polynomial regression experiments must be conducted with uniformly spaced points. Therefore, one could apply polynomial regression to the set of observations (x_i,y_i) from a golden section search, thus given a prediction equation over the interval within which the estimated optimum solution $(\hat{x},\hat{y})$ lies. This approach is particularly useful where there is experimental error in the observed values of the

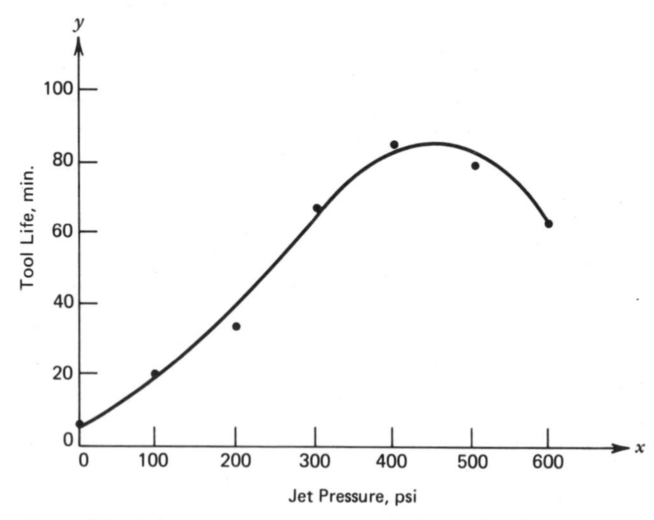

Figure 6.2. Polynomial regression to optimize cutting fluid pressure.

response y. If we apply polynomial regression to six of the seven observations in the golden section search, eliminating $x = 229$ psi because it lies well removed from the remaining six points, the quadratic prediction equation is

$$y = -41.74 + 0.5924x - 0.00069x^2$$

The predicted optimum based on this estimated equation is $\hat{x} = 429$ psi, $\hat{y} = 85.4$ min.

Certain advantages accrue to approaching the polynomial regression approach to optimization through a golden section search. The principal advantage lies in the fact that golden section search points are placed more densely near the optimum, owing to the process of interval elimination, thus providing a smaller confidence interval about the predicted optimum response $\hat{y}$. The main disadvantage of golden section search lies in the sequential nature of the experimentation. If a *block* of experiments requires t time units, regardless of the number of experiments conducted simultaneously, then a polynomial regression approach to experimental optimization using k experiments would require t time units. A golden section search involving k experiments would require kt time units. Thus, polynomial regression is preferred to golden section search when experimental time is an important factor.

EXAMPLE 2. OPTIMIZING YIELD IN A CHEMICAL PROCESS

The previous example illustrated two procedures for experimentally optimizing a single response y as a function of a single independent variable x. In the present example we consider a single response, process yield y (%), as a function of *three* independent variables.

Lind et al. [6] reported a response surface approach to maximizing the yield y (%) from a solvent extraction stage in a fermentation process for producing an antibiotic pharmaceutical. In the extraction operation, an organic solvent was used to extract the antibiotic from a filtrate. Reagents A and B combined with the antibiotic to form a complex molecule that was preferentially soluble in the organic solvent. Usage levels of the two complexing agents, and the pH at which the extraction took place, were known to have an important effect on the yield from the extraction process. Thus, the response variable and three independent variables for this process were

y = yield from the extraction process, %

x_1 = level of complexing agent A, %

x_2 = level of complexing agent B, %

x_3 = pH

The ranges of the independent variables evaluated in this study were

$$1.0 \leqslant x_1 \leqslant 2.0$$

$$1.0 \leqslant x_2 \leqslant 2.0$$

$$4.5 \leqslant x_3 \leqslant 5.5$$

These ranges were well known due to previous laboratory studies. In this section we compare the augmented 2^3 factorial design approach taken by Lind et al. [6] to a central composite design approach [1], and to a Box's Complex Search approach. We also see how a second-order response surface fitted to a set of experimental observations obtained via Box's Complex Search, as described in Chapter 5, is also an efficient and economical approach to experimental optimization.

Augmented 2^3 Factorial Design

Lind et al. [6] performed an augmented 2^3 factorial experimental design with three independent variables. As discussed in Chapter 3, the design points were first coded, employing the following coding scheme:

$$x_1 = \frac{\%A - 0.5}{0.5}$$

$$x_2 = \frac{\%B - 0.5}{0.5}$$

$$x_3 = \frac{\text{pH} - 5.0}{0.5}$$

This produced the experimental design shown in Table 6.2. The extraction yield y was measured three times at each corner point, three times at each face point, and nine times at the center point. Thus, Lind et al. employed 51 experiments in their study. The *average* response y at each design point is included in Table 6.2.

The second-order response surface model fitted to these 51 experimental design points is

$$y = 65.05 + 1.63x_1 + 3.28x_2 + 0.93x_3 - 2.93x_1^2 - 2.02x_2^2$$
$$- 1.07x_3^2 - 0.53x_1x_2 - 0.68x_1x_3 - 1.44x_2x_3$$

The maximum yield $\hat{y}$ can be obtained by taking the three first partial derivatives of this expression, equating the partial derivatives to zero, and

Table 6.2. Coded experimental design points for an extraction yield experiment [6] using an augmented 2^3 factorial design

Design point	x_1	x_2	x_3	y
1	-1	-1	-1	48.9
2	-1	-1	1	56.6
3	-1	1	-1	61.5
4	1	-1	-1	55.2
5	1	1	-1	64.0
6	1	-1	1	58.5
7	-1	1	1	63.3
8	1	1	1	61.8
9	1	0	0	65.5
10	0	1	0	65.0
11	0	0	1	64.6
12	0	0	-1	65.9
13	0	-1	0	63.6
14	-1	0	0	61.3
15	0	0	0	63.4

Note: Design points 1 through 14 were each repeated three times. Design point 15 was repeated nine times. The response y represents the average of all repeated observations.

solving for x_1^*, x_2^*, and x_3^*. Taking the three first partial derivatives, we get

$$\frac{\partial y}{\partial x_1} = 1.63 - 5.86x_1 - 0.53x_2 - 0.680x_3 = 0$$

$$\frac{\partial y}{\partial x_2} = 3.28 - 0.53x_1 - 4.04x_2 - 1.44x_3 = 0$$

$$\frac{\partial y}{\partial x_3} = 0.93 - 0.68x_1 - 1.44x_2 - 2.14x_3 = 0$$

This gives three linear algebraic equations in three unknowns, the solution of which in terms of the coded variables is $x_1 = 0.226$, $x_2 = 0.859$, $x_3 = -0.215$. In terms of the real physical variables, the solution giving maximum extraction yield is

$$x_1^* = 0.61\%A \quad x_2^* = 0.93\%B \quad x_3^* = 4.89 \text{ pH} \quad y^* = 66.54\% \text{ yield}$$

Central Composite Design

The 51 experiments employed to obtain this solution leaves much room for improved experimental efficiency. It is worthwhile to consider how a more conventional second-order response-surface design would perform with this problem. Using a uniform precision central-composite design, as recommended by Montgomery [7] and Box and Hunter [1], with the same coding scheme adopted by Lind et al. [6], eight factorial (corner) points, six axial points, and six replicates of the center point are needed. These 20 design points and their yield responses are shown in Table 6.3. The yield responses were obtained by using the second-order response surface found above and adding a random error term ε from the distribution $N(0, 1)$. Lind et al. reported an error variance of approximately $s^2 = 1$ for their experiment.)

It should be noted that the coded axial points -1.682 and 1.682 represent a larger range of experimentation than that conducted by Lind et al. The corresponding values for the real variables are $0.66 \leqslant \%A \leqslant 2.34\%$,

Table 6.3. Coded experimental design points for an extraction yield experiment [6] using a uniform precision central composite design

Design point	x_1	x_2	x_3	y
1	-1	-1	-1	49.9
2	-1	-1	1	56.7
3	-1	1	-1	62.0
4	1	-1	-1	54.2
5	1	1	-1	63.8
6	1	-1	1	60.2
7	-1	1	1	59.6
8	1	1	1	62.3
9	-1.682	0	0	55.1
10	1.682	0	0	59.7
11	0	-1.682	0	54.4
12	0	1.682	0	66.0
13	0	0	-1.682	61.3
14	0	0	1.682	64.4
15	0	0	0	65.0
16	0	0	0	63.3
17	0	0	0	65.8
18	0	0	0	65.4
19	0	0	0	66.3
20	0	0	0	65.3

$0.66 \leqslant \%B \leqslant 2.34\%$, and $4.16 \leqslant \text{pH} \leqslant 5.84$. It is assumed for the purpose of this illustrative problem that these somewhat wider ranges of values for the independent variables x_1, x_2, x_3 do not produce physically infeasible values.

The second-order response surface fitted to the 20 design points in Table 6.3 is

$$y = 65.25 + 1.62x_1 - 4.11x_2 + 0.885x_3 - 3.28x_1^2 - 1.345x_3^2 + 0.40x_1x_2 - 1.84x_2x_3$$

The three partial derivatives of this equation are

$$\frac{\partial y}{\partial x_1} = 1.62 - 6.56x_1 + 0.40x_2 = 0$$

$$\frac{\partial y}{\partial x_2} = -4.11 + 0.40x_1 - 4.58x_2 - 1.84x_3 = 0$$

$$\frac{\partial y}{\partial x_3} = 0.885 - 1.84x_2 - 2.69x_3 = 0$$

Solving these three equations gives the solution

$$x_1 = 0.313, \qquad x_2 = 1.093, \qquad x_3 = -0.416$$

in terms of the coded variables. For the original physical variables, this optimum solution is

$$x_1^* = 0.66\%A, \qquad x_2^* = 1.05\%B, \qquad x_3^* = 4.79 \text{ pH}, \qquad y^* = 67.5\% \text{ yield}$$

Assuming that the response equation obtained by Lind et al. is the true surface, the above solution would give a true yield of 65.7%, only 1.2% lower than that predicted by Lind et al. But we have used only 20 design points as compared to their 51. The economy derived by saving 31 design points would seem to justify a 1.2% lower value for the solution.

Box's Complex Search

It is worthwhile to compare the central composite design approach to Box's complex search from the standpoint of experimental efficiency. To make this comparison, we conduct a complex seach that terminates at 20 search points. Table 6.4 shows 20 Box's complex search points. The estimated optimum solution is $x_1^* = 1.65\%A$, $x_2^* = 2.15\%B$, $x_3^* = 4.88$ pH, and $y^* = 68.9\%$ yield. The true yield at this solution is 66.1%. It should be noted that the estimated optimum was found with the 12th search point.

As was indicated with the first example, in which polynomial regression was applied to the observations in a golden section search, it can sometimes be beneficial to apply response surface methodology to the sequence

Table 6.4. A Box's complex search approach to maximizing yield for an antibiotic extraction process [6]

Search point		x_1*	x_2*	x_3*	y
1	Initial Complex	0.89	1.12	5.24	55.2
2	Initial Complex	0.51	1.86	5.04	53.4
3	Initial Complex	1.94	1.09	5.58	59.8
4	Initial Complex	0.54	0.88	4.91	41.1†
5	Initial Complex	0.73	1.77	4.57	56.4
6		1.49	2.04	5.30	63.5
7		2.01	1.15	5.31	61.3
8		2.20	1.90	5.14	63.3
9		2.50	1.43	5.70	53.3
10		2.20	1.49	5.50	61.3
11		2.01	2.20	5.06	63.6
12		1.65	2.15	5.88	68.9‡
13		1.79	2.31	5.04	63.6
14		1.28	2.45	5.00	63.0
15		1.51	2.31	5.03	63.4
16		1.97	2.04	5.11	65.5
17		2.22	2.32	4.74	60.8
18		2.04	2.25	4.88	65.0
19		1.71	2.17	4.90	65.4
20		1.89	2.00	4.84	63.1

*Uncoded values.
‡Best point found in search.
†Worst point in initial complex.

of Box's complex search points. That is, we can use the least-squares method to fit a second-order response surface model to the complex search points as described in Chapter 5. The regression equation found by fitting a second-order model to the 20 search points given in Table 6.4 is

$$y = -833.6 - 15.6x_1 - 165.8x_2 + 302.6x_3 - 11.2x_1^2 - 12.5x_2^2 - 27.3x_3^2 + 1.53x_1x_2 + 9.66x_1x_3 - 24.0x_2x_3$$

The set of first partial derivatives equated to zero yields

$$\frac{\partial y}{\partial x_1} = -15.6 - 22.4x_1 + 1.53x_2 - 9.66x_3 = 0$$

$$\frac{\partial y}{\partial x_2} = 165.8 + 1.53x_1 - 25.0x_2 - 24.0x_3 = 0$$

$$\frac{\partial y}{\partial x_3} = 302.6 + 9.66x_1 - 24.0x_2 - 54.6x_3 = 0$$

The solution of these simultaneous linear algebraic equations is

$$x_1^* = 0.57\% A, \qquad x_2^* = 0.97\% B, \qquad x_3^* = 4.95 \text{ pH}, \qquad y^* = 67.1\% \text{ yield}$$

The true yield at these values of the process variables, which in coded form are $x_1 = 0.14$, $x_2 = 0.94$, and $x_3 = -0.1$, is 66.5%. This solution is almost exactly the same as that found by Lind et al. [6].

Thus Box's complex search, terminated at the same number of search points as contained in a uniform precision central composite design and fitted to a second-order response surface, yields effective optima.

EXAMPLE 3. MULTIPLE INDEPENDENT VARIABLES, MULTIPLE RESPONSES WITH A MACHINING PARAMETERS PROBLEM

In the two previous examples, we saw how to optimize a single response as a function of a single independent variables x, and a single response y as a function of several independent variables $x_1, \ldots, x_n$. In this example, we consider how to treat multiple responses $y_1, \ldots, y_m$ where each response y_j is a function of the same set of independent variables $x_1, \ldots, x_n$.

The selected example is taken from the paper by Taraman [12, 13] dealing with metal machining. The purpose of Taraman's work was to develop a methodology that would allow determination of the cutting conditions (cutting speed, feed rate, and depth of cut) such that a specified criterion for each of several machining parameters (surface finish, cutting force, and tool life) could be achieved simultaneously.

Central Composite Design

Taraman employed a central composite design with 24 design points. Three independent variables were involved:

v = cutting speed, fpm

f = feed rate, ipm.

d = depth of cut, in.

Each of three responses were measured:

R = surface finish, CLA microinches

F = cutting force, lb_f.

T = tool life, min.

A function relationship between each of these machining responses and the independent variables can be represented by

$$Y = CV^p f^m d^n$$

where Y represents the response (surface roughness, cutting force, or tool life) and V, f, and d are the cutting speed, feed, and depth of cut, respectively. This model can be written as

$$\ln Y = \ln C + p \ln V + m \ln f + n \ln d$$

which after a suitable transformation can be expressed as

$$y = b_0 + b_1 x_1 + b_2 x_2 + b_3 x_3$$

Using the coding scheme,

$$x_1 = \frac{\ln V - \ln 440}{\ln 570 - \ln 440}$$

$$x_2 = \frac{\ln f - \ln 0.00905}{\ln 0.01416 - \ln 0.00905}$$

$$x_3 = \frac{\ln d - \ln 0.029}{\ln 0.040 - \ln 0.029}$$

the first twelve design points in Table 6.5 were conducted. Utilizing the least-squares method, the following first-order response-surface models were developed

$$y_1 = 4.933 - 0.094 x_1 + 0.509 x_2 + 0.059 x_3$$

$$y_2 = 4.447 - 0.030 x_1 + 0.338 x_2 + 0.214 x_3$$

$$y_3 = 3.655 - 0.364 x_1 - 0.111 x_2 - 0.057 x_3$$

Transforming these equations by taking the exponential e of each side, the following models resulted:

$$R = 872{,}467\, V^{-0.366} f^{1.258} d^{0.185}$$

$$F = 90{,}780\, V^{-0.117} f^{0.835} d^{0.667}$$

$$T = 30{,}500\, V^{-1.408} f^{-0.273} d^{-0.1777}$$

These equations indicate that a reduction in all the investigated outputs (surface roughness, cutting force, and tool life) can be achieved by increasing cutting speed. However, as the feed rate increases, surface roughness and cutting force increase while tool life is reduced. An increase in the depth of cut reduces tool life, and increases surface roughness and cutting force.

Taraman [12, 12] performed an analysis of variance on each of the first-order models. He found that they were adequate, but that a second-order model might provide a better fit. He then performed the next twelve experiments in Table 6.5 and fitted second-order models to the data. Taraman chose to perform two replicates of axial points located at the

Table 6.5. Second-order response surface design for a multivariable, multiresponse machining experiment [12, 13]

Trial number	Performance sequence number	Conditions Speed (sfpm)	Feed (ipr)	Depth (in.)	Coding x_1	x_2	x_2	R (CLA)	T min	Fe (lb_f)
1	2	340	0.00630	0.0210	−1	−1	−1	88	70	53
2	7	570	0.00630	0.0210	1	−1	−1	776	29	48
3	10	340	0.01416	0.0210	−1	1	−1	259	60	100
4	6	570	0.01416	0.0210	1	1	−1	194	28	92
5	11	340	0.00630	0.0400	−1	−1	1	105	64	76
6	1	570	0.00630	0.0400	1	−1	1	82	32	74
7	3	340	0.01416	0.0400	−1	1	1	270	44	155
8	12	570	0.01416	0.0400	1	1	1	250	24	150
9	8	440	0.00905	0.0290	0	0	0	123	35	82
10	9	440	0.00905	0.0290	0	0	0	136	31	85
11	4	440	0.00905	0.0290	0	0	0	130	38	83
12	5	440	0.00905	0.0290	0	0	0	121	35	85
13	18	305	0.00905	0.0290	$-\sqrt{2}$	0	0	159	52	88
14	17	635	0.00905	0.0290	$\sqrt{2}$	0	0	115	23	80
15	15	440	0.00472	0.0290	0	$-\sqrt{2}$	0	77	40	50
16	14	440	0.01732	0.0290	0	$\sqrt{2}$	0	324	28	129
17	16	440	0.00905	0.0135	0	0	$-\sqrt{2}$	114	46	68
18	13	440	0.00905	0.0455	0	0	$\sqrt{2}$	215	33	124
19	20	305	0.00905	0.0290	$-\sqrt{2}$	0	0	139	46	87
20	19	635	0.00905	0.0290	$\sqrt{2}$	0	0	111	27	78
21	22	440	0.00472	0.0290	0	$-\sqrt{2}$	0	61	37	49
22	23	440	0.01732	0.0290	0	$\sqrt{2}$	0	340	34	130
23	24	440	0.00905	0.0135	0	0	$-\sqrt{2}$	128	41	71
24	21	440	0.00905	0.0455	0	0	$\sqrt{2}$	232	28	123

coded values $x_i = \pm 2$. The second-order models obtained by a least-squares fit of the data in Table 6.5 were

$$y_1 = 4.877 - 0.096x_1 + 0.535x_2 + 0.143x_3 - 0.022x_1^2 + 0.054x_2^2 + 0.097x_3^2$$

$$y_2 = 4.426 - 0.034x_1 + 0.339x_2 + 0.209x_3 - 0.002x_1^2 - 0.21x_2^2 + 0.052x_3^3$$

$$y_3 = 3.508 - 0.301x_1 - 0.094x_2 - 0.092x_3 + 0.049x_1^2 + 0.041x_2^2 + 0.069x_2^2$$

In each of these models, the interaction terms $b_{ij}x_ix_j$ were found to be inadequate and were dropped from the models.

Taraman illustrated the response surface approach to multivariable optimization of machining parameters by finding the maximum tool life L, subject to bounds on the machining parameters V, f, and d and on the responses surface roughness R and cutting force F. In terms of these variables and responses, the optimization problem is maximize L subject to

$$R \leqslant 80 \text{ CLA } \mu\text{in. (surface roughness)}$$

$$F \leqslant 100 \text{ lb}_f \text{ (cutting force)}$$

$$305 \leqslant V \leqslant 635 \text{ sfpm (cutting speed)}$$

$$0.00472 \leqslant f \leqslant 0.006 \text{ ipr (feed)}$$

$$0.021 \leqslant d \leqslant 0.06 \text{ in. (cutting depth)}$$

In terms of the transformed variables, this optimization problem becomes

$$\max y_1 = \ln L$$

subject to

$$y_2 = \ln R \leqslant 4.38$$

$$y_3 = \ln F \leqslant 4.60$$

$$-1.42 \leqslant x_1 \leqslant 1.42$$

$$-1.45 \leqslant x_2 \leqslant -0.918$$

$$-1.00 \leqslant x_3 \leqslant 2.26$$

where x_1, x_2, x_3 are as defined previously. Taraman approached this optimization graphically by plotting the contours of L and R as functions of f and V at constant values of d, with cutting force F taking any value. He found as an optimal solution $L^* = 75$ min., $R^* = 80$ CLA microinches, $f^* = 0.006$ ipr, $V^* = 305$ sfpm, at $d = 0.021$ inches.

If we approach the problem of optimizing exactly the same problem by applying Box's complex search in a purely computational manner, the following solution in terms of the transformed variables is obtained after

37 iterations: $x_1^* = -1.400$, $x_2^* = 1.236$, $x_3^* = -0.979$, $y_1^* = 4.36$, $y_2^* = 4.34$, $y_3^* = 3.86$. In terms of the actual machining variables, this solution is $L^* = 78.3$ min., $R^* = 76.9$ CLA μin., $F^* = 47.6$ lb_f, $V^* = 306$ sfpm, $f^* = 0.052$ ipr, $d^* = 0.021$ in.

This solution is somewhat better than Taraman's, in terms of both tool life (78.3–75 min.) and surface finish (76.9–80 CLA μin.). Presumably this improvement is due to a computational approach to optimization rather than a graphical approach.

It is also important to note that, if depth of cut d is allowed to take on values smaller than 0.021 in., superior solutions in terms of tool life can be obtained. If the lower bound on d is relaxed to 0.014 in., Box's complex search produces the following solution after 43 iterations:

$$L^* = 99.4 \text{ min.}, \quad R^* = 79.6 \text{ CLA } \mu\text{in.}, \quad F^* = 42.6 \text{ lb}_f,$$
$$V^* = 307 \text{ scfm}, \quad f^* = 0.00487 \text{ ipr}, \quad d^* = 0.0165 \text{ in.}$$

However, the small depth of cut prolongs cutting time, thus adding to the cost of the machined part, so that little is gained by this relaxation.

EXAMPLE 4. MULTIPLE INDEPENDENT VARIABLES, MULTIPLE RESPONSES WITH CHEMICAL PROCESSES

Hendrix [5] illustrated several experimental approaches to chemical process and product development using as an example a process by the Tanglefoot Flypaper Company for producing the insecticide dibutylfutile (which sounds suspiciously hypothetical) by reacting difutilic acid and butanol. As Hendrix described the problem, the process was put into production without a thorough laboratory investigation of the effects of the process variables upon the yield. Although the plant-scale process is operating smoothly, the yield is substantially less than that measured in some laboratory trials. It is desired to perform plant-scale experimentation to optimize process yield.

The process produces an undesirable by-product, futraldehyde, which must not exceed 2.5% based on pure dibutlyfutile in the product. Any product containing excessive amounts of futraldehyde is a total loss.

The process variables are as follows:

x_1 = catalyst concentration, ppm.

x_2 = reaction temperature, °C.

x_3 = agitator speed, rpm.

x_4 = feed ratio (butanol/difutilic acid).

The values of these process variables for the current production process are 16 ppm, 110°C, 400 rpm and 2.0, respectively. The feasible ranges of these process variables, based on equipment capability and past laboratory results, are as follows: $0 \leqslant x_1 \leqslant 25$ ppm; $50 \leqslant x_2 \leqslant 150$°C; $x_3 = 400$ rpm, 600 rpm or 800 rpm; and $1.5 \leqslant x_4 \leqslant 2.5$.

The last twelve production batches of dibutylfutile were made at the conditions $x_1 = 12$ ppm catalyst, $x_2 = 110$°C reaction temperature, $x_3 = 400$ rpm agitator speed, and $x_4 = 1.95$ feed ratio. The mean process yield over these 12 runs was 67.4%, with a standard deviation of 2.1%. The mean futraldehyde level was 1.95%, with a standard deviation of 0.1%.

The plant manager wishes to turn the production process over to the Process Development Department for 7 days. Including a "turn around" time to flush out the reactor and recharge it, a maximum of 35 experiments can be conducted during the allotted time.

Another important consideration in this experimental program is the necessity to have a low probability of a production batch exceeding 2.5% futraldehyde. If the experimental program should indicate an optimum solution that yields a mean futraldehyde level at the bounding value of 2.5%, about one-half of all batches would exceed this value. But the standard deviation of the futraldehyde level at the current production conditions is 0.1%. Assuming that this variable is normally distributed about its mean, if the predicted optimum is such that the resulting mean futraldehyde level is maintained at 2.3% (two standard deviations away from the bounding value of 2.5%) then only approximately 2.5% of all production batches would be discarded. This strategy was adopted for the experimental program.

The experimental optimization problem can be stated as follows:

$$\text{maximize } y_1 = g_1(x_1, x_2, x_3, x_4), \text{ process yield (\%)}$$

subject to

$$y_2 = g_2(x_1, x_2, x_3, x_4) \leqslant 2.3\% \text{ futraldehyde}$$

and

$$0 \leqslant x_1 \leqslant 25 \text{ ppm catalyst}$$

$$50 \leqslant x_2 \leqslant 150\text{°C reactor temperature}$$

$$x_3 = (400; 600; 800) \text{ rpm agitator speed}$$

$$1.5 \leqslant x_4 \leqslant 2.5 \text{ feed ratio}$$

$$N \leqslant 35 \text{ experiments}$$

Central Composite Design

One possible approach to maximizing process yield is to perform a second-order response surface experimental design over the known ranges of the process variables, employ regression analysis to estimate quadratic models for $y_1=g_1(x_1,x_2,x_3,x_4)$ and $y_2=g_2(x_1,x_2,x_3,x_4)$, and apply a mathematical programming method to find a predicted optimum solution. Confirmation experiments would then be performed in the near-neighborhood of the predicted solution. A central composite design would require at least $N=2^4+2\times4+1=25$ design points. It should be noted, however, that neither a uniform precision design nor an orthogonal design is possible since only three levels of agitator speed (400; 600; 800 rpm) are feasible instead of five.

An alternative experimental approach is to conduct three separate 3-variable central composite designs, one at each level of agitator speed, optimizing each such process separately and choosing the one "optimum" that maximizes process yield subject to the known restrictions. But this approach would require at least $N=3(2^3+2\times3+1)=45$ experiments, significantly exceeding the allotted 7 days. Moreover, no measure could be gained of the possible interactions of agitator speed with other process variables. This approach is clearly a poor idea.

We adopt the approach of performing $N=2^4+2\times4+4$ experiments arrayed in a nonstandard central composite design; that is, three of the process variables will be examined at five levels, but agitator speed will be given only its three feasible levels. Table 6.6 gives the values of these 28 design points and their responses.

Fitting second-order response surface models to these 28 design points, we obtain the following fitted equations:

$$\begin{aligned}\hat{y}_1 = &-2379.25-22.44x_1-3.180x_2-0.0938x_3+3029x_4\\ &-0.2426x_1^2-0.0285x_2^2+0.00006355x_3^2-954.0x_4^2\\ &+0.1816x_1x_2-0.00104x_1x_3+3.683x_1x_4+0.000568x_2x_3\\ &+3.969x_2x_4-0.01463x_3x_4\end{aligned}$$

$$\begin{aligned}\hat{y}_2 = &\,9.657-0.07267x_1-0.09107x_2+0.002531x_3-4.190x_4\\ &-0.006134x_1^2-0.0001317x_2^2+0.000001013x_3^2-0.707x_4^2\\ &-0.0007814x_1x_2+0.000005466x_1x_3+0.1075x_1x_4\\ &-0.00000187x_2x_3+0.07358x_2x_4-0.002031x_3x_4\end{aligned}$$

Table 6.6. Central composite design

Batch number	Randomization sequence	Catalyst concentration	Reaction temperature	Agitator speed	Feed ratio	Yield	Impurity
1	6	8	105	400	1.8	80.07	2.32
2	17	16	105	400	1.8	53.11	1.31
3	9	8	115	400	1.8	68.56	2.37
4	21	16	115	400	1.8	62.43	1.53
5	3	8	105	800	1.8	80.23	2.37
6	11	16	105	800	1.8	55.36	1.43
7	8	8	115	800	1.8	80.13	2.18
8	24	16	115	800	1.8	62.47	1.56
9	18	8	105	400	2.0	46.18	2.34
10	27	16	105	400	2.0	27.58	1.92
11	22	8	115	400	2.0	48.77	2.67
12	2	16	115	400	2.0	43.27	1.84
13	13	8	105	800	2.0	50.42	2.18
14	19	16	105	800	2.0	28.09	1.61
15	23	8	115	800	2.0	51.03	2.62
16	4	16	115	800	2.0	45.52	1.72
17	7	6	110	600	1.9	76.56	2.46
18	12	18	110	600	1.9	57.06	1.17
19	10	12	102	600	1.9	66.40	2.04
20	26	12	118	600	1.9	81.01	2.02
21	5	12	110	400	1.9	80.45	2.18
22	1	12	110	800	1.9	74.11	1.90
23	14	12	110	600	1.75	63.78	1.73
24	16	12	110	600	2.05	29.35	2.27
25	28	12	110	600	1.9	81.21	2.28
26	20	12	110	600	1.9	77.88	2.05
27	15	12	110	600	1.9	75.32	2.00
28	25	12	110	600	1.9	74.24	2.12

The F values for these two fitted models are $F(\hat{y}_1)=50.8$ and $F(\hat{y}_2)=13.1$, as compared to a tabulated F value of 2.04 at $\nu_1=14$, $\nu_2=13$, and $\alpha=0.05$. Therefore, each of these fitted models can be regarded as quite satisfactory representations of the respective relationships $\eta_1(x_1, x_2, x_3, x_4)$ and $\eta_2(x_1, x_2, x_3, x_4)$. Complete statistical summaries for these two responses, using the IBM Scientific Subroutine Package version of the multiple linear regression program, are given in Table 6.7 and 6.8, respectively.

The first step in finding an optimum solution is to hold the agitator speed x_3 at one of its fixed levels and optimize the three remaining process variables. Of course, this procedure must be repeated at the other fixed levels of x_3. Setting $x_3=800$ rpm yields the equations

$$\hat{y}_1=-2413.4-23.27x_1-2.726x_2+3017x_4-0.2426x_1^2$$
$$-0.0285x_2^2-954.0x_4^2+0.1816x_1x_2+3.683x_1x_4$$
$$+3.969x_2x_4$$

$$\hat{y}_2=12.33-0.0683x_1-0.0926x_2-5.813x_4-0.006134x_1^2$$
$$-0.0001317x_2^2-0.707x_4^2-0.0007814x_1x_2$$
$$+0.1075x_1x_4+0.07358x_2x_4$$

Now equating the set of first partial derivatives of $\hat{y}_1$ to zero yields

$$\frac{\partial \hat{y}_1}{\partial x_1}=-23.27-0.4852x_1+0.1816x_2+3.683x_4=0$$

$$\frac{\partial \hat{y}_1}{\partial x_2}=-2.726+0.1816x_1-0.057x_2+3.969x_4=0$$

$$\frac{\partial \hat{y}_1}{\partial x_4}=3017+3.683x_1+3.969x_2-1908x_4=0$$

Solving these three equations for x_1, x_2, and x_4 yields $x_1=14.57$ ppm, $x_2=127.7°C$, and $x_4=1.875$ feed ratio. The process yield at this solution is $y_1=74.98\%$, whereas at the impurity level it is $y_2=1.78\%$. Although this is certainly a feasible solution, it would hardly appear to be worthwhile given that several of the actual experimental trials gave higher yields.

At $x_3=600$ rpm agitator speed, the following equations hold:

$$\hat{y}_1=-2412.6-23.06x_1-3.52x_2+3020x_4-0.2426x_1^2$$
$$-0.0285x_2^2-954.0x_4^2+0.1816x_1x_2+3.683x_1x_4$$
$$+3.969x_2x_4$$

$$\hat{y}_2=11.54-0.0694x_1-0.0922x_2-5.409x_4-0.006134x_1^2$$
$$-0.0001317x_2^2-0.707x_4^2-0.0007814x_1x_2+0.1075x_1x_4$$
$$+0.07358x_2x_4$$

	VARIABLE NO.	MEAN	STANDARD DEVIATION	CORRELATION X VS Y	REGRESSION COEFFICIENT	STD. ERROR OF REG.COFF.	COMPUTED T VALUE
x_1	1	0.12000E 02	0.34854E 01	-0.38974E 00	-0.22439E 02	0.62399E 01	-0.35961E 01
x_2	2	0.11000E 03	0.44221E 01	0.15770E 00	-0.31804E 01	0.89432E 01	-0.35561E 00
x_3	3	0.60000E 03	0.16329E 03	0.45051E-01	-0.93786E-01	0.13052E 00	-0.71854E 00
x_{4_2}	4	0.18999E 01	0.87135E-01	-0.62851E 00	0.30286E 04	0.32850E 03	0.92195E 01
x_{1_2}	5	0.15571E 03	0.84277E 02	-0.41805E 00	-0.24261E 00	0.65938E-01	-0.36794E 01
x_{2_2}	6	0.12118E 05	0.97303E 03	0.15561E 00	-0.28499E-01	0.37882E-01	-0.75231E 00
x_{3_2}	7	0.38571E 06	0.19692E 06	0.14798E-01	0.63551E-04	0.41069E-04	0.15474E 01
x_4	8	0.36135E 01	0.33128E 00	-0.64709E 00	-0.95368E 03	0.71376E 02	-0.13361E 02
$x_1 x_2$	9	0.13200E 04	0.38735E 03	-0.35766E 00	0.18158E 00	0.41324E-01	0.43941E 01
$x_1 x_3$	10	0.72000E 04	0.29313E 04	-0.25578E 00	-0.10387E-02	0.10330E-02	-0.10055E 01
$x_1 x_4$	11	0.22796E 02	0.67119E 01	-0.47975E 00	0.36826E 01	0.20662E 01	0.17822E 01
$x_2 x_3$	12	0.66000E 05	0.18174E 05	0.68635E-01	0.56823E-03	0.82665E-03	0.68739E 00
$x_2 x_4$	13	0.20899E 03	0.12751E 02	-0.36578E 00	0.39688E 01	0.16535E 01	0.24002E 01
$x_3 x_4$	14	0.11400E 04	0.31501E 03	-0.60582E-01	-0.14632E-01	0.41325E-01	-0.35408E 00
y_1	DEPENDENT 15	0.61407E 02	0.17116E 02				

INTERCEPT -2379.25781

MULTIPLE CORRELATION , 0.99097

STD. ERROR OF ESTIMATE 3.30588

ANALYSIS OF VARIANCE FOR THE REGRESSION

SOURCE OF VARIATION	DEGREES OF FREEDOM	SUM OF SQUARES	MEAN SQUARES	F VALUE
ATTRIBUTABLE TO REGRESSION	14	0.77683E 04	0.55488E 03	0.50772E 02
DEVIATION FROM REGRESSION	13	0.14207E 03	0.10928E 02	
TOTAL	27	7910.47071		

	VARIABLE NO.	MEAN	STANDARD DEVIATION	CORRELATION X VS Y	REGRESSION COEFFICIENT	STD. ERROR OF REG.COFF.	COMPUTED T VALUE
x_1	1	0.12000E 02	0.34854E 01	-0.88483E 00	-0.72567E-01	0.27146E 00	-0.26768E 00
x_2	2	0.11000E 03	0.44221E 01	0.10571E 00	-0.91073E-01	0.38907E 00	-0.23407E 00
x_3	3	0.60000E 03	0.16329E 03	-0.10654E 00	0.25509E-02	0.56783E-02	0.44571E 00
x_{4_2}	4	0.18999E 01	0.87135E-01	0.28964E 00	-0.41899E 01	0.14291E 02	-0.29317E 00
x_{1_2}	5	0.15571E 03	0.84277E 02	-0.89641E 00	-0.61542E-02	0.28685E-02	-0.21384E 01
x_{2_2}	6	0.12118E 05	0.97303E 03	0.10564E 00	-0.13174E-03	0.16480E-02	-0.79937E-01
x_{3_2}	7	0.38571E 06	0.19692E 06	-0.10742E 00	0.10132E-05	0.17867E-05	0.56712E 00
x_4	8	0.36135E 01	0.33128E 00	0.28857E 00	-0.70690E 00	0.31051E 01	-0.22765E 00
$x_1 x_2$	9	0.13200E 04	0.38735E 03	-0.86253E 00	-0.78140E-03	0.17977E-02	-0.43464E 00
$x_1 x_3$	10	0.72000E 04	0.29313E 04	-0.70065E 00	0.54662E-05	0.44943E-04	0.12162E 00
$x_1 x_4$	11	0.22796E 02	0.67119E 01	-0.82413E 00	0.10746E 00	0.89891E-01	0.11954E 01
$x_2 x_3$	12	0.66000E 05	0.18174E 05	-0.90032E-01	-0.18694E-05	0.35962E-04	-0.51983E-01
$x_2 x_4$	13	0.20899E 03	0.12751E 02	0.28954E 00	0.73583E-01	0.71934E-01	0.10229E 01
$x_3 x_4$	14	0.11400E 04	0.31501E 03	-0.60815E-01	-0.20513E-02	0.17978E-02	-0.11298E 01
y_2	DEPENDENT 16	0.20067E 01	0.38742E 00				

INTERCEPT 9.65700

MULTIPLE CORRELATION . 0.96625

STD. ERROR OF ESTIMATE 0.14381

ANALYSIS OF VARIANCE FOR THE REGRESSION

SOURCE OF VARIATION	DEGREES OF FREEDOM	SUM OF SQUARES	MEAN SQUARES	F VALUE
ATTRIBUTABLE TO REGRESSION	14	0.37837E 01	0.27026E 00	0.13066E 02
DEVI&TION FROM REGRESSION	13	0.26889E 00	0.20684E-01	
TOTAL	27	4.05260		

The set of first partial derivatives of $\hat{y}_1$ are

$$\frac{\partial \hat{y}_1}{\partial x_1} = -23.06 - 0.4852x_1 + 0.1816x_2 + 3.683x_4 = 0$$

$$\frac{\partial \hat{y}_1}{\partial x_2} = -3.52 + 0.1816x_1 - 0.057x_2 + 3.969x_4 = 0$$

$$\frac{\partial \hat{y}_1}{\partial x_4} = -3020 + 3.683x_1 + 3.969x_2 - 1908x_4 = 0$$

The solution to these equations is $x_1 = 25.8$ ppm, $x_2 = 156°C$, and $x_4 = 1.95$. The values for x_1 and x_2 lie beyond the acceptable ranges. We can easily see that setting the agitator speed to 400 rpm would produce an even more infeasible solution.

We have a current solution at $x_1 = 14.57$, $x_2 = 127.7$, $x_3 = 800$, and $x_4 = 1.875$, giving $y_1 = 75\%$ yield and $y_2 = 1.78\%$ impurity. But we have seven experiments remaining. Since we cannot obtain a second-order response surface design with only seven design points, we should employ a local search around the current solution. If we set $x_3 = 800$ rpm, a sequential simplex search with x_1, x_2, and x_4 would involve four points to initiate the search. A Box's complex search would require $n + 2 = 5$ initial search points and two additional search points. We shall employ the sequential simplex search with a fairly wide point spacing, performing an experiment at each simplex point. The simplex of four points is as follows:

$$\mathbf{x}_1 = (x_1 = 10 \text{ ppm}, x_2 = 120°\text{C}, x_4 = 1.90)$$

$$\mathbf{x}_2 = (x_1 = 10 \text{ ppm}, x_2 = 120°\text{C}, x_4 = 1.85)$$

$$\mathbf{x}_3 = (x_1 = 10 \text{ ppm}, x_2 = 125°\text{C}, x_4 = 1.90)$$

$$\mathbf{x}_4 = (x_1 = 15 \text{ ppm}, x_2 = 120°\text{C}, x_4 = 1.90)$$

The responses at these four simplex points are

$$\mathbf{y}_1 = (y_1 = 77.2, y_2 = 2.1)$$

$$\mathbf{y}_2 = (y_1 = 83.6, y_2 = 2.0)$$

$$\mathbf{y}_3 = (y_1 = 76.1, y_2 = 2.4)$$

$$\mathbf{y}_4 = (y_1 = 80.7, y_2 = 1.7)$$

Thus point $\mathbf{x}_3$ gives both the lowest yield and the highest impurity level and is thus the worst point in the simplex. Designating $\mathbf{x}_3$ as $\mathbf{x}^w$, we compute the centroid of the three remaining points as

$$\mathbf{x}^c = (x_1 = 11.67, x_2 = 120, x_4 = 1.88)$$

The reflected or "image" point for $\mathbf{x}^w$ is therefore

$$\mathbf{x}_3' = (x_1 = 13.34, x_2 = 120, x_4 = 1.86)$$

Performing an experiment at this image point we get

$$\mathbf{y}(\mathbf{x}_3') = (y_1 = 77.3, y_2 = 2.1)$$

Our simplex now consists of the following points: $\mathbf{x}_1$, $\mathbf{x}_2$, $\mathbf{x}_4$, and $\mathbf{x}_3'$, with $\mathbf{x}_1$ now the worst point. Reflecting to a new image point $\mathbf{x}_1'$ we obtain

$$\mathbf{x}_1' = (x_1 = 15.56, x_2 = 120, x_4 = 1.84)$$

at which

$$\mathbf{y}(\mathbf{x}_1') = (y_1 = 67.5, y_2 = 1.8)$$

This point is the worst point in its new simplex, but its image point is $\mathbf{x}_1$, which we have already evaluated. In such an event we typically take the image of the *next* worst point, but $\mathbf{x}_3'$ has already had its image point $\mathbf{x}_3$ evaluated. So we choose to deploy our one remaining experiment by performing a confirmation trial at the best result we have obtained in our experimental program, which is $\mathbf{x}_2$ in the sequential simplex experiments. Performing another trial at $\mathbf{x}_2$ gives $\mathbf{y} = (y_1 = 82.8, y_2 = 2.0)$. Thus, we take $\mathbf{x}_2$ as our optimum solution, or

$$\hat{\mathbf{x}}^* = (x_1 = 10 \text{ ppm}, x_2 = 120°\text{C}, x_3 = 800 \text{ rpm}, x_4 = 1.85)$$

$$\hat{\mathbf{y}}^* \cong (y_1 \cong 83\% \text{ yield}, y_2 \cong 2\% \text{ impurity})$$

SUMMARY

This chapter has illustrated a few of the experimental optimization procedures described in Chapters 4 and 5. These examples point out the necessity for the experimenter to intervene in the experimental sequence, particularly where search procedures are being employed. The statistician is often horrified at subjective intervention in experimentation, but as a practical matter managers will usually demand that an experimental activity be conducted at as low a cost as possible. Search procedures can afford the experimenter the opportunity to obtain "good" solutions within the budgetary constraints to which he is confined.

REFERENCES

1. Box, G. E. P., and J. S. Hunter, "Multifactor Experimental Designs for Exploring Response Surfaces," *Annals of Mathematical Statistics*, vol. 28, 1957, pp. 195–242.
2. Box, G. E. P., and K. B. Wilson, "On the Experimental Attainment of Optimum Conditions," *Journal of the Royal Statistical Society*, series B, vol. 13, pp. 1–45, 1951.
3. Carr, J. M., Jr., and E. A. McCracken, "Statistical Program Planning for Process Development," *Chemical Engineering Progress*, vol. 56, no. 11, 1960, pp. 56–61.
4. Claycombe, W. W., and W. G. Sullivan, "Use of Response Surface Methodology to Select

a Cutting Tool to Maximize Profit," *American Society of Mechanical Engineers Transactions*, vol. 98, series B, no. 1, 1976, pp. 63–65.

5. Hendrix, C. D., "Experimental Design—An Efficient Route to Process and Product Development," Technical Report, Union Carbide Corp., S. Charleston, WV, 1970.
6. Lind, E. E., J. Goldin, and J. B. Hickman, "Fitting Yield and Cost Response Surfaces," *Chemical Engineering Progress*, vol. 45, no. 11, 1960, pp. 62–68.
7. Montgomery, D. C., *Design and Analysis of Experiments*, John Wiley and Sons, New York, 1976.
8. Myers, R. H., *Response Surface Methodology*, Allyn and Bacon, Boston, 1971.
9. Nagpal, B. K., and C. S. Sharma, "Cutting Fluids Performance: Part I—Optimization of Pressure for Hi-Jet Method of Cutting Fluid Application," *American Society of Mechanical Engineers Transactions*, series B, vol. 95, no. 3, 1973, pp. 881–889.
10. Pigott, R. J. S., and A. T. Colwell, "Hi-Jet System for Increasing Tool Life," *Society of Automotive Engineers Quarterly Transactions*, vol. 6, 1952, pp. 547–566.
11. Read, D. R., "The Design of Chemical Experiments," *Biometrics*, March 1954, pp. 1–15.
12. Taraman, K. S., "Multi Machining Output—Multi Independent Variable Turning Research by Response Surface Methodology," *International Journal of Production Research*, vol. 12, no. 2, 1974, pp. 233–245.
13. Taraman, K. S., and B. K. Lambert, "Application of Response Surface Methodology to the Selection of Machining Variables," *American Institute of Industrial Engineers Transactions*, vol. 4, no. 2, 1972, pp. 111–115.
14. Wu, S. M., "Tool Life Testing by Response Surface Methodology—Part I," *American Society of Mechanical Engineers Transactions*, series B, vol. 86, no. 2, 1964, pp. 105–110; "Part 2," pp. 111–116.
15. Wu, S. M., and R. N. Meyer, "Cutting Tool Temperature—Predicting Equation by Response—Surface Methodology," *American Society of Mechanical Engineers Transactions*, series B, vol. 86, no. 2, 1964, pp. 150–156.

Chapter 7

Optimization and Computer Simulation Experiments

In the previous chapters of this book, we have examined means by which the concepts and techniques of optimization can be applied to experimentation. Throughout these chapters, particularly in Chapter 6, we have focused primarily on *physical experimentation*; that is, we considered experimentation in the physical or real-world realm, even though that experimentation might well be performed at a *subscale* level due to reasons of practicality and economy.

In many instances, physical experimentation, even on a subscale level, is neither practical nor economical. We may be dealing with physical systems that are so large, complex, or expensive that actual experimentation is prohibitive. In these cases, it becomes necessary to approach the problem by defining and analyzing a *model* of the real system. If the model mimics the behavior of the real system on a digital computer, it is called a *computer simulation model*. Computer simulation provides a fast, convenient, and inexpensive means of evaluating systems

- Without constructing them, if they are proposed systems.
- Without interfering with their operation, if they are costly, complex operational systems.
- Without damaging them, if the purpose of experimentation is to assess the limits of strength or endurance.

This chapter discusses the basic concepts involved in computer simulation, the design of simulation models, the use of statistical methodology in simulation, and how optimization techniques are applied to computer simulation experiments. The concepts discussed here are illustrated in several examples in which optimization techniques are applied to real computer simulation models.

BASIC CONCEPTS

Introduction

Simulation is a problem-solving procedure for defining and analyzing a *model* of a system. Simulation can take several forms, including electrical analog, mechanical analog, fluid analog, or digital computer simulation. Of these, electrical analog and digital computer simulation are the most common, with *hybrid simulation* encompassing the combination of these two modes of simulation.

Electrical analog simulation uses electrical quantities (current, voltage, resistance, capacitance, etc.) to represent the physical quantities in real systems. It is based on the notion of an *analog model*, which substitutes a physical quantity of one species (e.g., electrical) for an analogous quantity of another species (e.g., mechanical). For example, we could use voltage to represent the volume of a certain chemical in a storage tank. We might scale this representation as follows: 1 V = 1000 gal. The behavior of this particular voltage over simulated time in a specially structured electrical circuit corresponds to the variation of the volume of the chemical in the storage tank over real time. Electrical analog simulation involves observing the dynamic behavior of the electrical circuit *model* over time, perhaps using an oscilloscope to view the model behavior. Experimentation with an electrical analog simulation model involves (1) structuring electrical circuits to represent the mathematical relationships that model the real system, and (2) manipulating the electrical input quantities (voltage and current) as the simulation progresses. Some of the characteristics of electrical analog simulation are

- Continuous representation of variables over time.
- High speed operation.
- Increasing problem size requires increasing computer size.
- Accuracy and dynamic range limited by physical measurement capability.
- Continuous integration.
- Difficult to handle and store discrete data.
- Programmed by hand patching of electrical components.

Analog simulation is restricted to *continuous* systems, which are modeled by differential equations. It is used largely by electrical, mechanical, and chemical engineers to model physical processes that change continuously over time.

In contrast to electrical analog computers, digital computers are capable of performing arithmetic and logical operations at very high speeds and storing many programs and large volumes of data. They are superior to electrical analog computers when these specific characteristics are required, but they can be very inefficient for problems involving partial or ordinary differential equations. Some of the characteristics of digital computer simulation are as follows:

- Variables represented by discrete numbers.
- Operations performed in sequence.
- Slower operating speed than analog computers.
- Increasing problem size requires increase in memory size and solution time.
- Accuracy and dynamic range can be extended as desired.
- Integration by finite-difference calculus.
- Easy to handle and store discrete data.
- Programmed by computer language.

Digital simulation is used in almost all areas of science, engineering, business, industry and government. A digital simulation model consists of a set of program steps that carry out the several functions involved in the simulation process.

In this book we focus on digital computer simulation, because it has the greatest popularity and most widespread use of any of the simulation modes, and because it is an extremely important experimental tool.

Models, Systems, and Simulation

Given the rather restrictive focus on digital computer simulation in this book, simulation will be defined as follows (see Pritsker [26]): "*Simulation is the establishment of a mathematical-logical model of a system and the experimental manipulation of the model on a digital computer*." This definition contains several concepts, the discussion of which will help to elucidate the simulation process.

Computer simulation offers a convenient means of studying the behavior of a system. By *system*, we mean some circumscribed sector of reality upon which we focus analysis for the purpose of accomplishing some logical end. Simulation is the process of capturing the salient features of the system without fully duplicating it. A system is a collection of related *entities*, or objects. Each entity is characterized by a set of descriptors which, in the fashion of Pritsker [26], we shall call *attributes*. Entities engage in *activities* that elapse over time. Activities culminate in *events*, which alter the *state* of the system. An activity may last some known or precisely predictable time, or its duration may be a random variable. We refer to the former as *deterministic*, and the latter as *probabilistic* or *stochastic*. ("Stochastic" simply means probabilistic relative to time). The occurrence of an event, usually at the termination of an activity, generally alters the state of the system by changing the values of certain attributes of one or more system entities. We shall retain this conceptual structure of a system in describing the construction and manipulation of a simulation model of the system.

We say that simulation is the action of performing experimentation on a model of the system. A *model* is a mathematical-logical representation of the system based on physical theory, empirical observation, or a combination of both. Thus a computer simulation model must accomplish the following:

- Represent the entities in the system.
- Maintain values for the attributes of these entities.
- Cause the entities to engage in activities.
- Schedule the occurrence of the events that mark the completion of these activities.
- Represent the passage of time in the appropriate units (seconds, minutes, hours, days, etc.), expanding or compressing time as necessary to achieve the desired *macro* or *micro* representation of system operation.
- Control sources of variation, eliminating unwanted variation.
- Stop the simulation to review results to date, maintaining all values and states at the time simulation is stopped, and then resume the simulation at these values and states.
- Perform an analysis of simulation results.
- Generate reports on simulation results.

Digital computer simulation allows us to do all these things. The simulation model provides the *experimental process* shown in the "black-box" representation in Figure 1.1. In this chapter we examine the particular concepts that pertain to simulation experimentation uniquely.

DESIGN OF SIMULATION MODELS

This section describes the process of designing a digital computer simulation model of a system. It discusses and compares various *simulation languages* that can be employed in the development of a simulation model. The different mechanisms for *time control* in the operation of a simulation model are examined. The very important simulation functions (1) random number generation and (2) random process generation are discussed.

Simulation Languages

In the previous section, we saw that digital simulation modeling involves a number of different functions. These include time and event control, state variable updating, information storage and retrieval, system state initialization, system performance data collection, random process generation, statistical computations, and simulation report generation. Any given simulation will typically involve some or all of these functions. To provide all these functional capabilities each time we developed a particular computer simulation model would entail a tedious and burdensome task. If we were to engage in a number of different modeling activities over a period of time, perhaps using a general purpose computer language such as FORTRAN, we would doubtlessly find ourselves repeating many model components over and over again: indeed, we would probably learn to reuse certain model components from one modeling activity to the next.

Fortunately, experts came to that very conclusion some time ago, and have developed highly structured *simulation languages* that provide those functional capabilities that are used consistently in simulation modeling. Several programming languages exist for the specific purpose of computer simulation. In addition to providing the various functional capabilities needed in simulation modeling, each of these simulation languages also affords a particular modeling convention that may be the same as or somewhat different from the entity, attribute, activity, event, system state structure employed in this discussion. These simulation languages typically require far less programming than the general purpose, problem-oriented languages such as FORTRAN - IV, ALGOL, COBOL, and PL/1.

The principal means by which simulation languages are classified is according to a discrete versus continuous dichotomy based on the manner in which the state variables of the model change value. *Discrete simulation* involves changing the values of the state variables discretely at specified points in simulated time. The time variable may be either discrete or continuous in such a model, depending on whether time is advanced in discrete increments or moves directly from one event to the next (next-event simulation). For example, we could consider a simulation model of a finished product inventory policy in which time is advanced one day at a time; a variable number of orders for the product are received and processed each day and inventory levels vary accordingly, with purchase or production orders for new stock placed if the stock level drops below some established reorder point. Alternatively, we could move from event to event, with the arrival of customer orders, inventory level reviews, and the placement of purchase or production orders forming individual events that occur randomly (or deterministically) through simulated time. GPSS-111 [12, 30], GASP-11 [25], SIMSCRIPT [15, 17] and SIMULA [4] are among the most widely used discrete simulation languages.

In *continuous simulation* the state variables of the model may change continuously over simulated time. The time variable may be either discrete or continuous, depending on the mechanism for advancing time, but the values of the state variables change in a manner typically modeled through the use of differential equations. CSMP [13], MIMIC [22], and CSSL [31] are the most important continuous simulation languages.

Certain simulation languages developed recently afford *both* continuous and discrete simulation modeling. Called *combined simulation*, the state variables of the model can change discretely, continuously, or continuously with discrete jumps superimposed. The time variable may be discrete or continuous. GASP-IV [26], GASP-PL/I [27], and SIMSCRIPT 2.5 [15] are each fully documented, combined discrete-continuous simulation languages. These languages represent the most capable computer simulation languages available for general simulation modeling.

Certain highly specialized simulation languages are gaining wide acceptance and use in business, industry, and government. Notable among these are the so-called *network modeling techniques*, such as GERT and GERTS-111 [35] and Q-GERT [28]. GERT [35] is an analytical modeling technique, based on signal flowgraph theory, which allows a system to be modeled in terms of a network consisting of *nodes* and *directed branches*. The nodes represent events or decision points; the branches either denote activities or simply connect certain nodes. GERT is very restrictive in terms of the node structure that it can accommodate, so a simulation version (GERTS) was developed to incorporate greater modeling capability; GERTS-111 [35] represents one of the most widespread of the GERTS modeling languages. The codeveloper of GERT and GERTS, A. A. B. Pritsker, has evolved an even more powerful network modeling technique called Q-GERT [28]. As with GERTS-111, Q-GERT models can be simulated with only data inputs to the program package to describe the characteristics of the nodes and branches comprising the network representation of the system. Although these GERT-based modeling techniques require a highly specialized modeling structure to obtain an analysis of a system, the variety of systems that have been modeled using network techniques is truly awesome. They include manufacturing plants, chemical processes, health care units, educational systems, courts, military operations, and many more. Thus the highly specialized modeling structure exacted by the network modeling techniques does not limit the areas of simulation modeling application for these easy-to-use tools.

Time Control

We have several times referred to the mechanism for advancing time in a digital computer simulation. The two typical types of time control in computer simulation are *uniform time flow* and *variable time flow*. With uniform time flow, simulation time is advanced through each and every time period in fixed steps or intervals. Variable flow causes time to be advanced immediately to the next event, resulting in variable time increments through the simulation. Figure 7.1 illustrates the difference between uniform time flow and variable time flow.

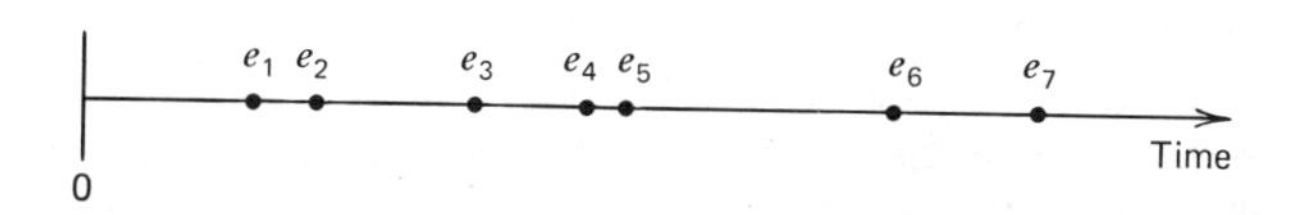

(a) Variable or Next-Event Time Flow

(b) Uniform Time Flow in Discrete-Event Simulation

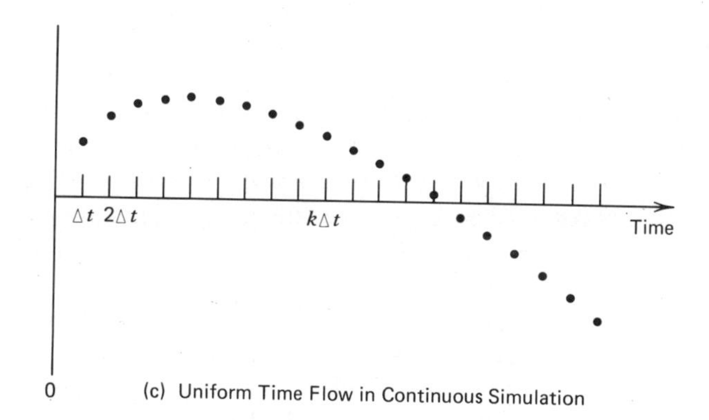

(c) Uniform Time Flow in Continuous Simulation

Figure 7.1. Time-flow mechanisms in digital computer simulation.

Uniform time flow is used most often with continuous simulation, where the time increment Δt approximates the denominator in the derivative dy/dt. More accurate results are obtained when Δt is small, but computer run time is increased. Variable, or next-event, time control is used most often with discrete simulation. The combined simulation languages employ both methods of time advance. In the GASP-IV language [26], for example, the discrete component of the model flows from event-to-event, while the continuous component increments time in a user-selected Δt.

Monte Carlo Sampling

One of the most basic techniques employed in digital computer simulation, especially with systems involving probabilistic or stochastic elements, is that called *Monte Carlo sampling*. This technique exploits the nature of the cumulative distribution function $F(y)$ of a random variable y to generated a value of y. Recall from Chapter 2 that the cumulative distribution function gives the probability of a value of y less than or equal to a specified value a; that is

$$P(y \leqslant a) = F(a) = \begin{cases} \sum_{y=0}^{a} f(y) & Y \text{ discrete} \\ \int_{-\infty}^{a} f(y)\,dy & Y \text{ continuous} \end{cases} \tag{7.1}$$

Note that the range of the quantity $F(y)$ is $0 \leqslant F(y) \leqslant 1$. In the Monte Carlo sampling technique, a *random number* r is taken from the interval $0 \leqslant R \leqslant 1$ and equated to $F(y)$. The value of y corresponding to this particular value of $F(y) = r$ is the desired value of the random variable Y. In this manner we can generate a sequence of values for Y, a procedure called *random variate generation* or *random process generation*.

Figure 7.2 illustrates the procedure employed in Monte Carlo sampling. We must first identify, either from theory or empirically, the probability mass function or probability density function $f(Y)$ for the random variable Y. This is often accomplished by using statistical goodness-of-fit tests [23] to compare the sample distribution to a theoretical distribution. If a suitable theoretical distribution for Y cannot be established using goodness-of-fit tests, then regression techniques can be employed to fit an empirical model to the sample data. Shannon [32] and Schmidt and Taylor [29] give quite lucid descriptions of both procedures. Having identified the mass or density function $f(Y)$, as shown in Figure 7.2(a), we must next form the cumulative distribution function $F(Y)$, as seen in Figure 7.2(b). Then we find the *inverse cumulative distribution function* $\phi[F(Y)]$, shown in

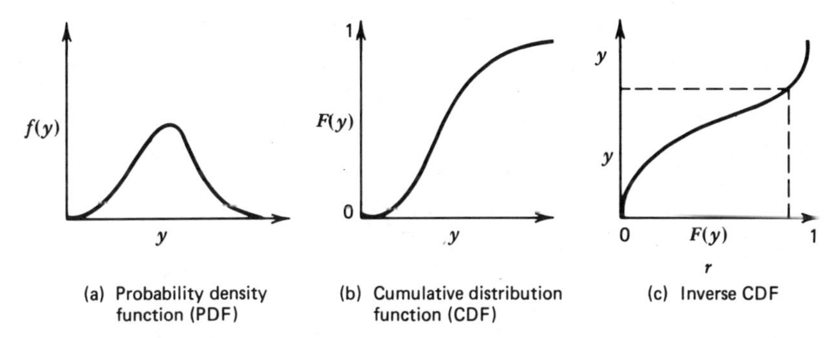

Figure 7.2. Monte Carlo procedure for random process generation.

Figure 7.2(c). This function simply provides a value y of the random variable Y corresponding to a specific value of $F(Y)$. To capsulize the Monte Carlo sampling procedure:

1. Generate a random number r, using a table or a random number generator. Then $F(Y)=r$.
2. For $F(Y)=r$, find the corresponding value of y.

To illustrate the Monte Carlo sampling technique, consider a random process in which pallets of grocery goods arrive at the shipping dock of a grocery warehouse according to a Poisson distribution with mean rate $\mu=0.4$ pallets/min. Then the random variable Y, the time between successive pallet arrivals, follows an *exponential distribution* with parameter $\lambda=1/\mu$ min. Thus the probability density function is

$$f(y)=\lambda e^{-\lambda y}, \qquad 0 \leqslant y \leqslant \infty$$

and the cumulative distribution function is

$$F(y)=1-e^{-\lambda y}$$

Note that the parameter $\lambda=2.5$ min is the mean time between arrivals. Solving for y in terms of $F(y)$, we get

$$e^{-\lambda y}=1-F(y)$$

which, upon taking the natural logarithm of both sides, yields

$$-\lambda y=\ln[1-F(y)]$$

or

$$y=-\frac{1}{\lambda}\ln[1-F(y)]$$

But $1/\lambda$ is simply μ, and we know that $F(y)=r$, so

$$y=-\mu\ln(1-r)$$

Table 7.1. A Monte Carlo sequence of exponentiate variates ($\lambda = 2.5$)

Sequence number	Random number, r	Exponential variate, y	Sequential time, t (min)
1	0.3273	2.79	2.79
2	0.1520	4.71	7.50
3	0.5401	1.54	9.04
4	0.5940	1.30	10.34
5	0.8403	0.43	10.77
6	0.2950	3.05	13.82
7	0.7789	0.62	14.44
8	0.8866	0.30	14.74

Now if r is a uniformly distributed quantity $0 \leqslant R \leqslant 1$, then $1-r$ has exactly the same distribution. So we can use the simpler expression

$$y = -\mu \ln r$$

to generate a deviate from the exponential distribution. Note that, because r is a decimal fraction, $\ln r$ is a negative quantity; hence y is a positive quantity. Since we are probably using the deviate y to establish the time of the next pallet arrival according to the relation

$$t_{\text{next}} = t_{\text{now}} + y$$

we would certainly want y to be nonnegative. Table 7.1 gives a sequence of values for R, Y, and T using the exponential generator with $\lambda = 2.5$ min.

In digital computer simulation, we must have a means of creating sequences of random variates such as the values of y shown in Table 7.1. Since the Monte Carlo technique uses a random number r to generate a value of y, we must also have a means of generating random numbers. In the following sections, we shall discuss random number generation and random variate generation as they apply to the simulation of probabilistic or stochastic processes.

Random Number Generation

In digital computer simulation we make frequent use of so-called *random numbers*; that is, values r of a random variable R that has the continuous uniform distribution over the interval $0 \leqslant R \leqslant 1$. Figure 7.3 shows the probability density function $f(R)$, which has mean $\mu = 1/2$ and variance $\sigma^2 = 1/12$.

There are two available mechanisms by which we could generate random numbers with the computer:

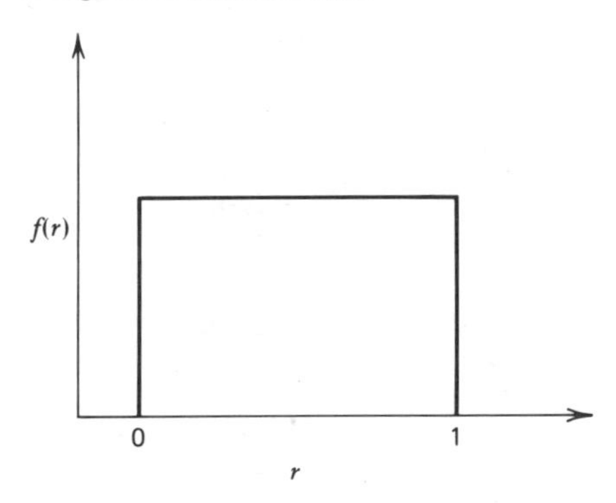

Figure 7.3. The uniform probability density function $0 \leqslant r \leqslant 1$.

1. We could store a very large *table* of random numbers in memory, and use them in sequence through a systematic access procedure.
2. We could store a program that algorithmically uses the ith value in the random number sequence to generate the $(i+1)$st value

A program would require far less core storage than would a table of random numbers. Unfortunately, because an algorithm exploits a dependency on previous values, it is not truly "random"; hence, we call them "pseudorandom number" generators. Because of their many desirable characteristics, we usually employ pseudorandom number generators in digital computer simulation. But hereafter in this book, we refer to them simply as random number generators.

What properties must a random number generator possess to make it attractive for use in computer simulation modeling?

- The algorithm should be *fast*, to keep computer run time as low as possible.
- The program should be short, to minimize core storage.
- The generator should have a long *period*; that is, it should produce a long sequence of random numbers before the same sequence reappears.
- It should generate the same sequence of random numbers at will, so that we can duplicate the experiment as needed.
- The algorithm should not *degenerate* (degeneracy is the condition in which the same number is continuously produced by the generator).
- The generator should be *algorithmic*, so that the ith value in the sequence is used to produce the $(i+1)$st value, and so on.

We examine several techniques that have been used to generate sequences of random numbers, including the mid-square, mid-product, multiplicative congruential, and the mixed congruential technique. Of these procedures, the mixed congruential technique is most widely used.

Table 7.2. Random number sequence by the mid-square technique

Sequence i	Seed s_i	Mid-square quantity, s_i^2	Random number, r_i
0	1367	01868689	0.1367
1	8686	75446596	0.8686
2	4465	19936225	0.4465
3	9362	87647044	0.9362
4	6470	41860900	0.6470

With the *mid-square* technique, we initialize the generator with an n-digit integer *seed*, s_0. This seed is squared, zeros are added to the left if necessary to produce a number having an even number of digits, and the middle n digits are taken as s_1. This procedure is repeated throughout the sequence. Table 7.2 illustrates a sequence of random numbers generated by a 4-digit mid-square technique.

Note that the random number r_i is simply the seed s_i normalized to a decimal fraction. The mid-square technique is seldom used today due to its tendency to degenerate. It should be apparent that the occurrence of a zero produces a degenerate sequence of zeroes. (The probability that any given n-digit number is zero is $1/10^n$).

The mid-product technique is very similar to the mid-square technique, except that a successive number is obtained by multiplying the current number by a constant k; that is,

$$s_{i+1} = ks_i$$

The mid-product technique has a longer period than the mid-square procedure, but suffers the same tendency to degenerate.

The random number generators in common used today are those based on the *congruence* relation. Let m be a fixed integer greater than 1. If a and b are integers such that $a-b$ is divisible by m, we say that "a is congruent to b modulo m." This relationship is written

$$a \equiv b \pmod{m} \tag{7.2}$$

If we rewrite (7.2) in the form

$$s_{i+1} \equiv s_i \pmod{m} \tag{7.3}$$

we have a *congruential generator*. That is, if we divide the random number seed s_i by a fixed integer m, s_{i+1} is the remainder. For example, suppose $s_0 = 37$ and $m = 4$. Then

$$s_1 \equiv 37 \pmod{4} = 1$$

We see that the very next step will produce a degenerate sequence of zeroes.

The *multiplicative congruential* generator has the form

$$s_{i+1} \equiv as_i \pmod{m} \tag{7.4}$$

where the product as_i is divided by m and the remainder taken as s_{i+1}. Another effective random number generator is the *mixed congruential method*, which employs the relation

$$s_{i+1} \equiv (as_i + c) \pmod{m} \tag{7.5}$$

Shannon [32] gives the following sample program to guide one toward creating a FORTRAN subroutine to produce uniformly distributed random numbers RN, given a seed IX, where a and m are the parameters in relation (7.4).

```
      SUBROUTINE RANDN (IX,IY,RN)
1 IY=IX*a
2 IF (IY) 3,4,4
3 IY=IY+m
4 RN=IY
5 RN=RN*m^-1
6 IX=IY
7 RETURN
8 END
```

Actual values for a, m, and m^{-1} must be supplied by the simulation analyst.

Having generated a sequence of pseudorandom, rather than truly random numbers, it is appropriate to perform certain statistical tests to ascertain the "randomness" of the sequence. Shannon [32] describes the application of a number of different tests to a random number generator. The most crucial tests are (1) a goodness-of-fit test to compare the distribution of the generated set to a uniform distribution, and (2) a runs test to evaluate the significance of sequences of random numbers above or below the mean 0.5.

Random Variate Generation

We have already discussed the *Monte Carlo sampling technique* for generating values of a random variable Y in a digital computer simulation. This technique employs the *inverse transformation method* in which the relationship between a random variable Y and its inverse cumulative distribution function

$$y=\phi[F(Y)]=F^{-1}(X) \tag{7.6}$$

is exploited. Recall that we simply substitute a uniformly distributed random number r for $F(Y)$ in the inverse relationship (7.6) to produce a value y of the random variable Y. We illustrated this Monte Carlo technique with a simple example involving an exponentially distributed random variable.

Another distribution for which the inverse transformation method is used to generate a value of the random variable is the *Weibull distribution*, which is a family of probability density functions used in modeling the failure rate characteristics in reliability analysis. The probability density function is

$$f(y)=\alpha\beta y^{\beta-1}e^{-\alpha y^{\beta}}, \qquad y\geqslant 0 \tag{7.7}$$

where $y>0$, $\alpha>0$, $\beta>0$. The Weibull probability density function comprises a family of curves as α and β take on different values. The cumulative distribution function for the Weibull distribution is

$$F(y)=1-e^{-\alpha y^{\beta}} \tag{7.8}$$

Equating this to a uniformly distributed random number r, we obtain

$$F(y)=r=1-e^{-\alpha y^{\beta}}$$

Rearranging we get

$$1-r=e^{-\alpha y^{\beta}}$$

But if r is uniformly distributed, so must be $1-r$, so we can write the above simply as

$$r=e^{-\alpha y^{\beta}}$$

Taking the natural logarithm of both sides and rearranging yields a random value from the Weibull distribution,

$$y=\left[-\frac{1}{\alpha}\ln r\right]^{1/\beta} \tag{7.9}$$

For example, for a Weibull distribution with $\alpha=1$ and $\beta=2$, if $r=0.356$, $y=1.016$ time units.

The examples with the exponential and Weibull distributions illustrate how to generate variates by exploiting the inverse transformation scheme from Monte Carlo sampling. There are other techniques for random variate generation, including the rejection technique, the composition method, a mathematical derivation approach, approximation techniques, and other specialized procedures. Phillips et al. [24] give an excellent description of these techniques and the probability distributions to which they apply. Fishman [7] provides a comprehensive treatment of the entire subject of random variate generation. Shannon [32] presents an easily readable discussion on this subject and gives FORTRAN subroutines by which random variates from several distributions are generated.

STATISTICAL TECHNIQUES IN COMPUTER SIMULATION

Just as statistical analysis is the backbone of objective experimental analysis, it underlies our entire approach to digital computer simulation. Statistical analysis is an integral part of simulation modeling. There are four principal areas into which the statistical techniques in computer simulation can be regarded. They are:

- *Input analysis*, including the selection of probability laws that appropriately characterize the input variables to a particular simulation and the assignment of numerical values to the parameters of distributions corresponding to these laws.
- *Estimation*, which involves the development of point estimates and interval estimates based on samples of input data.
- *Output analysis*, which involves the estimation of population parameters and the identification of distributions for the output responses, as well as hypothesis testing and confidence interval estimation for the population parameters.
- *Design of computer simulation experiments*, including the selection of values for the input variables, the choice of random number seeds on corresponding replications, determining the simulation run length, and incorporating an appropriate variance reduction technique.

We discuss each of these issues here.

Input Analysis

An expression that anyone associated with digital computer simulation hears from time to time is "garbage in!—garbage out!" Although this statement is true to some extent for any experimental endeavor, it is of paramount concern for simulation. We must always bear in mind that, as a *model* of the real system, a simulation must be carefully constructed if meaningful results are to be obtained from its use. In order that a computer simulation model serve its intended function as an experimental tool, its *credibility* must be clearly established. Credibility connotes two aspects: (1) the *verification* of the model to ensure that it behaves as the analyst has intended; and (2) the *validation* of the model to ascertain that it reproduces the real-world system to a reasonable degree of accuracy. The time and money invested in establishing a credible simulation model should be a function of the potential savings that might accrue from its use: the expenditure of $50,000, say, over a period of 6 to 12 months is easily justified when a capital expenditure of several million dollars hangs in the balance, but even $1000 over a 2-week period might not be worthwhile when the potential return is only several thousand dollars over the next few years.

A significant factor in the credibility of a digital computer simulation model is the acquisition and analysis of available data from the real system so as to construct the probability laws. We typically confront two situations here.

- In simulating *existing* systems, we must collect and analyze historical data that portray the behavior of the system in as wide an array of circumstances as possible.
- In simulating *proposed* systems, we must either characterize the inputs of similar systems or conjecture the parameters of the system inputs if similar configurations and operations are not known to exist.

In either case, the simulation analyst bears the responsibility of carefully representing the extent of the input data analysis so decision-makers can exercise proper caution in reaching conclusions based on simulation results.

Estimation

The most basic tool in input analysis is the *estimation* of population parameters based on samples of n observations. Since the basic concepts of estimation were discussed in Chapter 2, they will not be reiterated here except to identify the most important *point estimates* and *interval estimates*. The *point estimates* most often found useful in input analysis include:

- The measures of central tendency, such as the mean, median and mode.
- The measures of dispersion, such as the variance, standard deviation and range.
- The *quantile* measures that provide the subranges of the random variable x for equiprobable intervals.
- The *cumulative distribution function* for the sample data.
- *Histograms* that give the relative frequency of occurrence of values of the

random variable X, at specific values for discrete random variables and within specified ranges for continuous random variables.

The latter three estimates enable the analyst to make a proper choice of a *distribution* for the probability law represented by the random variable X, and the central tendency and dispersion quantities enable an estimate of the population parameters of that distribution.

Interval estimators, also described in Chapter 2, enable the analyst to find the confidence intervals of a population parameter and in so doing gain an assessment of how representative the sample estimators are. Confidence intervals serve as a guide in designing the computer simulation experiments; that is, one would typically plan to place experimental points within, say, the 95% confidence interval for a given population parameter. Thus interval estimators are an important part of the input analysis.

Output Analysis

In analyzing the output of a simulation, it is necessary to first distinguish between *static* response variables and *dynamic* response variables. The term "static" does not imply that the random variable has a single value, but connotes the concept that the *parameters* of its probability distribution are invariant with time. For example, the time to repair failed equipment might be a gamma distributed random variable whose parameters remain constant throughout the simulation, hence, it is a static response variable. In contrast to that, the length of the queue of failed equipment awaiting repair could well be a dynamic response variable having a *transient* period and a *steady-state* period of behavior. We usually want to estimate the steady-state parameters for a dynamic process.

The analysis of a static response variable is typically carried out in exactly the same way as that of the input data for a simulation model; that is, given n independent observations $y_1,\ldots,y_n$ generated in a particular simulation trial, we would compute the standard point and interval estimators, and develop quartile estimators, histograms, and cumulative distribution plots for each such static response variable. We would then perform tests of hypothesis to assist us in reaching objective conclusions about the system under study. Static output analysis differs little from a standard statistical analysis of experimental data.

Dynamic output analysis, however, is somewhat different and more formidable. Consider that the queue length of failed equipment awaiting repair (in fact the failed equipment may remain in place in the plant, but only *figuratively* joins a queue), denoted $\{Y(t)\}$, is a *stochastic process*; that is, it evolves over simulated time from an *initial value* $Y_0(0)$ to a *final value*

$Y_T(T)$. There are two principal characteristics of dynamic response variables that we might take into account in an output analysis of such a random variable:

- The *steady-state* behavior of the variable $Y_t(t)$, $S<t\leqslant T$, may be quite different from its *transient* behavior Y_t, $0\leqslant t\leqslant S$. In some instances it is necessary to measure the mean of the entire process, $0\leqslant t\leqslant T$, whereas in other cases the steady-state mean is desired.
- The value of the response $Y_t(t)$ is often *dependent* upon its value at a previous point $Y_r(r)$, $r\leqslant t$. That is, we seek the conditional mean $E[Y(t)|Y(r)]$, $r<t$. In particular, we must be attentive to the *initial* condition $Y_0(0)$.

Fishman [7] and Kleijnen [16] give excellent treatments of the statistical analysis of simulation output.

Design of Computer Simulation Experiments

The design of simulation experiments embodies all that is involved in designing physical experiments and more. Just as with the physical experiment, a simulation trial is conducted at a point $\mathbf{x}^k$, where

$$\mathbf{x}^k=\{x_i^k, i=1,\ldots,n\} \tag{7.10}$$

In addition, the analyst must establish the *simulation run length* T and the *initial set of random number seeds* $\mathbf{s}$ at each point $\mathbf{x}^k$ in the experimental design. Just as with the classical experimental design, the independent variables x_i, $i=1,\ldots,n$ can be varied over a set of *qualitative* levels or over a range of *quantitative* values. We shall concentrate on the latter, since they are more pertinent to the merger of optimization and simulation experimentation.

By analyzing the simulation output for a dependent or response variable y at each of the levels of the input or independent variables $\mathbf{x}^k$, $k=1,\ldots,K$, where K is the number of simulation trials or *design points* at which the model is run, usually at constant run time T or a constant sample size N for the response variable $\mathbf{y}$, and at either (1) the same set of initial random number seeds $\mathbf{s}$ or (2) at randomly selected sets of seeds $\mathbf{s}^k$, $k=1,\ldots,K$, it is possible to estimate a *response surface* that relates y and $\mathbf{x}$. Such response surfaces, which may be either linear, quadratic, or some special function, were discussed in Chapter 3. Several classes of experimental designs, such as *full factorials*, *fractional factorials*, *simplex* designs for first-order models, and *central composite* and extended simplex designs for higher-order models can be applied, just as with physical experimentation. *Linear*, *curvilinear*, and *multiple linear* regression are employed to estimate the prediction equation $\hat{y}=f(\mathbf{x})$. By employing *analysis of variance*, together with *tests of hypothesis of the regression coefficients*, we are able to determine the extent to which each of the independent variables x_i, $i=1,\ldots,n$ influences the simulation response. This kind of analysis is also utilized as an integral part of the "simulation/optimization" schemes we shall discuss later.

Variance Reduction Techniques

In physical experimentation we have to *replicate* an experimental point $\mathbf{x}^k$ to gain a better estimate of the response $y(\mathbf{x}^k)$. In computer simulation, however, we obtain numerous observations of the response variable y at a given set of input conditions $\mathbf{x}^k$. This mechanism affords us the opportunity to reduce the variance of the simulation response y *internal* to the simulation. Such measures are termed *variance reduction techniques*. Because variance reduction is unique to computer simulation, it is important that the reader understand how to apply them in his or her modeling. Since these techniques tend to become rather sophisticated, we shall limit the discussion here to a general treatment of these methods, leaving the reader to gain a more detailed view of the subject by referring to Fishman [7] and Kleijnen [16]. We shall briefly discuss four methods of variance reduction: (1) stratified sampling, (2) importance sampling, (3) antithetic variates, and (4) control variates.

Stratified sampling involves partitioning the cumulative distribution function $F(y)$ into segments, sampling separately from each segment, and combining the results into a single estimate. The determination of the size of the segments, or *strata*, is based on prior knowledge of the characteristics of the population to be sampled. We seek to have the elements within the resulting strata show greater homogeneity (less variation) than the elements in the actual population. One way to stratify the sampling distribution is to divide the total interval into subintervals of equal size. An alternative means of selecting strata is to choose segments such that the variance is equal for all subintervals. As an example of a stratified sampling scheme, suppose we perform two simulation trials, at the same set of input conditions, involving a sample size of k observations of a particular random variable y. We might run the first trial using the k random numbers $r_1,\ldots,r_k$ and the second trial using the correlated set of random number $r_1',\ldots,r_k'$, where

$$r_j'=\begin{cases} r_j+\frac{1}{2} & 0\leqslant r_j<\frac{1}{2} \\ r_j-\frac{1}{2} & \frac{1}{2}\leqslant r_j\leqslant 1 \end{cases} \tag{7.11}$$

where r_j and r_j' are uniformly distribution in the interval zero to one with mean $\frac{1}{2}$ and variance $\frac{1}{12}$. Suppose we use these random numbers to generate exponentially distributed random variates y with parameter λ.

$$y_j = -\frac{1}{\lambda}\ln r_j$$

$$y_j' = \begin{cases} -\frac{1}{\lambda}\ln\left(r_j + \frac{1}{2}\right) & 0 \leqslant r_j < \frac{1}{2} \\ -\frac{1}{\lambda}\ln\left(r_j - \frac{1}{2}\right) & \frac{1}{2} \leqslant r_j \leqslant 1 \end{cases} \tag{7.12}$$

One can use expressions (2.38) and (2.39) to show that the correlation coefficient for the variables r_j and r_j' is -0.5, and for y_j and y_j' is approximately 0.303. We could of course develop alternate strata for r_j, and correspondingly for y_j, but the above example illustrates the concept of stratified sampling. Stratified sampling has been shown to effect variance reduction ratios in excess of 10 to 1.

Importance sampling is a variance reduction scheme that concentrates the sampling effort on those values or ranges of a random variable y that are *most likely* to occur or, in some instances, those that have a small probability of occurrence but are the values of real concern. For example, suppose that we are concerned about machine failures that occur according to an exponential failure distribution with a mean time between failures of $1/\lambda$ hr. The variance of this distribution is $1/\lambda^2$. In importance sampling we would attempt to increase the probability of occurrence of a machine failure. We might choose to sample from another exponential distribution with parameter λ' such that the mean of the sampled distribution is greater than $1/\lambda$, thus increasing the probability of a machine failure in the simulated time interval. We must then apply a weighting factor W to correct the sample result. W is the ratio of the true probability $P(E)$ to the distorted probability $P'(E)$, or

$$W = \frac{P(E)}{P'(E)} \tag{7.13}$$

where $P(E)$ is the probability of the event E. In our example, E might be the occurrence of a machine failure in a specified time interval Δt.

One of the most popular variance reduction techniques is the method of *antithetic variates*. The objective of antithetic sampling is to have two estimators, x_1 and x_2, of a random variable y, such that x_1 is negatively correlated with x_2. If the random variable is

$$y = \tfrac{1}{2}(x_1 + x_2) \tag{7.14}$$

then its mean is

$$\mu_Y = \tfrac{1}{2}(\mu_1 + \mu_2) \tag{7.15}$$

and its variance is

$$\sigma_Y^2 = \tfrac{1}{4}(\sigma_1^2 + \sigma_2^2) + \tfrac{1}{2}\text{cov}(x_1, x_2) \tag{7.16}$$

Because x_1 and x_2 are negatively correlated, $\text{cov}(x_1,x_2)$ is negative, and σ_Y^2, is even smaller than it would be if x_1 and x_2 were independent replications [in which case $\text{cov}(x_1,x_2)=0$]. One way to obtain negatively correlated variates x_1 and x_2 is to use complementary uniform variates in generating them. For instance, if the deviates $r_1,\ldots,r_k$ are used to generate $x_{11},\ldots,x_{1k}$, then $r'_1,\ldots,r'_k$ would be used to generate $x_{21},\ldots,x_{2k}$ where

$$r'_j = 1 - r_j \tag{7.17}$$

In this case, r and r' are perfectly negatively correlated.

OPTIMIZATION OF SIMULATION PARAMETERS

A "black-box" view of the interaction between optimization and simulation is seen in Figure 7.4. To elaborate on that interaction, we shall assign the following characteristics to the *optimizer*:

- It contains the optimization algorithm, including any search techniques, mathematical programming methods, experimental design generators, least-square regression procedures, and statistical techniques for hypothesis testing.
- It possesses the means of assigning values to those independent or *input* variables x_i, $i=1,\ldots,n$, as well as initial random number seeds **s**, simulation run length T, or sample size K.
- It accepts as input the results of one or more simulation trials. If p trials are conducted, each generating m simulation responses y_j, $j=1,\ldots,m$, the

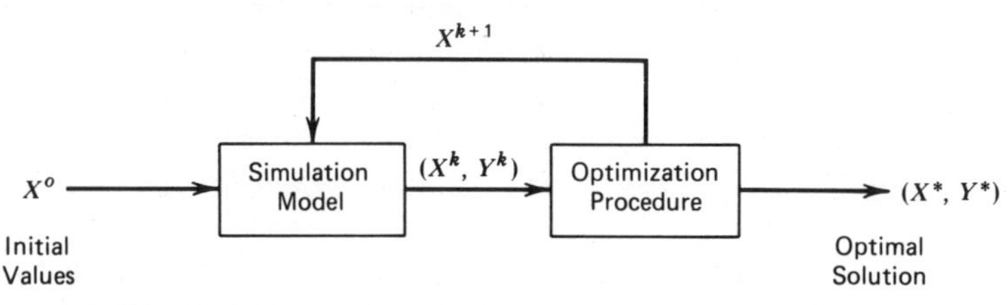

Figure 7.4. Interaction between optimization and simulation.

optimizer receives as input the response matrix

$$Y=\begin{bmatrix} y_{11} & y_{12} & \cdots & y_{1p} \\ y_{21} & y_{22} & \cdots & y_{2p} \\ \vdots & \vdots & & \vdots \\ y_{m1} & y_{m2} & \cdots & y_{mp} \end{bmatrix} \tag{7.18}$$

where y_{jl} is the jth simulation response observed in the lth simulation trial.

- It contains the stopping rule for the experimental procedure, typically either the "best" responses $\hat{y}_j$, $j=1,\ldots,m$ or the expenditure of a maximum allotted computer running time.

The optimizer can either be a completely algorithmic procedure, which could be computer programmed to be fully automatic, or it could be completely analyst-controlled, or it could involve a combination of these.

The simulation model typically requires as input,

- Values for the controllable values x_i, $i=1,\ldots,n$.
- Initial values for some or all of the response variables y_j, $j=1,\ldots,m$.
- Initial random number seeds.
- A means of terminating the simulation run, such as a required sample size k_j for each response variable, or a simulated time T.

In this section, we present a survey of optimization applied to computer simulation and relate how the techniques described in Chapter 5 would apply in a simulation situation.

Simulation and Optimization: A Historical Survey

Although many simulationists regard computer simulation as primarily a tool for systems *analysis*, there are numerous references in the literature of operations research and management science that describe efforts to use simulation models as vehicles for systems *optimization*. For example, Hunter and Naylor [11] employed a response surface approach to finding the optimum values of x_1=reorder point and x_2=order quantity to minimize the average daily cost y of an inventory system. Nelson and Krisbergh [22] applied an optimum-seeking procedure called *razor search* to large-scale simulation models using Forresters *Urban Dynamics* simulation language [8]. Farrell et al. [6] reviewed available research and applications associated with the optimization of computer simulations, and list one of the most exhaustive bibliographies available on the state-of-the-art. The work cited in Farrell et al. [6] pertained exclusively to the optimization of a single simulation response y as an unknown function of several independent variables x_i, $i=1,\ldots,m$.

Smith [33, 34] described several *external* optimization procedures that are applicable to computer simulation. "External" procedures or those which possess the characteristics of the *optimizer* as outlined above. Smith focused on an optimization procedure, called "Optimizer", based on response surface methodology. His recommended procedure actually consists of a *strategy*, and combined a search technique based on the method of steepest ascent, a screening method for eliminating those independent variables that contribute only marginally to the behavior of the response, and variance reduction measures. Smith's "Optimizer" was tested with selected computer simulation models of naval operations.

Biles [1] reported a sequential, first-order response surface method for seeking the solution to a constrained optimization formulation of the multivariable (x_i, $i=1,\ldots,n$), multiresponse (y_j, $j=1,\ldots,m$) simulation problem. Biles procedure is initiated at a feasible starting point $\mathbf{x}^0$, employs first-order response surface experimental designs along with multiple linear regression in pursuing one or more gradient directions, and when a boundary of the feasible region is encountered resorts to a *gradient projection* direction to continue the search. A generalized version of this procedure was described in Chapter 5, as it also has excellent potential for application to physical experimentation. One drawback to Biles so-called *gradient-regression* procedure is its dependence on a constrained formulation of the simulation/optimization problem.

Montgomery and Bettencourt [19] described the application of a second-order response surface approach to the optimization of a multiple response simulation model of a tank duel. The problem was formulated as a multiple-objective optimization. In one version of the tank combat model, two independent variables were $x_1=$ time for Blue tank to fire the first round and $x_2=$ time between rounds. Two responses were $y_1=$ probability of Blue victory and $y_2=$ expected Blue rounds fired. One Blue tank was opposed by one Red tank. A detailed description of this problem is given later in this chapter. The principal development in the Montgomery and Bettencourt work was the multiple objective problem formulation. Biles [2] described a nonlinear goal programming approach to the problem formulation, thus extending the Montgomery and Bettencourt results.

Biles and Swain [3] gave a generalized approach to optimizing the multivariable, multiresponse computer simulation model. They described several problem formulations, including:

- Constrained optimization.
- Weighted objective function.

- Multiple objective optimization.
- Goal programming.

and outline several general optimization procedures which could be applied to any of these formulations. These procedures included:

- Complex search.
- Sequential first-order respones surface methods.
- Second-order response surface procedures.

These are essentially the same procedures described in Chapter 5, beyond those that apply only to functions of a single independent variable x.

A General Procedure

This brief narrative has not attempted to provide an exhaustive discussion of optimization in computer simulation. But it serves to bring us to the point of recommending a general procedure by which one would approach such a problem. This procedure focuses all of the concepts developed in this book on the simulation/optimization problem, including optimization, experimental design, and statistical analysis. The steps in this procedure are as follows:

1. Identify the controllable input variables x_i and their respective ranges $a_i \leqslant x_i \leqslant c_i$, $i=1,\ldots,n$. This represents the a priori experimental region for the simulation/optimization process.
2. Identify the simulation responses y_j, $j=1,\ldots,m$.
3. Formulate an optimization problem in one of the forms outlined above and in Chapter 5, or in any specialized format that might pertain to the problem at hand. In general, the optimization problem must establish a function that is to be optimized, usually in terms of one or more of the simulation responses y_j, $j=1,\ldots,m$, expressed in a dependency upon the controllable variables x_i, $i=1,\ldots,n$.
4. Choose an appropriate optimization procedure from among those described in Chapter 5. Proceed with the optimization until (1) a chosen stopping criterion is met, or (2) available computer time is exhausted.

SIMULATION EXAMPLES

Each of the simulation optimization procedures, complex search, sequential first-order optimization, and second-order response-surface optimization, was evaluated with two simulation models, as follows:

1. A discrete-event simulation model of a stochastic production inventory system in which daily demand and order lead time are random variables with known probability density (mass) functions. It is desired to minimize the mean daily cost, y, which is the sum of carrying, setup, and shortage costs. The two controllable input variables are x_1 = reorder point and x_2 = economic order quantity. The corresponding optimization problem is

$$\min y = \eta(x_1, x_2)$$

subject to

$$-5 \leqslant x_1 \leqslant 90 \qquad 50 \leqslant x_2 \leqslant 250$$

The known solution is $x_1^* = 45$, $x_2^* = 175$, and $y^* = \$76.00$.

2. A tank duel model as described by Montgomery and Bettencourt [19] in which a brief fire engagement between two armored vehicles is simulated. A stationary defending vehicle (Blue Tank) fires first at a fully exposed attacking vehicle (Red Tank). The engagement ends when a kill occurs or a predetermined time limit of 120 sec expires. The two input variables to the model were x_1 = mean time to fire the first round (sec), and x_2 = mean time between rounds (sec). The simulation responses are y_1 = probability of Blue victory, and y_2 = expected number of rounds fired by the Blue Tank. The optimization problem was framed as one of constrained optimization.

$$\max y_1 = g_1(x_1, x_2)$$

subject to

$$5 \leqslant x_1 \leqslant 25 \qquad 5 \leqslant x_2 \leqslant 25 \qquad 0 \leqslant y_2 \leqslant 2$$

The known solution for this test problem is $x_1^* = 8.2$ sec., $x_2^* = 12.5$ sec., $y_1^* = 0.61$, and $y_2^* = 2$ rounds.

The results of evaluation trials with these three simulation optimization procedures with the stochastic production inventory model are shown in Table 7.3. The two sets of complex search results correspond to two sets of starting points, derived by using two different random number seeds for the random placement of the four initial points. In terms of optimization efficiency, the second-order central composite design gave the best results (9 simulation trials), followed by the complex search (13 and 14 trials), with the first-order response surface approaches faring poorly by comparison. In terms of the precision of the predicted solution, errors as large as 5% were incurred.

Table 7.4 gives the results of the comparison of the same three optimization procedures with the tank duel model. Again the second-order response-surface model was most efficient, followed by complex search and

Table 7.3. Summary of simulation/optimization results for the stochastic inventory model

Optimization method	Starting seed	Estimated solution x_1	x_2	y	Number of trials
Complex search	12471	50	177	\$76.20	13
	21437	43	245	\$81.00	14
First-order factorial design	17332	70	125	\$79.00	40
First-order simplex design	17332	42	159	\$76.78	41
Second-order central composite design	35188	53	183	\$70.32 (Predicted) \$79.00 (Actual)	9

Table 7.4. Summary of simulation/optimization results for the tank duel model

Optimization method	Estimated Solution x_1	x_2	y_1	y_2	Number of trials
Complex search	12.9	10.1	0.59	1.97	20
First-order simplex design	8.2	12.4	0.60	1.97	55
Second-order central composite	8.0	12.5	0.61	2.00	9

the first-order response-surface method. Greater accuracy was obtained with each procedure as compared to the previous simulation model, owing to a smaller variance in the responses.

Optimization procedures have also been successfully coupled with a FORTRAN simulation model of material handling in automated manufacturing, and with a naval minefield effectiveness model. One observation that resulted from those efforts is that the interface between the computer program for the optimization procedure and that for the simulation model must often be "custom fitted."

REFERENCES

1. Biles, W. E., "A Gradient-Regression Search Procedure for Simulation Experimentation," *Proceedings of the 1974 Winter Simulation Conference*, 1974, pp. 491–497.
2. Biles, W. E., "Optimization of Multiple-Objective Computer Simulations: A Nonlinear Goal Programming Approach," Proceedings of the 1977 AIDS Conference, 1977, pp. 134–142.

3. Biles, W. E., and J. J. Swain, "Strategies for Optimization of Multiple Response Simulation Models," *Proceedings of the 1977 Winter Simulation Conference*, 1977, pp. 134–142.
4. Dahl, O., and K. Nygaard, "SIMULA-an ALGOL-Based Simulation Language," *Communications of the ACM*, vol. 9, no. 9, 1966, pp. 671–628.
5. Emshoff, J. R. and R. L. Sisson, *Design and Use of Computer Simulation Models*, MacMillan, New York, 1970.
6. Farrell, W., C. H. McCall and E. C. Russell, *Optimization Techniques for Computerized Simulation Models*, Technical Report 1200-4-75, CACI, Inc., 1975.
7. Fishman, G. S., *Concepts and Methods in Discrete Event Digital Simulation*, Wiley-Interscience, New York, 1973.
8. Forrester, J., *Industrial Dynamics*, MIT Press, Cambridge, MA, 1961.
9. Gordon, G., *System Simulation*, Prentice-Hall, Englewood Cliffs, NJ, 1969.
10. Hammersly, J. M., and D. C. Hanscomb, *Monte Carlo Methods*, John Wiley and Sons, New York, 1964.
11. Hunter, J. S., and T. H. Naylor, "Experimental Designs for Computer Simulation Experiments," *Management Science*, vol. 16, no. 7, 1970.
12. IBM Corporation, *General Purpose Simulation System/360 OS and D05 Version 2 User's Manual* SH20-0694-0, White Plains, NY, 1969.
13. IBM Corporation, "Introduction to 1130 Continuous Systems Modeling Program II," GH20-0848-1, White Plains, NY, 1970.
14. Jacoby, J. E., and S. Harrison, "Multi-Variable Experimentation and Simulation Models," *Naval Research Logistics Quarterly*, vol. 9, 1962, pp. 121–136.
15. Kiviat, P. J., R. Villanueva, and H. Markowitz, *The SIMSCRIPT 11 Programming Language*, Prentice-Hall, Englewood Cliffs, NJ, 1969.
16. Kleihnen, J. P. C., *Statistical Techniques in Simulation*, vol. I and vol. II, Marcel Dekker, Inc., New York, 1974/1975.
17. Markowitz, H. M., H. W. Karr and B. Hausner, *SIMSCRIPT: A Simulation Programming Language*, Prentice-Hall, Englewood Cliffs, NJ, 1963.
18. Mihram, G. A., *Simulation: Statistical Foundations and Methodology*, Academic Press, New York, 1972.
19. Montgomery, D. C., and V. M. Bettencourt, "Multiple Response Surface Methods in Computer Simulation," *SIMULATION*, October 1977, pp. 113–121.
20. Naylor, T. H., *Computer Simulation Experiments with Models of Economic Systems*, John Wiley and Sons, New York, 1971.
21. Nelson, C. W., and H. M. Krisbergh, "A Search Procedure for Policy Oriented Simulations: Applications to Urban Dynamics," *Management Science*, vol. 20, no. 8, 1974, pp. 1164–1174.
22. Petersen, N. D., "MIMIC, An Alternative Programming Language for Industrial Dynamics," *Journal of Socio-Economic Planning Science*, vol. 6, 1972, pp. 319–327.
23. Phillips, D. T., *Applied Goodness of Fit Testing*, AIIE Monograph Series, AIIE-OR-72-1, Atlanta, GA, 1972.
24. Phillips, D. T., A. Ravindran and J. J. Solberg, *Operations Research: Principles and Practice*, John Wiley and Sons, New York, 1976.
25. Pritsker, A. A. B., and P. J. Kiviat, *Simulation With GASP-11*, Prentice-Hall, Englewood Cliffs, NJ, 1969.

26. Pritsker, A. A. B., *The GASP-IV Simulation Language*, Wiley-Interscience, New York, 1974.
27. Pritsker, A. A. B., and R. E. Young, *Simulation With GASP-PL/1*, John Wiley and Sons, New York, 1975.
28. Pritsker, A. A. B., *Modeling and Analysis Using Q-GERT Networks*, Wiley-Halsted, New York, 1977.
29. Schmidt, J. W. and R. E. Taylor, *Simulation and Analysis of Industrial Systems*, Richard D. Irwin, Homewood, IL, 1970.
30. Schriber, T., *A GPSS Primer*, Ulrich's Books, Ann Arbor, MI, 1972.
31. SCi Simulation Software Committee, "The SCi Continuous System Simulation Language (CSSL)," *SIMULATION*, vol. 9, 1967, pp. 281–303.
32. Shannon, R. E., *Systems Simulation: The Art and Science*, Prentice-Hall, Englewood Cliffs, NJ, 1975.
33. Smith, D. E., "Requirements of an Optimizer for Computer Simulations," *Naval Research Logistics Quarterly*, vol. 20, no. 1, 1973.
34. Smith, D. E., "An Empirical Investigation of Optimum-Seeking in the Computer Simulation Situation," *Operations Research*, vol. 21, 1973, pp. 475–497.
35. Whitehouse, G. E., *Systems Analysis and Design Using Network Techniques*, Prentice-Hall, Englewood Cliffs, NJ, 1973.
36. Williams, D. R. and D. L. Weeks, "Technique for Designing and Augmenting Simulation Experiments," *Management Science*, vol. 20, 1974, pp. 1385–1392.

Appendix A

Selected Statistical Tables

z	.00	.01	.02	.03	.04	z
.0	.500 00	.503 99	.507 98	.511 97	.515 95	.0
.1	.539 83	.543 79	.547 76	.551 72	.555 67	.1
.2	.579 26	.583 17	.587 06	.590 95	.594 83	.2
.3	.617 91	.621 72	.625 51	.629 30	.633 07	.3
.4	.655 42	.659 10	.662 76	.666 40	.670 03	.4
.5	.691 46	.694 97	.698 47	.701 94	.705 40	.5
.6	.725 75	.729 07	.732 37	.735 65	.738 91	.6
.7	.758 03	.761 15	.764 24	.767 30	.770 35	.7
.8	.788 14	.791 03	.793 89	.796 73	.799 54	.8
.9	.815 94	.818 59	.821 21	.823 81	.826 39	.9
1.0	.841 34	.843 75	.846 13	.848 49	.850 83	1.0
1.1	.864 33	.866 50	.868 64	.870 76	.872 85	1.1
1.2	.884 93	.886 86	.888 77	.890 65	.892 51	1.2
1.3	.903 20	.904 90	.906 58	.908 24	.909 88	1.3
1.4	.919 24	.920 73	.922 19	.923 64	.925 06	1.4
1.5	.933 19	.934 48	.935 74	.936 99	.938 22	1.5
1.6	.945 20	.946 30	.947 38	.948 45	.949 50	1.6
1.7	.955 43	.956 37	.957 28	.958 18	.959 07	1.7
1.8	.964 07	.964 85	.965 62	.966 37	.967 11	1.8
1.9	.971 28	.971 93	.972 57	.973 20	.973 81	1.9
2.0	.977 25	.977 78	.978 31	.978 82	.979 32	2.0
2.1	.982 14	.982 57	.983 00	.983 41	.938 82	2.1
2.2	.986 10	.986 45	.986 79	.987 13	.987 45	2.2
2.3	.989 28	.989 56	.989 83	.990 10	.990 36	2.3
2.4	.991 80	.992 02	.992 24	.992 45	.992 66	2.4
2.5	.993 79	.993 96	.994 13	.994 30	.994 46	2.5
2.6	.995 34	.995 47	.995 60	.995 73	.995 85	2.6
2.7	.996 53	.996 64	.996 74	.996 83	.996 93	2.7
2.8	.997 44	.997 52	.997 60	.997 67	.997 74	2.8
2.9	.998 13	.998 19	.998 25	.998 31	.998 36	2.9
3.0	.998 65	.998 69	.998 74	.998 78	.998 82	3.0
3.1	.999 03	.999 06	.999 10	.999 13	.999 16	3.1
3.2	.999 31	.999 34	.999 36	.999 38	.999 40	3.2
3.3	.999 52	.999 53	.999 55	.999 57	.999 58	3.3
3.4	.999 66	.999 68	.999 69	.999 70	.999 71	3.4
3.5	.999 77	.999 78	.999 78	.999 79	.999 80	3.5
3.6	.999 84	.999 85	.999 85	.999 86	.999 86	3.6
3.7	.999 89	.999 90	.999 90	.999 90	.999 91	3.7
3.8	.999 93	.999 93	.999 93	.999 94	.999 94	3.8
3.9	.999 95	.999 95	.999 96	.999 96	.999 96	3.9

[a]Reproduced with permission from *Probability and Statistics in Engineering and Management Science* by W. W. Hines and D. C. Montgomery, The Ronald Press, New York, 1972.

$$\Phi(z) = \int_{-\infty}^{z} \frac{1}{\sqrt{2\pi}} e^{-u^2/2} \, du$$

z	.05	.06	.07	.08	.09	z
.0	.519 94	.523 92	.527 90	.531 88	.535 86	.0
.1	.559 62	.563 56	.567 49	.571 42	.575 34	.1
.2	.598 71	.602 57	.606 42	.610 26	.614 09	.2
.3	.636 83	.640 58	.644 31	.648 03	.651 73	.3
.4	.673 64	.677 24	.680 82	.684 38	.687 93	.4
.5	.708 84	.712 26	.715 66	.719 04	.722 40	.5
.6	.742 15	.745 37	.748 57	.751 75	.754 90	.6
.7	.773 37	.776 37	.779 35	.782 30	.785 23	.7
.8	.802 34	.805 10	.807 85	.810 57	.813 27	.8
.9	.828 94	.831 47	.833 97	.836 46	.838 91	.9
1.0	.853 14	.855 43	.857 69	.859 93	.862 14	1.0
1.1	.874 93	.876 97	.879 00	.881 00	.882 97	1.1
1.2	.894 35	.896 16	.897 96	.899 73	.901 47	1.2
1.3	.911 49	.913 08	.914 65	.916 21	.917 73	1.3
1.4	.926 47	.927 85	.929 22	.930 56	.931 89	1.4
1.5	.939 43	.940 62	.941 79	.942 95	.944 08	1.5
1.6	.950 53	.951 54	.952 54	.953 52	.954 48	1.6
1.7	.959 94	.960 80	.961 64	.962 46	.963 27	1.7
1.8	.967 84	.968 56	.969 26	.969 95	.970 62	1.8
1.9	.974 41	.975 00	.975 58	.976 15	.976 70	1.9
2.0	.979 82	.980 30	.980 77	.981 24	.981 69	2.0
2.1	.984 22	.984 61	.985 00	.985 37	.985 74	2.1
2.2	.987 78	.988 09	.988 40	.988 70	.988 99	2.2
2.3	.990 61	.990 86	.991 11	.991 34	.991 58	2.3
2.4	.992 86	.993 05	.993 24	.993 43	.993 61	2.4
2.5	.994 61	.994 77	.994 92	.995 06	.995 20	2.5
2.6	.995 98	.996 09	.996 21	.996 32	.996 43	2.6
2.7	.997 02	.997 11	.997 20	.997 28	.997 36	2.7
2.8	.997 81	.997 88	.997 95	.998 01	.998 07	2.8
2.9	.998 41	.998 46	.998 51	.998 56	.998 61	2.9
3.0	.998 86	.998 89	.998 93	.998 97	.999 00	3.0
3.1	.999 18	.999 21	.999 24	.999 26	.999 29	3.1
3.2	.999 42	.999 44	.999 46	.999 48	.999 50	3.2
3.3	.999 60	.999 61	.999 62	.999 64	.999 65	3.3
3.4	.999 72	.999 73	.999 74	.999 75	.999 76	3.4
3.5	.999 81	.999 81	.999 82	.999 83	.999 83	3.5
3.6	.999 87	.999 87	.999 88	.999 88	.999 89	3.6
3.7	.999 91	.999 92	.999 92	.999 92	.999 92	3.7
3.8	.999 94	.999 94	.999 95	.999 95	.999 95	3.8
3.9	.999 96	.999 96	.999 96	.999 97	.999 97	3.9

α / ν	.25	.1	.05	.025	.01	.005
1	1.000	3.078	6.314	12.706	31.821	63.657
2	.816	1.886	2.920	4.303	6.965	9.925
3	.765	1.638	2.353	3.182	4.541	5.841
4	.741	1.533	2.132	2.776	3.747	4.604
5	.727	1.476	2.015	2.571	3.365	4.032
6	.718	1.440	1.943	2.447	3.143	3.707
7	.711	1.415	1.895	2.365	2.998	3.499
8	.706	1.397	1.860	2.306	2.896	3.355
9	.703	1.383	1.833	2.262	2.821	3.250
10	.700	1.372	1.812	2.228	2.764	3.169
11	.697	1.363	1.796	2.201	2.718	3.106
12	.695	1.356	1.782	2.179	2.681	3.055
13	.694	1.350	1.771	2.160	2.650	3.012
14	.692	1.345	1.761	2.145	2.624	2.977
15	.691	1.341	1.753	2.131	2.602	2.947
16	.690	1.337	1.746	2.120	2.583	2.921
17	.689	1.333	1.740	2.110	2.567	2.898
18	.688	1.330	1.734	2.101	2.552	2.878
19	.688	1.328	1.729	2.093	2.539	2.861
20	.687	1.325	1.725	2.086	2,528	2.845
21	.686	1.323	1.721	2.080	2.518	2.831
22	.686	1.321	1.717	2.074	2.508	2.819
23	.685	1.319	1.714	2.069	2.500	2.807
24	.685	1.318	1.711	2.064	2.492	2.797
25	.684	1.316	1.708	2.060	2.485	2.787
26	.684	1.315	1.706	2.056	2.479	2.779
27	.684	1.314	1.703	2.052	2.473	2.771
28	.683	1.313	1.701	2.048	2.467	2.763
29	.683	1.311	1.699	2.045	2.462	2.756
30	.683	1.310	1.697	2.042	2.457	2.750
40	.681	1.303	1.684	2.021	2.423	2.704
60	.679	1.296	1.671	2.000	2.390	2.660
120	.677	1.289	1.658	1.980	2.358	2.617
∞	.674	1.282	1.645	1.960	2.326	2.576

ν \ α	.995	.990	.975	.950	.050	.025	.010	.005
1	392704×10^{-10}	157088×10^{-9}	982069×10^{-9}	393214×10^{-8}	3.84146	5.02389	6.63490	7.87944
2	.0100251	.0201007	.0506356	.102587	5.99147	7.37776	9.21034	10.5966
3	.0717212	.114832	.215795	.351846	7.81473	9.34840	11.3449	12.8381
4	.206990	.297110	.484419	.710721	9.48773	11.1433	13.2767	14.8602
5	.411740	.554300	.831211	1.145476	11.0705	12.8325	15.0863	16.7496
6	.675727	.872085	1.237347	1.63539	12.5916	14.4494	16.8119	18.5476
7	.989265	1.239043	1.68987	2.16735	14.0671	16.0128	18.4753	20.2777
8	1.344419	1.646482	2.17973	2.73264	15.5073	17.5346	20.0902	21.9550
9	1.734926	2.087912	2.70039	3.32511	16.9190	19.0228	21.6660	23.5893
10	2.15585	2.55821	3.24697	3.94030	18.3070	20.4831	23.2093	25.1882
11	2.60321	3.05347	3.81575	4.57481	19.6751	21.9200	24.7250	26.7569
12	3.07382	3.57056	4.40379	5.22603	21.0261	23.3367	26.2170	28.2995
13	3.56503	4.10691	5.00874	5.89186	22.3621	24.7356	27.6883	29.8194
14	4.07468	4.66043	5.62872	6.57063	23.6848	26.1190	29.1413	31.3193
15	4.60094	5.22935	6.26214	7.26094	24.9958	27.4884	30.5779	32.8013
16	5.14224	5.81221	6.90766	7.96164	26.2962	28.8454	31.9999	34.2672
17	5.69724	6.40776	7.56418	8.67176	27.5871	30.1910	33.4087	35.7185
18	6.26481	7.01491	8.23075	9.39046	28.8693	31.5264	34.8053	37.1564
19	6.84398	7.63273	8.90655	10.1170	30.1435	32.8523	36.1908	38.5822

20	7.43386	8.26040	9.59083	10.8508	31.4104	34.1696	37.5662	39.9968
21	8.03366	8.89720	10.28293	11.5913	32.6705	35.4789	38.9321	41.4010
22	8.64272	9.54249	10.9823	12.3380	33.9244	36.7807	40.2894	42.7956
23	9.26042	10.19567	11.6885	13.0905	35.1725	38.0757	41.6384	44.1813
24	9.88623	10.8564	12.4011	13.8484	36.4151	39.3641	42.9798	45.5585
25	10.5197	11.5240	13.1197	14.6114	37.6525	40.6465	44.3141	46.9278
26	11.1603	12.1981	13.8439	15.3791	38.8852	41.9232	45.6417	48.2899
27	11.8076	12.8786	14.5733	16.1513	40.1133	43.1944	46.9630	49.6449
28	12.4613	13.5648	15.3079	16.9279	41.3372	44.4607	48.2782	50.9933
29	13.1211	14.2565	16.0471	17.7083	42.5569	45.7222	49.5879	52.3356
30	13.7867	14.9535	16.7908	18.4926	43.7729	46.9792	50.8922	53.6720
40	20.7065	22.1643	24.4331	26.5093	55.7585	59.3417	63.6907	66.7659
50	27.9907	29.7067	32.3574	34.7642	67.5048	71.4202	76.1539	79.4900
60	35.5346	37.4848	40.4817	43.1879	79.0819	83.2976	88.3794	91.9517
70	43.2752	45.4418	48.7576	51.7393	90.5312	95.0231	100.425	104.215
80	51.1720	53.5400	57.1532	60.3915	101.879	106.629	112.329	116.321
90	59.1963	61.7541	65.6466	69.1260	113.145	118.136	124.116	128.299
100	67.3276	70.0648	74.2219	77.9295	124.342	129.561	135.807	140.169

† From *Biometrika Tables for Statisticians*, Vol. 1 (2nd edition), Cambridge University Press (1958); edited by E. S. Pearson and H. O. Hartley; reproduced by permission of the publishers.

$\alpha = .10$

ν_2 \ ν_1	1	2	3	4	5	6	7	8	9
1	39.864	49.500	53.593	55.833	57.241	58.204	58.906	59.439	59.858
2	8.5263	9.0000	9.1618	9.2434	9.2926	9.3255	9.3491	9.3668	9.3805
3	5.5383	5.4624	5.3908	5.3427	5.3092	5.2847	5.2662	5.2517	5.2400
4	4.5448	4.3246	4.1908	4.1073	4.0506	4.0098	3.9790	3.9549	3.9357
5	4.0604	3.7797	3.6195	3.5202	3.4530	3.4045	3.3679	3.3393	3.3163
6	3.7760	3.4633	3.2888	3.1808	3.1075	3.0546	3.0145	2.9830	2.9577
7	3.5894	3.2574	3.0741	2.9605	2.8833	2.8274	2.7849	2.7516	2.7247
8	3.4579	3.1131	2.9238	2.8064	2.7265	2.6683	2.6241	2.5893	2.5612
9	3.3603	3.0065	2.8129	2.6927	2.6106	2.5509	2.5053	2.4694	2.4403
10	3.2850	2.9245	2.7277	2.6053	2.5216	2.4606	2.4140	2.3772	2.3473
11	3.2252	2.8595	2.6602	2.5362	2.4512	2.3891	2.3416	2.3040	2.2735
12	3.1765	2.8068	2.6055	2.4801	2.3940	2.3310	2.2828	2.2446	2.2135
13	3.1362	2.7632	2.5603	2.4337	2.3467	2.2830	2.2341	2.1953	2.1638
14	3.1022	2.7265	2.5222	2.3947	2.3069	2.2426	2.1931	2.1539	2.1220
15	3.0732	2.6952	2.4898	2.3614	2.2730	2.2081	2.1582	2.1185	2.0862
16	3.0481	2.6682	2.4618	2.3327	2.2438	2.1783	2.1280	2.0880	2.0553
17	3.0262	2.6446	2.4374	2.3077	2.2183	2.1524	2.1017	2.0613	2.0284
18	3.0070	2.6239	2.4160	2.2858	2.1958	2.1296	2.0785	2.0379	2.0047
19	2.9899	2.6056	2.3970	2.2663	2.1760	2.1094	2.0580	2.0171	1.9836
20	2.9747	2.5893	2.3801	2.2489	2.1582	2.0913	2.0397	1.9985	1.9649
21	2.9609	2.5746	2.3649	2.2333	2.1423	2.0751	2.0232	1.9819	1.9480
22	2.9486	2.5613	2.3512	2.2193	2.1279	2.0605	2.0084	1.9668	1.9327
23	2.9374	2.5493	2.3387	2.2065	2.1149	2.0472	1.9949	1.9531	1.9189
24	2.9271	2.5383	2.3274	2.1949	2.1030	2.0351	1.9826	1.9407	1.9063
25	2.9177	2.5283	2.3170	2.1843	2.0922	2.0241	1.9714	1.9292	1.8947
26	2.9091	2.5191	2.3075	2.1745	2.0822	2.0139	1.9610	1.9188	1.8841
27	2.9012	2.5106	2.2987	2.1655	2.0730	2.0045	1.9515	1.9091	1.8743
28	2.8939	2.5028	2.2906	2.1571	2.0645	1.9959	1.9427	1.9001	1.8652
29	2.8871	2.4955	2.2831	2.1494	2.0566	1.9878	1.9345	1.8918	1.8560
30	2.8807	2.4887	2.2761	2.1422	2.0492	1.9803	1.9269	1.8841	1.8498
40	2.8354	2.4404	2.2261	2.0909	1.9968	1.9269	1.8725	1.8289	1.7929
60	2.7914	2.3932	2.1774	2.0410	1.9457	1.8747	1.8194	1.7748	1.7380
120	2.7478	2.3473	2.1300	1.9923	1.8959	1.8238	1.7675	1.7220	1.6843
∞	2.7055	2.3026	2.0838	1.9449	1.8473	1.7741	1.7167	1.6702	1.6315

$\alpha = .10$

10	12	15	20	24	30	40	60	120	∞
60.195	60.705	61.220	61.740	62.002	62.265	62.529	62.794	63.061	63.328
9.3916	9.4081	9.4247	9.4413	9.4496	9.4579	9.4663	9.4746	9.4829	9.4913
5.2304	5.2156	5.2003	5.1845	5.1764	5.1681	5.1597	5.1512	5.1425	5.1337
3.9199	3.8955	3.8689	3.8443	3.8310	3.8174	3.8036	3.7896	3.7753	3.7607
3.2974	3.2682	3.2380	3.2067	3.1905	3.1741	3.1573	3.1402	3.1228	3.1050
2.9369	2.9047	2.8712	2.8363	2.8183	2.8000	2.7812	2.7620	2.7423	2.7222
2.7025	2.6681	2.6322	2.5947	2.5753	2.5555	2.5351	2.5142	2.4928	2.4708
2.5380	2.5020	2.4642	2.4246	2.4041	2.3830	2.3614	2.3391	2.3162	2.2926
2.4163	2.3789	2.3396	2.2983	2.2768	2.2547	2.2320	2.2085	2.1843	2.1592
2.3226	2.2841	2.2435	2.2007	2.1784	2.1554	2.1317	2.1072	2.0818	2.0554
2.2482	2.2087	2.1671	2.1230	2.1000	2.0762	2.0516	2.0261	1.9997	1.9721
2.1878	2.1474	2.1049	2.0597	2.0360	2.0115	1.9861	1.9597	1.9323	1.9036
2.1376	2.0966	2.0532	2.0070	1.9827	1.9576	1.9315	1.9043	1.8759	1.8462
2.0954	2.0537	2.0095	1.9625	1.9377	1.9119	1.8852	1.8572	1.8280	1.7973
2.0593	2.0171	1.9722	1.9243	1.8990	1.8728	1.8454	1.8168	1.7867	1.7551
2.0281	1.9854	1.9399	1.8913	1.8656	1.8388	1.8108	1.7816	1.7507	1.7182
2.0009	1.9577	1.9117	1.8624	1.8362	1.8090	1.7805	1.7506	1.7191	1.6856
1.9770	1.9333	1.8868	1.8368	1.8103	1.7827	1.7537	1.7232	1.6910	1.6567
1.9557	1.9117	1.8647	1.8142	1.7873	1.7592	1.7298	1.6988	1.6659	1.6308
1.9367	1.8924	1.8449	1.7938	1.7667	1.7382	1.7083	1.6768	1.6433	1.6074
1.9197	1.8750	1.8272	1.7756	1.7481	1.7193	1.6890	1.6569	1.6228	1.5862
1.9043	1.8593	1.8111	1.7590	1.7312	1.7021	1.6714	1.6389	1.6042	1.5668
1.8903	1.8450	1.7964	1.7439	1.7159	1.6864	1.6554	1.6224	1.5871	1.5490
1.8775	1.8319	1.7831	1.7302	1.7019	1.6721	1.6407	1.6073	1.5715	1.5327
1.8658	1.8200	1.7708	1.7175	1.6890	1.6589	1.6272	1.5934	1.5570	1.5176
1.8550	1.8090	1.7596	1.7059	1.6771	1.6468	1.6147	1.5805	1.5437	1.5036
1.8451	1.7989	1.7492	1.6951	1.6662	1.6356	1.6032	1.5686	1.5313	1.4906
1.8359	1.7895	1.7395	1.6852	1.6560	1.6252	1.5925	1.5575	1.5198	1.4784
1.8274	1.7808	1.7306	1.6759	1.6465	1.6155	1.5825	1.5472	1.5090	1.4670
1.8195	1.7727	1.7223	1.6673	1.6377	1.6065	1.5732	1.5376	1.4989	1.4564
1.7627	1.7146	1.6624	1.6052	1.5741	1.5411	1.5056	1.4672	1.4248	1.3769
1.7070	1.6574	1.6034	1.5435	1.5107	1.4755	1.4373	1.3952	1.3476	1.2915
1.6524	1.6012	1.5450	1.4821	1.4472	1.4094	1.3676	1.3203	1.2646	1.1926
1.5987	1.5458	1.4871	1.4206	1.3832	1.3419	1.2951	1.2400	1.1686	1.0000

$\alpha = .05$

ν_2 \ ν_1	1	2	3	4	5	6	7	8	9
1	161.45	199.50	215.71	224.58	230.16	233.99	236.77	238.88	240.54
2	18.513	19.000	19.164	19.247	19.296	19.330	19.353	19.371	19.385
3	10.128	9.5521	9.2766	9.1172	9.0135	8.9406	8.8868	8.8452	8.8123
4	7.7086	6.9443	6.5914	6.3883	6.2560	6.1631	6.0942	6.0410	5.9988
5	6.6079	5.7861	5.4095	5.1922	5.0503	4.9503	4.8759	4.8183	4.7725
6	5.9874	5.1433	4.7571	4.5337	4.3874	4.2839	4.2066	4.1468	4.0990
7	5.5914	4.7374	4.3468	4.1203	3.9715	3.8660	3.7870	3.7257	3.6767
8	5.3177	4.4590	4.0662	3.8378	3.6875	3.5806	3.5005	3.4381	3.3881
9	5.1174	4.2565	3.8626	3.6331	3.4817	3.3738	3.2927	3.2296	3.1789
10	4.9646	4.1028	3.7083	3.4780	3.3258	3.2172	3.1355	3.0717	3.0204
11	4.8443	3.9823	3.5874	3.3567	3.2039	3.0946	3.0123	2.9480	2.8962
12	4.7472	3.8853	3.4903	3.2592	3.1059	2.9961	2.9134	2.8486	2.7964
13	4.6672	3.8056	3.4105	3.1791	3.0254	2.9153	2.8321	2.7669	2.7144
14	4.6001	3.7389	3.3439	3.1122	2.9582	2.8477	2.7642	2.6987	2.6458
15	4.5431	3.6823	3.2874	3.0556	2.9013	2.7905	2.7066	2.6408	2.5876
16	4.4940	3.6337	3.2389	3.0069	2.8524	2.7413	2.6572	2.5911	2.5377
17	4.4513	3.5915	3.1968	2.9647	2.8100	2.6987	2.6143	2.5480	2.4943
18	4.4139	3.5546	3.1599	2.9277	2.7729	2.6613	2.5767	2.5102	2.4563
19	4.3808	3.5219	3.1274	2.8951	2.7401	2.6283	2.5435	2.4768	2.4227
20	4.3513	3.4928	3.0984	2.8661	2.7109	2.5990	2.5140	2.4471	2.3928
21	4.3248	3.4668	3.0725	2.8401	2.6848	2.5727	2.4876	2.4205	2.3661
22	4.3009	3.4434	3.0491	2.8167	2.6613	2.5491	2.4638	2.3965	2.3419
23	4.2793	3.4221	3.0280	2.7955	2.6400	2.5277	2.4422	2.3748	2.3201
24	4.2597	3.4028	3.0088	2.7763	2.6207	2.5082	2.4226	2.3551	2.3002
25	4.2417	3.3852	2.9912	2.7587	2.6030	2.4904	2.4047	2.3371	2.2821
26	4.2252	3.3690	2.9751	2.7426	2.5868	2.4741	2.3883	2.3205	2.2655
27	4.2100	3.3541	2.9604	2.7278	2.5719	2.4591	2.3732	2.3053	2.2501
28	4.1960	3.3404	2.9467	2.7141	2.5581	2.4453	2.3593	2.2913	2.2360
29	4.1830	3.3277	2.9340	2.7014	2.5454	2.4324	2.3463	2.2782	2.2229
30	4.1709	3.3158	2.9223	2.6896	2.5336	2.4205	2.3343	2.2662	2.2107
40	4.0848	3.2317	2.8387	2.6060	2.4495	2.3359	2.2490	2.1802	2.1240
60	4.0012	3.1504	2.7581	2.5252	2.3683	2.2540	2.1665	2.0970	2.0401
120	3.9201	3.0718	2.6802	2.4472	2.2900	2.1750	2.0867	2.0164	1.9588
∞	3.8415	2.9957	2.6049	2.3719	2.2141	2.0986	2.0096	1.9384	1.8799

α = .05

10	12	15	20	24	30	40	60	120	∞
241.88	243.91	245.95	248.01	249.05	250.09	251.14	252.20	253.25	254.32
19.396	19.413	19.429	19.446	19.454	19.462	19.471	19.479	19.487	19.496
8.7855	8.7446	8.7029	8.6602	8.6385	8.6166	8.5944	8.5720	8.5494	8.5265
5.9644	5.9117	5.8578	5.8025	5.7744	5.7459	5.7170	5.6878	5.6581	5.6281
4.7351	4.6777	4.6188	4.5581	4.5272	4.4957	4.4638	4.4314	4.3984	4.3650
4.0600	3.9999	3.9381	3.8742	3.8415	3.8082	3.7743	3.7398	3.7047	3.6688
3.6365	3.5747	3.5108	3.4445	3.4105	3.3758	3.3404	3.3043	3.2674	3.2298
3.3472	3.2840	3.2184	3.1503	3.1152	3.0794	3.0428	3.0053	2.9669	2.9276
3.1373	3.0729	3.0061	2.9365	2.9005	2.8637	2.8259	2.7872	2.7475	2.7067
2.9782	2.9130	2.8450	2.7740	2.7372	2.6996	2.6609	2.6211	2.5801	2.5379
2.8536	2.7876	2.7186	2.6464	2.6090	2.5705	2.5309	2.4901	2.4480	2.4045
2.7534	2.6866	2.6169	2.5436	2.5055	2.4663	2.4259	2.3842	2.3410	2.2962
2.6710	2.6037	2.5331	2.4589	2.4202	2.3803	2.3392	2.2966	2.2524	2.2064
2.6021	2.5342	2.4630	2.3879	2.3487	2.3082	2.2664	2.2230	2.1778	2.1307
2.5437	2.4753	2.4035	2.3275	2.2878	2.2468	2.2043	2.1601	2.1141	2.0658
2.4935	2.4247	2.3522	2.2756	2.2354	2.1938	2.1507	2.1058	2.0589	2.0096
2.4499	2.3807	2.3077	2.2304	2.1898	2.1477	2.1040	2.0584	2.0107	1.9604
2.4117	2.3421	2.2686	2.1906	2.1497	2.1071	2.0629	2.0166	1.9681	1.9168
2.3779	2.3080	2.2341	2.1555	2.1141	2.0712	2.0264	1.9796	1.9302	1.8780
2.3479	2.2776	2.2033	2.1242	2.0825	2.0391	1.9938	1.9464	1.8963	1.8432
2.3210	2.2504	2.1757	2.0960	2.0540	2.0102	1.9645	1.9165	1.8657	1.8117
2.2967	2.2258	2.1508	2.0707	2.0283	1.9842	1.9380	1.8895	1.8380	1.7831
2.2747	2.2036	2.1282	2.0476	2.0050	1.9605	1.9139	1.8649	1.8128	1.7570
2.2547	2.1834	2.1077	2.0267	1.9838	1.9390	1.8920	1.8424	1.7897	1.7331
2.2365	2.1649	2.0889	2.0075	1.9643	1.9192	1.8718	1.8217	1.7684	1.7110
2.2197	2.1479	2.0716	1.9898	1.9464	1.9010	1.8533	1.8027	1.7488	1.6906
2.2043	2.1323	2.0558	1.9736	1.9299	1.8842	1.8361	1.7851	1.7307	1.6717
2.1900	2.1179	2.0411	1.9586	1.9147	1.8687	1.8203	1.7689	1.7138	1.6541
2.1768	2.1045	2.0275	1.9446	1.9005	1.8543	1.8055	1.7537	1.6981	1.6377
2.1646	2.0921	2.0148	1.9317	1.8874	1.8409	1.7918	1.7396	1.6835	1.6223
2.0772	2.0035	1.9245	1.8389	1.7929	1.7444	1.6928	1.6373	1.5766	1.5089
1.9926	1.9174	1.8364	1.7480	1.7001	1.6491	1.5943	1.5343	1.4673	1.3893
1.9105	1.8337	1.7505	1.6587	1.6084	1.5543	1.4952	1.4290	1.3519	1.2539
1.8307	1.7522	1.6664	1.5705	1.5173	1.4591	1.3940	1.3180	1.2214	1.0000

$\alpha = .025$

ν_2 \ ν_1	1	2	3	4	5	6	7	8	9
1	647.79	799.50	864.16	899.58	921.85	937.11	948.22	956.66	963.28
2	38.506	39.000	39.165	39.248	39.298	39.331	39.355	39.373	39.387
3	17.443	16.044	15.439	15.101	14.885	14.735	14.624	14.540	14.473
4	12.218	10.649	9.9792	9.6045	9.3645	9.1973	9.0741	8.9796	8.9047
5	10.007	8.4336	7.7636	7.3879	7.1464	6.9777	6.8531	6.7572	6.6810
6	8.8131	7.2598	6.5988	6.2272	5.9876	5.8197	5.6955	5.5996	5.5234
7	8.0727	6.5415	5.8898	5.5226	5.2852	5.1186	4.9949	4.8994	4.8232
8	7.5709	6.0595	5.4160	5.0526	4.8173	4.6517	4.5286	4.4332	4.3572
9	7.2093	5.7147	5.0781	4.7181	4.4844	4.3197	4.1971	4.1020	4.0260
10	6.9367	5.4564	4.8256	4.4683	4.2361	4.0721	3.9498	3.8549	3.7790
11	6.7241	5.2559	4.6300	4.2751	4.0440	3.8807	3.7586	3.6638	3.5879
12	6.5538	5.0959	4.4742	4.1212	3.8911	3.7283	3.6065	3.5118	3.4358
13	6.4143	4.9653	4.3472	3.9959	3.7667	3.6043	3.4827	3.3880	3.3120
14	6.2979	4.8567	4.2417	3.8919	3.6634	3.5014	3.3799	3.2853	3.2093
15	6.1995	4.7650	4.1528	3.8043	3.5764	3.4147	3.2934	3.1987	3.1227
16	6.1151	4.6867	4.0768	3.7294	3.5021	3.3406	3.2194	3.1248	3.0488
17	6.0420	4.6189	4.0112	3.6648	3.4379	3.2767	3.1556	3.0610	2.9849
18	5.9781	4.5597	3.9539	3.6083	3.3820	3.2209	3.0999	3.0053	2.9291
19	5.9216	4.5075	3.9034	3.5587	3.3327	3.1718	3.0509	2.9563	2.8800
20	5.8715	4.4613	3.8587	3.5147	3.2891	3.1283	3.0074	2.9128	2.8365
21	5.8266	4.4199	3.8188	3.4754	3.2501	3.0895	2.9686	2.8740	2.7977
22	5.7863	4.3828	3.7829	3.4401	3.2151	3.0546	2.9338	2.8392	2.7628
23	5.7498	4.3492	3.7505	3.4083	3.1835	3.0232	2.9024	2.8077	2.7313
24	5.7167	4.3187	3.7211	3.3794	3.1548	2.9946	2.8738	2.7791	2.7027
25	5.6864	4.2909	3.6943	3.3530	3.1287	2.9685	2.8478	2.7531	2.6766
26	5.6586	4.2655	3.6697	3.3289	3.1048	2.9447	2.8240	2.7293	2.6528
27	5.6331	4.2421	3.6472	3.3067	3.0828	2.9228	2.8021	2.7074	2.6309
28	5.6096	4.2205	3.6264	3.2863	3.0625	2.9027	2.7820	2.6872	2.6106
29	5.5878	4.2006	3.6072	3.2674	3.0438	2.8840	2.7633	2.6686	2.5919
30	5.5675	4.1821	3.5894	3.2499	3.0265	2.8667	2.7460	2.6513	2.5746
40	5.4239	4.0510	3.4633	3.1261	2.9037	2.7444	2.6238	2.5289	2.4519
60	5.2857	3.9253	3.3425	3.0077	2.7863	2.6274	2.5068	2.4117	2.3344
120	5.1524	3.8046	3.2270	2.8943	2.6740	2.5154	2.3948	2.2994	2.2217
∞	5.0239	3.6889	3.1161	2.7858	2.5665	2.4082	2.2875	2.1918	2.1136

α = .025

10	12	15	20	24	30	40	60	120	∞
968.63	976.71	984.87	993.10	997.25	1001.4	1005.6	1009.8	1014.0	1018.3
39.398	39.415	39.431	39.448	39.456	39.465	39.473	39.481	39.490	39.498
14.419	14.337	14.253	14.167	14.124	14.081	14.037	13.992	13.947	13.902
8.8439	8.7512	8.6565	8.5599	8.5109	8.4613	8.4111	8.3604	8.3092	8.2573
6.6192	6.5246	6.4277	6.3285	6.2780	6.2269	6.1751	6.1225	6.0693	6.0153
5.4613	5.3662	5.2687	5.1684	5.1172	5.0652	5.0125	5.9589	4.9045	4.8491
4.7611	4.6658	4.5678	4.4667	4.4150	4.3624	4.3089	4.2544	4.1989	4.1423
4.2951	4.1997	4.1012	3.9995	3.9472	3.8940	3.8398	3.7844	3.7279	3.6702
3.9639	3.8682	3.7694	3.6669	3.6142	3.5604	3.5055	3.4493	3.3918	3.3329
3.7168	3.6209	3.5217	3.4186	3.3654	3.3110	3.2554	3.1984	3.1399	3.0798
3.5257	3.4296	3.3299	3.2261	3.1725	3.1176	3.0613	3.0035	2.9441	2.8828
3.3736	3.2773	3.1772	3.0728	3.0187	2.9633	2.9063	2.8478	2.7874	2.7249
3.2497	3.1532	3.0527	2.9477	2.8932	2.8373	2.7797	2.7204	2.6590	2.5955
3.1469	3.0501	2.9493	2.8437	2.7888	2.7324	2.6742	2.6142	2.5519	2.4872
3.0602	2.9633	2.8621	2.7559	2.7006	2.6437	2.5850	2.5242	2.4611	2.3953
2.9862	2.8890	2.7875	2.6808	2.6252	2.5678	2.5085	2.4471	2.3831	2.3163
2.9222	2.8249	2.7230	2.6158	2.5598	2.5021	2.4422	2.3801	2.3153	2.2474
2.8664	2.7689	2.6667	2.5590	2.5027	2.4445	2.3842	2.3214	2.2558	2.1869
2.8173	2.7196	2.6171	2.5089	2.4523	2.3937	2.3329	2.2695	2.2032	2.1333
2.7737	2.6758	2.5731	2.4645	2.4076	2.3486	2.2873	2.2234	2.1562	2.0853
2.7348	2.6368	2.5338	2.4247	2.3675	2.3082	2.2465	2.1819	2.1141	2.0422
2.6998	2.6017	2.4984	2.3890	2.3315	2.2718	2.2097	2.1446	2.0760	2.0032
2.6682	2.5699	2.4665	2.3567	2.2989	2.2389	2.1763	2.1107	2.0415	1.9677
2.6396	2.5412	2.4374	2.3273	2.2693	2.2090	2.1460	2.0799	2.0099	1.9353
2.6135	2.5149	2.4110	2.3005	2.2422	2.1816	2.1183	2.0517	1.9811	1.9055
2.5895	2.4909	2.3867	2.2759	2.2174	2.1565	2.0928	2.0257	1.9545	1.8781
2.5676	2.4688	2.3644	2.2533	2.1946	2.1334	2.0693	2.0018	1.9299	1.8527
2.5473	2.4484	2.3438	2.2324	2.1735	2.1121	2.0477	1.9796	1.9072	1.8291
2.5286	2.4295	2.3248	2.2131	2.1540	2.0923	2.0276	1.9591	1.8861	1.8072
2.5112	2.4120	2.3072	2.1952	2.1359	2.0739	2.0089	1.9400	1.8664	1.7867
2.3882	2.2882	2.1819	2.0677	2.0069	1.9429	1.8752	1.8028	1.7242	1.6371
2.2702	2.1692	2.0613	1.9445	1.8817	1.8152	1.7440	1.6668	1.5810	1.4822
2.1570	2.0548	1.9450	1.8249	1.7597	1.6899	1.6141	1.5299	1.4327	1.3104
2.0483	1.9447	1.8326	1.7085	1.6402	1.5660	1.4835	1.3883	1.2684	1.0000

$\alpha = .01$

ν_2 \ ν_1	1	2	3	4	5	6	7	8	9
1	4052.2	4999.5	5403.3	5624.6	5763.7	5859.0	5928.3	5981.6	6022.5
2	98.503	99.000	99.166	99.249	99.299	99.332	99.356	99.374	99.388
3	34.116	30.817	29.457	28.710	28.237	27.911	27.672	27.489	27.345
4	21.198	18.000	16.694	15.977	15.522	15.207	14.976	14.799	14.659
5	16.258	13.274	12.060	11.392	10.967	10.672	10.456	10.289	10.158
6	13.745	10.925	9.7795	9.1483	8.7459	8.4661	8.2600	8.1016	7.9761
7	12.246	9.5466	8.4513	7.8467	7.4604	7.1914	6.9928	6.8401	6.7188
8	11.259	8.6491	7.5910	7.0060	6.6318	6.3707	6.1776	6.0289	5.9106
9	10.561	8.0215	6.9919	6.4221	6.0569	5.8018	5.6129	5.4671	5.3511
10	10.044	7.5594	6.5523	5.9943	5.6363	5.3858	5.2001	5.0567	4.9424
11	9.6460	7.2057	6.2167	5.6683	5.3160	5.0692	4.8861	4.7445	4.6315
12	9.3302	6.9266	5.9526	5.4119	5.0643	4.8206	4.6395	4.4994	4.3875
13	9.0738	6.7010	5.7394	5.2053	4.8616	4.6204	4.4410	4.3021	4.1911
14	8.8616	6.5149	5.5639	5.0354	4.6950	4.4558	4.2779	4.1399	4.0297
15	8.6831	6.3589	5.4170	4.8932	4.5556	4.3183	4.1415	4.0045	3.8948
16	8.5310	6.2262	5.2922	4.7726	4.4374	4.2016	4.0259	3.8896	3.7804
17	8.3997	6.1121	5.1850	4.6690	4.3359	4.1015	3.9267	3.7910	3.6822
18	8.2854	6.0129	5.0919	4.5790	4.2479	4.0146	3.8406	3.7054	3.5971
19	8.1850	5.9259	5.0103	4.5003	4.1708	3.9386	3.7653	3.6305	3.5225
20	8.0960	5.8489	4.9382	4.4307	4.1027	3.8714	3.6987	3.5644	3.4567
21	8.0166	5.7804	4.8740	4.3688	4.0421	3.8117	3.6396	3.5056	3.3981
22	7.9454	5.7190	4.8166	4.3134	3.9880	3.7583	3.5867	3.4530	3.3458
23	7.8811	5.6637	4.7649	4.2635	3.9392	3.7102	3.5390	3.4057	3.2986
24	7.8229	5.6136	4.7181	4.2184	3.8951	3.6667	3.4959	3.3629	3.2560
25	7.7698	5.5680	4.6755	4.1774	3.8550	3.6272	3.4568	3.3239	3.2172
26	7.7213	5.5263	4.6366	4.1400	3.8183	3.5911	3.4210	3.2884	3.1818
27	7.6767	5.4881	4.6009	4.1056	3.7848	3.5580	3.3882	3.2558	3.1494
28	7.6356	5.4529	4.5681	4.0740	3.7539	3.5276	3.3581	3.2259	3.1195
29	7.5976	5.4205	4.5378	4.0449	3.7254	3.4995	3.3302	3.1982	3.0920
30	7.5625	5.3904	4.5097	4.0179	3.6990	3.4735	3.3045	3.1726	3.0665
40	7.3141	5.1785	4.3126	3.8283	3.5138	3.2910	3.1238	2.9930	2.8876
60	7.0771	4.9774	4.1259	3.6491	3.3389	3.1187	2.9530	2.8233	2.7185
120	6.8510	4.7865	3.9493	3.4796	3.1735	2.9559	2.7918	2.6629	2.5586
∞	6.6349	4.6052	3.7816	3.3192	3.0173	2.8020	2.6393	2.5113	2.4073

$\alpha = .01$

10	12	15	20	24	30	40	60	120	∞
6055.8	6106.3	6157.3	6208.7	6234.6	6260.7	6286.8	6313.0	6339.4	6366.0
99.399	99.416	99.432	99.449	99.458	99.466	99.474	99.483	99.491	99.501
27.229	27.052	26.872	26.690	26.598	26.505	26.411	26.316	26.221	26.125
14.546	14.374	14.198	14.020	13.929	13.838	13.745	13.652	13.558	13.463
10.051	9.8883	9.7222	9.5527	9.4665	9.3793	9.2912	9.2020	9.1118	9.0204
7.8741	7.7183	7.5590	7.3958	7.3127	7.2285	7.1432	7.0568	6.9690	6.8801
6.6201	6.4691	6.3143	6.1554	6.0743	5.9921	5.9084	5.8236	5.7372	5.6495
5.8143	5.6668	5.5151	5.3591	5.2793	5.1981	5.1156	5.0316	4.9460	4.8588
5.2565	5.1114	4.9621	4.8080	4.7290	4.6486	4.5667	4.4831	4.3978	4.3105
4.8492	4.7059	4.5582	4.4054	4.3269	4.2469	4.1653	4.0819	3.9965	3.9090
4.5393	4.3974	4.2509	4.0990	4.0209	3.9411	3.8596	3.7761	3.6904	3.6025
4.2961	4.1553	4.0096	3.8584	3.7805	3.7008	3.6192	3.5355	3.4494	3.3608
4.1003	3.9603	3.8154	3.6646	3.5868	3.5070	3.4253	3.3413	3.2548	3.1654
3.9394	3.8001	3.6557	3.5052	3.4274	3.3476	3.2656	3.1813	3.0942	3.0040
3.8049	3.6662	3.5222	3.3719	3.2940	3.2141	3.1319	3.0471	2.9595	2.8684
3.6909	3.5527	3.4089	3.2588	3.1808	3.1007	3.0182	2.9330	2.8447	2.7528
3.5931	3.4552	3.3117	3.1615	3.0835	3.0032	2.9205	2.8348	2.7459	2.6530
3.5082	3.3706	3.2273	3.0771	2.9990	2.9185	2.8354	2.7493	2.6597	2.5660
3.4338	3.2965	3.1533	3.0031	2.9249	2.8442	2.7608	2.6742	2.5839	2.4893
3.3682	3.2311	3.0880	2.9377	2.8594	2.7785	2.6947	2.6077	2.5168	2.4212
3.3098	3.1729	3.0299	2 8796	2.8011	2.7200	2.6359	2.5484	2.4568	2.3603
3.2576	3.1209	2.9780	2.8274	2.7488	2.6675	2.5831	2.4951	2.4029	2.3055
3.2106	3.0740	2.9311	2.7805	2.7017	2.6202	2.5355	2.4471	2.3542	2.2559
3.1681	3.0316	2.8887	2.7380	2.6591	2.5773	2.4923	2.4035	2.3099	2.2107
3.1294	2.9931	2.8502	2.6993	2.6203	2.5383	2.4530	2.3637	2.2695	2.1694
3.0941	2.9579	2.8150	2.6640	2.5848	2.5026	2.4170	2.3273	2.2325	2.1315
3.0618	2.9256	2.7827	2.6316	2.5522	2.4699	2.3840	2.2938	2.1984	2.0965
3.0320	2.8959	2.7530	2.6017	2.5223	2.4397	2.3535	2.2629	2.1670	2.0642
3.0045	2.8685	2.7256	2.5742	2.4946	2.4118	2.3253	2.2344	2.1378	2.0342
2.9791	2.8431	2.7002	2.5487	2.4689	2.3860	2.2992	2.2079	2.1107	2.0062
2.8005	2.6648	2.5216	2.3689	2.2880	2.2034	2.1142	2.0194	1.9172	1.8047
2.6318	2.4961	2.3523	2.1978	2.1154	2.0285	1.9360	1.8363	1.7263	1.6006
2.4721	2.3363	2.1915	2.0346	1.9500	1.8600	1.7628	1.6557	1.5330	1.3805
2.3209	2.1848	2.0385	1.8783	1.7908	1.6964	1.5923	1.4730	1.3246	1.0000

Appendix B

Review of Matrix Algebra

1. BASIC DEFINITIONS AND PROPERTIES OF MATRICES

Matrix notation is used extensively in experimental design, regression analysis, and optimization. It is a perceptive notation, which unites disparate problems with a single representation, a common method of analysis used throughout the text. Because of this, a review of the fundamentals of matrix notation is included here.

Matrices and Vectors

A real matrix is a rectangular array of nm real numbers arranged into n rows and m columns

$$A=(a_{ij})=\begin{bmatrix} a_{11} & a_{12} & \cdots & a_{1m} \\ a_{21} & a_{22} & \cdots & a_{2m} \\ \cdots & & & \\ a_{n1} & a_{n2} & \cdots & a_{nm} \end{bmatrix}$$

and denoted by a capital letter or array of elements a_{ij}. The set of $n \times m$ real matrices is given by $R^{n \times m}$.

Matrices with the same number of rows and columns are called square. The elements a_{ii} are called diagonal elements and lie on the main diagonal of the matrix. Matrices in which only the diagonal elements are nonzero are called diagonal. For instance, D is a diagonal matrix.

$$D=\begin{bmatrix} 1 & 0 & 0 \\ 0 & 3 & 0 \\ 0 & 0 & 4 \end{bmatrix}$$

A matrix with all elements above or below the main diagonal zero is called a triangular matrix. A triangular matrix with zeros above the main diagonal is called lower triangular

$$\begin{bmatrix} 1 & 0 & 0 \\ 2 & 1 & 0 \\ 4 & 6 & 15 \end{bmatrix}$$

and one with zeros below the main diagonal is upper triangular

$$\begin{bmatrix} 1 & 2 & 7 \\ 0 & -5 & 10 \\ 0 & 0 & 11 \end{bmatrix}$$

A diagonal matrix is thus both upper and lower triangular.

Matrices of a single row or column are called vectors. A column vector of dimension n consists of n elements arranged in a column and represented by a boldface lowercase letter

$$\mathbf{a} = \begin{bmatrix} a_1 \\ a_2 \\ \cdots \\ a_n \end{bmatrix}$$

The set of all n-dimensional real column vectors is denoted by $R^{n\times 1}$ or R^n. An n-dimensional row vector has n elements arranged in a row; the set of all n-dimensional row vectors is denoted by $R^{1\times n}$.

Matrix Transpose

The transpose of a matrix A, denoted by A' or A^T, is formed by interchanging the rows and columns of A. For example,

$$\begin{bmatrix} 1 & 2 & 3 \\ -5 & 7 & 4 \end{bmatrix}^T = \begin{bmatrix} 1 & -5 \\ 2 & 7 \\ 3 & 4 \end{bmatrix}$$

In this text all vectors are column vectors. Row vectors are designated by transposed column vectors, $\mathbf{a}^T$.

A matrix equal to its own transpose is called a symmetric matrix. For instance,

$$\begin{bmatrix} 1 & 2 & 5 \\ 2 & 3 & 4 \\ 5 & 4 & 6 \end{bmatrix}$$

is a symmetric matrix.

Matrix Addition and Multiplication

Two matrices can be added together if they have the same number of rows and columns. Addition is performed with like elements.

$$\begin{bmatrix} 1 & 0 \\ 2 & 4 \end{bmatrix} + \begin{bmatrix} 5 & 3 \\ 6 & 8 \end{bmatrix} = \begin{bmatrix} 1+5 & 0+3 \\ 2+6 & 4+8 \end{bmatrix} = \begin{bmatrix} 6 & 3 \\ 8 & 12 \end{bmatrix}$$

Matrix addition is both commutative and associative,

$$A + B = B + A$$

and

$$(A + B) + C = A + (B + C)$$

Any real number or scalar can multiply a matrix by multiplying all the elements of the matrix

$$5\begin{bmatrix} 0 & 1 \\ 10 & -5 \end{bmatrix} = \begin{bmatrix} 0 & 5 \\ 50 & -25 \end{bmatrix}$$

The matrix product AB between $A \in R^{n \times m}$ and $B \in R^{m \times p}$ is defined for the elements c_{il} by

$$(a_{ij})(b_{jl}) = \sum_{j=1}^{m} a_{ij} b_{jl} = (c_{il})$$

and belongs to $R^{n \times p}$. The product is defined when the number of columns of A and the number of rows of B are equal, and A and B are then called conformable for multiplication. The resultant matrix has the same number of rows as A and the same number of columns as B. For instance,

$$\begin{bmatrix} 1 & 4 \\ 2 & 2 \\ 3 & 1 \end{bmatrix} \begin{bmatrix} 5 & 2 \\ 0 & 6 \end{bmatrix} = \begin{bmatrix} 5 & 26 \\ 10 & 16 \\ 15 & 12 \end{bmatrix}$$

The vector product

$$\mathbf{a}^T \mathbf{b} = s$$

between $\mathbf{a}, \mathbf{b} \in R^n$ is a 1×1 matrix s which is treated like a scalar. Matrix multiplication is associative,

$$A(BC) = (AB)C$$

but is generally not commutative. For example,

$$\begin{bmatrix} 1 & 2 \\ 3 & 4 \end{bmatrix} \begin{bmatrix} 1 & 2 \\ 0 & 1 \end{bmatrix} = \begin{bmatrix} 1 & 4 \\ 3 & 10 \end{bmatrix}$$

does not equal

$$\begin{bmatrix} 1 & 2 \\ 0 & 1 \end{bmatrix} \begin{bmatrix} 1 & 2 \\ 3 & 4 \end{bmatrix} = \begin{bmatrix} 7 & 10 \\ 3 & 4 \end{bmatrix}$$

The diagonal matrix of order $n \times n$ of all ones (shown for $n=3$)

$$\begin{bmatrix} 1 & 0 & 0 \\ 0 & 1 & 0 \\ 0 & 0 & 1 \end{bmatrix}$$

is the multiplicative identity and denoted by I, and therefore

$$AI = IA = A$$

The transpose of a product of matrices is the product of transpose matrices taken in the reverse order

$$(ABC)^T = C^t B^t A^t$$

A special transpose product $A^T A$ results in a square symmetric matrix. A square matrix B is said to have an inverse B^{-1} if and only if

$$B^{-1}B = BB^{-1} = I$$

Determinants

A determinant $|A|$ or det(A) is a scalar function of a square matrix A with the following properties:

1. The determinant of a matrix product is equal to the product of the determinants

$$|AB| = |A| \cdot |B|$$

2. The determinants of a matrix and its transpose are equal.
3. The exchange of a row (column) with another row (column) changes the sign of the determinant.
4. The determinant is zero if any row or column is zero, or if any row or column can be expressed as a combination of the other rows or columns.
5. The determinant of a scalar multiple of an $n \times n$ matrix is

$$|\alpha A| = \alpha^n |A|$$

6. The determinant of a diagonal matrix $D = (d_{ii})$ is

$$|D| = \prod_{i=1}^{n} d_{ii}$$

7. If we define as M_{ij} the minor matrix created by deleting the ith row and jth column, the cofactor A_{ij} of the element a_{ij} is given by $(-1)^{i+j}|M_{ij}|$. The determinant of A can then be expressed as an expansion of

elements and cofactors along any row i or column j of A according to

$$|A| = \sum_{j=1}^{n} a_{ij}A_{ij}$$

$$= \sum_{i=1}^{n} a_{ij}A_{ij}$$

Eigenvalues and Eigenvectors

The roots of the characteristic polynomial

$$|A - \lambda I| = 0$$

are called the characteristic roots or eigenvalues of the matrix A, λ_i or $\lambda_i(A)$. For instance the square matrix

$$A = \begin{bmatrix} 3 & 2 \\ 2 & 1 \end{bmatrix}$$

has as its characteristic polynomial

$$\lambda^2 - 4\lambda - 1$$

whose roots are $\lambda_1 = 2 + \sqrt{5}$ and $\lambda_2 = 2 - \sqrt{5}$.

A vector solution $\boldsymbol{\varepsilon}_i$ to the equation

$$(A - \lambda_i I)\boldsymbol{\varepsilon}_i = \mathbf{0}$$

for an eigenvalue λ_i is called an eigenvector of the matrix A. Such a solution is not unique, since $\beta\boldsymbol{\varepsilon}_i$ is also an eigenvector to A.

The eigenvalues of real symmetric matrices are all real.

It can also be shown that the determinant of A is equal to the product of its eigenvalues,

$$|A| = \prod_{i=1}^{n} \lambda_i(A)$$

Quadratic Forms

A quadratic expression of the form

$$q = \sum_{i=1}^{n} a_{ii}x_i^2 + \sum_{i>j}\sum a_{ij}x_ix_j$$

can be expressed in terms of a symmetric matrix $A = (a_{ij})$ as

$$q = \mathbf{x}^T A \mathbf{x}$$

The matrix A is called positive definite if for any nonzero $\mathbf{x}$

$$\mathbf{x}^T A \mathbf{x} > 0$$

A is called positive semidefinite if for any nonzero $\mathbf{x}$

$$\mathbf{x}^T A \mathbf{x} \geqslant 0$$

A necessary and sufficient condition for A to be positive definite is for all of its eigenvalues to be positive. For A to be positive semidefinite it is necessary and sufficient for it's eigenvalues to be greater than or equal to zero.

Distance, Norms, and Inner Products

A distance function $d(\mathbf{x}, \mathbf{z})$ for vectors will embody certain intuitive notions about distance. We expect that the distance be a nonnegative number and that the distance can only be zero when the two vectors are identical. We expect that distance between $\mathbf{x}$ and $\mathbf{z}$ be the same as that between $\mathbf{z}$ and $\mathbf{x}$. Finally, the triangle inequality should hold, so that the distance from a point $\mathbf{x}$ to a point $\mathbf{z}$ is less than or equal to the sum of distances from $\mathbf{x}$ to $\mathbf{y}$ and from $\mathbf{y}$ to $\mathbf{z}$.

One way to define distance is in terms of a vector norm, $\|\mathbf{x}\|$. A norm is a function that assigns a scalar $\|\mathbf{x}\|$ to every vector $\mathbf{x} \in R^n$. The norm has the properties

1. $\|\mathbf{x}\| \geqslant 0$ and $\|\mathbf{x}\| = 0$ if and only if $\mathbf{x} = \mathbf{0}$
2. $\|\alpha\mathbf{x}\| = |\alpha| \, \|\mathbf{x}\|$ for all real α
3. $\|\mathbf{x} + \mathbf{z}\| \leqslant \|\mathbf{x}\| + \|\mathbf{z}\|$

There are an infinite number of p-norms of the form

$$\|\mathbf{x}\|_P = \left(\sum_{i=1}^{n} |x_i^p| \right)^{1/p}$$

which can be used as distance functions. The most familiar are the 1-norm or Chebyshev norm,

$$\|\mathbf{x}\|_1 = \sum_{i=1}^{n} |x_i|$$

and the Euclidean or two-norm,

$$\|\mathbf{x}\|_2 = \left(\sum_{i=1}^{n} x_i^2 \right)^{1/2}$$

and the infinity norm,

$$\|\mathbf{x}\|_\infty = \max_i |x_i|$$

The distance between two points $\mathbf{x}$ and $\mathbf{y}$ can then be defined by the vector norms

$$d(\mathbf{x},\mathbf{y}) = \|\mathbf{x}-\mathbf{y}\|$$

The vector inner product has certain properties similar to the vector norms. An inner product is also a scalar function $(\mathbf{x},\mathbf{z})$ of vectors with the properties

1. $(\mathbf{x},\mathbf{x}) \geqslant 0$
2. $(\mathbf{x},\mathbf{x}) = 0$ if and only if $\mathbf{x} = \mathbf{0}$
3. $(\mathbf{x},\mathbf{z}) = (\mathbf{z},\mathbf{x})$
4. $(\mathbf{x},\alpha\mathbf{z}+\beta\mathbf{w}) = \alpha(\mathbf{x},\mathbf{z}) + \beta(\mathbf{x},\mathbf{w})$

A general form of the inner product is

$$(\mathbf{x},\mathbf{z}) = \mathbf{x}^T Q \mathbf{x}$$

where Q is a real symmetric $n \times n$ positive definite matrix. When Q is the identity matrix I, $(\mathbf{x},\mathbf{x})$ is the square of the Euclidian norm.

Linear Independence and Matrix Rank

A set of n vectors $\mathbf{x}_i$ is said to be linearly independent if the equation

$$\alpha_1\mathbf{x}_1 + \alpha_2\mathbf{x}_2 + \cdots + \alpha_n\mathbf{x}_n = \mathbf{0}$$

is only satisfied for $\alpha_1 = \alpha_2 = \cdots = \alpha_n = 0$. If there are nonzero α_i the vectors are linearly dependent and at least one vector can be expressed as a combination of the others.

In R^n there can at most be n vectors in a set of independent vectors. Since any vector in R^n can be expressed as a combination of n independent vectors such a set is said to span or generate R^n.

The rank of a matrix is given by the number of independent vectors in its row or column space. The row and column rank of a matrix are always equal.

In a square $n \times n$ matrix A the following are equivalent

1. A is of rank n.
2. The determinant is nonzero.
3. The n rows of A are linearly independent.
4. The n columns of A are linearly independent.
5. All the eigenvalues of A are nonzero.
6. The inverse A^{-1} exists.

2. LINEAR EQUATIONS

A system of linear equations

$$a_{11}x_1 + a_{12}x_2 + \cdots + a_{1m}x_m = b_1$$
$$a_{21}x_1 + a_{22}x_2 + \cdots + a_{2m}x_m = b_2$$
$$\cdots$$
$$a_{n1}x_1 + a_{n2}x_2 + \cdots + a_{nm}x_m = b_n$$

can be represented by

$$A\mathbf{x} = \mathbf{b}$$

where $A = (a_{nm})$, $\mathbf{x} = (x_m)$ and $\mathbf{b} = (b_n)$.

When $n = m$, A is square and the number of equations is equal to the number of unknowns. A unique solution exists if the matrix is of full rank. The solution is formally designated by

$$\mathbf{x} = A^{-1}\mathbf{b}$$

If A is not of full rank an infinite number of solutions exist if $\mathbf{b}$ is in the space spanned by the independent column vectors of A. Otherwise no solution for the equations exist.

It is of theoretical importance that solutions to full rank problems can be designated by $A^{-1}\mathbf{b}$, but in terms of numerical computation, calculation of an inverse is less efficient and more prone to error than other approaches. For instance, the procedure of Gauss factorizes A into a lower triangular matrix L and an upper triangular matrix U so that

$$A = LU$$

and the equivalent system of equations

$$L\mathbf{c} = \mathbf{b}$$

and

$$U\mathbf{x} = \mathbf{c}$$

is solved instead, without the necessity of a matrix inversion. These procedures are used in all computation packages and are so efficient that A^{-1} is only calculated when it has significance of its own.

For systems of equations where $n > m$, there are more equations than unknowns. Such a system is called overdetermined, and in general no unique solution to the problem exists. Solutions are generally posed for these problems by minimizing the distance between the left and right sides of the equation using either the three norms denoted earlier or a quadratic inner product as the distance function.

Matrix Calculus

The solution of minimum distance and other optimization problems is frequently approached via the matrix calculus. In this section we note some of the derivatives of scalar and vector valued functions of vectors.

The derivative of a scalar valued function of a vector is a row vector of derivatives with respect to each component direction

$$\nabla f(\mathbf{x}) = \left(\frac{\partial f}{\partial x_1}, \frac{\partial f}{\partial x_2}, \ldots, \frac{\partial f}{\partial x_n} \right)$$

called the gradient of f, "del f" or "grad f". Each component gives the rate of change in the function with respect to that direction. The gradient of some common scalars is given by

$$\nabla \mathbf{a}^t\mathbf{x} = \mathbf{a}^t$$

$$\nabla \mathbf{x}^t\mathbf{x} = 2\mathbf{x}^t$$

$$\nabla \mathbf{x}^t B\mathbf{x} = \mathbf{x}^t B + \mathbf{x}^t B^t = \mathbf{x}^t(B + B^t)$$

when B is symmetric, $B = B^t$ so that

$$\nabla \mathbf{x}^t B\mathbf{x} = 2\mathbf{x}^t B$$

The Hessian matrix consists of the second derivatives of $f(\mathbf{x})$ and is designated by H_f or $\nabla^2 f(\mathbf{x})$

$$\nabla^2 f(\mathbf{x}) = \begin{bmatrix} \frac{\partial^2 f}{\partial x_1^2} & \frac{\partial^2 f}{\partial x_1 \partial x_2} & \cdots & \frac{\partial^2 f}{\partial x_1 \partial x_n} \\ \frac{\partial^2 f}{\partial x_2 \partial x_1} & \frac{\partial^2 f}{\partial x_2^2} & \cdots & \frac{\partial^2 f}{\partial x_2 \partial x_n} \\ & \cdots & & \\ \frac{\partial^2 f}{\partial x_n \partial x_1} & \frac{\partial^2 f}{\partial x_n \partial x_2} & \cdots & \frac{\partial^2 f}{\partial x_n^2} \end{bmatrix}$$

The derivative of a vector valued function for a vector is called the Jacobian or Jacobian matrix. The rows of the Jacobian consist of the individual gradients of each component of the vector:

$$J_g = \begin{bmatrix} \nabla g_1 \\ \nabla g_2 \\ \vdots \\ \nabla g_m \end{bmatrix} = \begin{bmatrix} \frac{\partial g_1}{\partial x_1} & \frac{\partial g_1}{\partial x_2} & \cdots & \frac{\partial g_1}{\partial x_n} \\ \frac{\partial g_2}{\partial x_1} & \frac{\partial g_2}{\partial x_2} & \cdots & \frac{\partial g_2}{\partial x_n} \\ & \cdots & & \\ \frac{\partial g_n}{\partial x_1} & \frac{\partial g_n}{\partial x_2} & \cdots & \frac{\partial g_n}{\partial x_n} \end{bmatrix}$$

Index

Acceleration, in searches, 194, 196
Adaptive random search, 192
Algorithm, 16
Aliasing, 108, 129
Alias matrix, 107
Analysis of variance, 89
  one-way, 90
  two-way, 94
ANOVA table, 93, 95, 103
  general regression, 104
  one-way, 93
  two-way, 95
Augmented simplex design, 150

Barrier function, 198
Bernoulli trial, 44, 48
Best feasible direction, 226, 245, 279, 281
Bias, 103, 106, 126, 256
Bias structure, 126, 128, 134
Binomial distribution, 45, 48
Blocking, 132, 152
Box-Behnkin design, 132, 147
Box's "complex" search, 17, 23, 95, 250, 297

Central composite designs, 137, 296, 299
Central limit theorem, 62
Chi-square distribution, 64
  tables of, 345
Chi-square test, 74
Classical optimization, 5
Coding, 116, 123
Complex search, 17, 23, 259, 297
Computer simulation, 313, 315
Concave function, 10
Conditional probability, 35
Conjugate gradient method, 208
Constrained optimization, 212, 248, 270
Continuity, 9
Contrast, 129
Convex function, 10
Correlation, 77, 87
  sample, 88
Covariance, 88
Curve fitting, 118
Curvilinear regression, 83

Defining contrast, 129
Design criteria, 153
Design matrix, 100, 115
  first-order, 124
  second-order, 137
Design moments, 141
Design resolution, 177
Determinant, 358, 361
Dichotomous search, 184
Direct search, 23, 189, 250
Distribution, 3, 29

Efficiency, 125
Eigenvalue, 118, 161, 359
Eigenvector, 161, 359
Equiradial designs, 148
Errors, 3, 102, 238
  distributions of, 105
  roundoff, 117, 161
Estimation, 66
  interval, 66
  point, 66
  simulation, 328
Estimators, 103
  unbiased, 103
Euclidean norm, 360
Event, 32, 34
Evolutionary operation, 23
Expectation, 41

Expected value, 41, 105, 109
Experiment, 1, 4
  random, 29
Experimental design, 121
Experimental matrix, 101, 107
Experimental optimization, 238, 289
Experimentation, 1, 21, 238, 289
Exponential distribution, 50, 56
Exterior penalty function, 199
Extreme point, 199

Factorial designs, 124, 131, 178, 257, 294
F-distribution, 65
  tables of, 347-354
Feasible direction methods, 17, 213, 225, 270
Fibonacci search, 186, 244
First-order designs, 124, 252
First-order response surfaces, 252
Fletcher-Reeves method, 208
Fractional factorial designs, 131, 176
Function, 7
  concave, 10
  convex, 10
  unimodal, 10

Gamma distribution, 51, 56
Geoffrion-Dyer algorithm, 234
Geometric programming, 20
Goal programming, 18, 238
Golden section search, 188, 244, 290
Goodness-of-fit, 76
Gradient, 122
Gradient projection method, 228, 245, 254, 271
Gradient search, 17, 199, 205, 276
Gradient vector, 200

Hessian matrix, 201, 216, 363
Hexagonal designs, 148, 155
Hooke-Jeeves, *see* Pattern search
Hypothesis, 68
  null, 68
  test of, 68

Icosahedral design, 150, 266
Inner product, 360
Input analysis, simulation, 327
Interior penalty function, 198
Interval estimation, 66, 67
Interval reduction, 184
Inverse matrix, 361

Jacobian matrix, 222, 363
Jamming, 227, 246

Lack-of-fit, 106, 257, 266
Lagrangian analysis, 216, 219
Least-squares, 77, 102, 114, 157, 160, 166, 221
Likelihood function, 168, 174, 190
Linear combination, 361
Linear equations, 362
Linear independence, 361
Linear programming, 18, 217, 226
Linear regression model, 101
Line search, 184, 206, 244. *See also* Dichotomous search; Fibonacci search; Golden section search; Polynomial regression
Local maximum, 12
Local minimum, 12

Mathematical programming, 17
Matrix, 355
  addition, 357
  calculus, 363
  defined, 355
  multiplication, 357
  rank, 361
  transpose, 356
Maximum, 12, 14, 201
  global, 12
  local, 12
Maximum likelihood methods, 168
Mean, 41
  tests of, 71
Methods of feasible directions, 19
Minimum, 12, 14, 201
  global, 12
  local, 12
Modality, 10
Monte Carlo technique, 320
Multiple-gradient-summation technique, 225
Multiple objective optimization, 229, 248, 274
Multiple responses, 247, 303

Newton method, 206

Newton-Raphson method, 208, 223
Nonlinear programming, 18
Nonlinear regression, 171
Norm, 360
  Chebyshev, 170, 360
  Euclidean, 360
  infinity, 170, 361
Normal distribution, 52, 56, 61
  bivariate normal, 54
  standard normal, 53, 342
Normal equations, 79, 117
  curvilinear regression, 83
  linear regression, 79
  multiple regression, 86
Normal probability paper, 110

Optimization, 5, 14, 21, 181, 182, 238, 289, 313
  classical, 5
  experimental, 21, 238
  multivariable, 189, 229
  simulation, 333
  single-variable, 183
  weighted, 230
Optimization methods, 249
  analytical, 15
  classical, 5
  derivative, 6, 14
  experimental, 238
  numerical, 6, 14, 15
Optimum, 1, 5, 12
Orthogonal polynomials, 120
Orthogonal vector, 103
Outcome, 29
Outliers, 111
Output analysis, simulation, 329

Parallel tangents (PARTAN), 211
Pattern search, 17, 195
Plackett-Burman designs, 176
Point estimation, 66
Poisson distribution, 47, 48
Polynomial regression,183, 115, 118, 259, 291
Population, 57, 62
Positive definite matrix, 359
Positive semidefinite matrix, 360
Probability, 28, 29, 31
  conditional, 35
Probability distribution, 38
  continuous, 49
  discrete, 43
Probability function, 33
  cumulative distribution, 40
  density function, 39
  mass function, 38
Projection method, 228

Q-orthogonal, 208
Quadratic form, 359
Quadratic programming, 19, 159, 218

Random experiment, 29
Randomization, 3
Random number generation, 322
Random sample, 57
Random search, 189
  adaptive, 192
  stratified, 191
Random variable, 37
Random variate generation, 325
Rank of a matrix, 361
Reduced gradient method, 227
Redundancy, 128
Regression, 77, 99, 155
  curvilinear, 83, 291
  linear, 78
  multiple, 86
  nonlinear, 171
  polynomial, 291
  ridge, 118, 164
Relaxation methods, 207
Replication, 3
Residual analysis, 109, 113
Response, 3, 7
  multiple, 7
  single, 7
Response surface methods, 24, 121, 241, 252
  first-order, 24, 124, 244, 252
  second-order, 24, 135, 247, 255
Ridge analysis, 203, 229
Ridge trace, 166
Rotatable designs, 140, 154

Saddle point, 14
Sample, 57, 62
  mean, 59
  variance, 60
Sample space, 29
  continuous, 30

discrete, 30
Screening experiments, 175, 242
Search procedures, 16, 184
complex search, 17, 195, 250
dichotomous search, 184
direct search, 16, 250
Fibonacci search, 186
golden section search, 188
gradient search, 17, 199
line search, 244
pattern search, 17, 195
random search, 189
sequential search, 16
simplex search, 17, 192
Second-order design, 135, 255, 264, 285
Simplex designs, 132, 262
Simplex search, 17, 192
Simulation, 315
continuous, 318
discrete, 317
model, 316
Standard error of estimate, 82
Stationary point, 12
Statistic, 62
Steepest ascent, 243, 257, 262
Student's *t*-distribution, 63

Taylor series, 122
t-distribution, defined, 63
table of, 344
Triangular matrix, 355

Unbiased estimate, 103
Uniform distribution, 49, 56, 110, 170
Uniform precision estimate, 141, 145
Unimodality, 10, 184
Utility function, 229, 234

Variable, 2
controllable, 2
dependent, 2
independent, 2
qualitative, 3
quantitative, 3
uncontrollable, 2
Variance, 42
matrix, 103, 125, 167
of parameters, 104
residuals, 114
test of, 73
Variance function, 141
Variance reduction, simulation, 331
Vector, 355
column or row, 356
defined, 355
distance, 360
inner product, 360
linear independence, 361
norm, 360
orthogonal, 103
Q-orthogonal, 208

Weighted least squares, 166
Weighted optimization, 230, 248
Winsorizing, 111

Zigzagging in gradient search, 206
Zoutendylc's methods, 270, 279, 281
Z-statistic, 62
Z-transformation, 53